CANCER VACCINATION AND CHALLENGES

Volume 1

Strategies for Therapeutic Cancer Vaccine Development

Cancer Vaccination and Challenges, 2-volume set:
ISBN: 978-1-77491-672-8 (hbk)
ISBN: 978-1-77491-673-5 (pbk)

Volume 1: Strategies for Therapeutic Cancer Vaccine Development
ISBN: 978-1-77491-674-2 (hbk)
ISBN: 978-1-77491-675-9 (pbk)
ISBN: 978-1-00350-171-8 (ebk)

Volume 2: Delivery Strategies for Cancer Vaccine and Immunotherapy in Management of Various Carcinomas
ISBN: 978-1-77491-676-6 (hbk)
ISBN: 978-1-77491-677-3 (pbk)
ISBN: 978-1-00350-307-1 (ebk)

CANCER VACCINATION AND CHALLENGES

Volume 1

Strategies for Therapeutic Cancer Vaccine Development

Edited by

Rishabha Malviya, PhD

Bhupendra Prajapati, PhD

Sonali Sundram, MPharm

AAP | APPLE ACADEMIC PRESS

First edition published 2025

Apple Academic Press Inc.
1265 Goldenrod Circle, NE,
Palm Bay, FL 32905 USA

760 Laurentian Drive, Unit 19,
Burlington, ON L7N 0A4, CANADA

CRC Press
2385 NW Executive Center Drive,
Suite 320, Boca Raton FL 33431

4 Park Square, Milton Park,
Abingdon, Oxon, OX14 4RN UK

Library and Archives Canada Cataloguing in Publication

CIP data on file with Canada Library and Archives

Library of Congress Cataloging-in-Publication Data

CIP data on file with US Library of Congress

ISBN: 978-1-77491-674-2 (hbk)
ISBN: 978-1-77491-675-9 (pbk)
ISBN: 978-1-00350-171-8 (ebk)

About the Editors

Rishabha Malviya, PhD
Associate Professor, Department of Pharmacy, School of Medical and Allied Sciences, Galgotias University, Greater Noida, Uttar Pradesh, India

Rishabha Malviya, PhD, completed a BPharm from Uttar Pradesh Technical University and an MPharm (Pharmaceutics) from Gautam Buddha Technical University, Lucknow Uttar Pradesh. His PhD (Pharmacy) work was in Novel formulation development techniques. He has 13 years of research experience and has been working as an Associate Professor in the Department of Pharmacy, School of Medical and Allied Sciences, Galgotias University for the past 11 years. His area of interest includes formulation optimization, Nano formulation, targeted drug delivery, localized drug delivery, and characterization of natural polymers as pharmaceutical excipients. He has authored more than 150 research/review papers for national/international journals of repute. He has 51 patents (12 grants, 38 published, 1 filed) and publications in reputed National and International journals with more than 300 cumulative impact factors. He has also received an Outstanding Reviewer award from Elsevier. He has authored/Edited/edited 70 books (Wily, Springer Nature, De Gruyter, CRC Press/Taylor and Francis, River Publisher, Apple Academic Press, Nova Science Publishers, and OMICS publication) and authored 125 book chapters. His name has been included in Word's top 2% scientist list for the years 2020, 2021 and 2022 by Elsevier BV and Stanford University. He is a Reviewer/Editor/Editorial board member of more than 50 national and international journals of repute.

Bhupendra Prajapati, PhD
*Professor, Department of Pharmaceutics,
Shree S. K. Patel College of Pharmaceutical Education and Research, Ganpat University, Gujarat, India*

Bhupendra Prajapati, PhD, is a Professor in the Department of Pharmaceutics, Shree S. K. Patel College of Pharmaceutical Education and Research, Ganpat University, Gujarat, India.

He has more than 20 years of academic and research experience and has published more than 100 research and review papers in international and national journals. He has edited three books and has authored 20 book chapters in the field of novel drug delivery. He has published two Indian patents and has three applications under evaluation. He is a reviewer for three high impact journals and is on the editorial boards of several scientific journals.

Sonali Sundram, MPharm
Assistant Professor, School of Medical & Allied Sciences, Galgotias University, Greater Noida, Uttar Pradesh India

Sonali Sundram, MPharm, completed BPharm and MPharm (pharmacology) from AKTU, Lucknow. She has worked as research scientist in project of ICMR in King George's Medical University, Lucknow after that she has joined BBDNIIT and currently she is working in Galgotias university, Greater Noida. Her PhD (Pharmacy) work is in the area of Neurodegeneration and Nanoformulation. Her area of interest is neurodegeneration, clinical research, artificial intelligence. She has authored/edited/editing more than 30 books (Wiley, CRC Press/Taylor and Francis, River Publisher, IOP Publishing, Apple academic press etc.) She has attended as well organized more than 15 national and international seminar/conferences/workshop. She has more than 8 patents national and international in her credit.

Contents

Contributors

B. Chaudhary Ankit
Department of Pharmaceutics, Saraswati Institute of Pharmaceutical Sciences, Dhanap, Gandhinagar, Gujarat, India

Sachin P. Bhatt
Gyanmanjari Pharmacy College, Bhavnagar, Gujarat, India

Phool Chandra
Teerthanker Mahaveer College of Pharmacy, Teerthanker Mahaveer University, Moradabad, Uttar Pradesh, India

Mehul R. Chorawala
Department of Pharmacology, L. M. College of Pharmacy, Opp. Gujarat University, Navrangpura, Ahmedabad, Gujarat, India

Shrikalp S. Deshpande
Department of Pharmacology and Pharmacy Practice, K. B. Institute of Pharmaceutical Education and Research, Gandhinagar, Gujarat, India

J. Patel Dhaval
Department of Pharmaceutics, Saraswati Institute of Pharmaceutical Sciences, Dhanap, Gandhinagar, Gujarat, India

Arde Gayatri
Research Scholar, Sanjivani College of Pharmaceutical Education and Research, Kopargaon, Maharashtra, India

Patel Gayatri
Professor, Department of Pharmaceutics and Pharmaceutical Technology, Ramanbhai Patel College of Pharmacy, Charotar University of Science and Technology (CHARUSAT), CHARUSAT Campus, Changa, Gujarat, India

Shivani N. Jani
Department of Pharmaceutical Quality Assurance, Smt. S. M. Shah Pharmacy College, Gujarat Technological University, Amsaran, Gujarat, India

Avani N. Joshi
Gyanmanjari Pharmacy College, Bhavnagar, Gujarat, India

Nimisha P. Kakadia
Department of Pharmacology and Pharmacy Practice, K. B. Institute of Pharmaceutical Education and Research, Gandhinagar, Gujarat, India

Hitesh B. Katariya
Department of Pharmaceutics, Smt. S. M. Shah Pharmacy College, Gujarat Technological University, Amsaran, Gujarat, India

Rishabha Malviya
Department of Pharmacy, School of Medical and Allied Science, Galgotias University, Greater Noida, Uttar Pradesh, India

Dhanalekshmi Unnikrishnan Meenakshi
College of Pharmacy, National University of Science and Technology, Muscat, Oman

Deepak Mishra
Amity Institute of Pharmacy, Amity University, Noida, Uttar Pradesh, India

Lucy Mohapatra
Amity Institute of Pharmacy, Amity University, Noida, Uttar Pradesh, India

Mohsen Aghabozorgi Nafchi
Department of Computer Science and Engineering, Shiraz University, Shiraz, Iran

Sagar Pandav
Research Scholar, Sanjivani College of Pharmaceutical Education and Research, Kopargaon, Maharashtra, India

Sambit Kumar Parida
Seth Vishambhar Nath Institute of Pharmacy, Barabanki, Uttar Pradesh, India

Vipul P. Patel
Professor, Sanjivani College of Pharmaceutical Education and Research, Kopargaon, Maharashtra, India

Vishvas N. Patel
Department of Pharmacology, L. M. College of Pharmacy, Opp. Gujarat University, Navrangpura, Ahmedabad, Gujarat, India

Rashmi Pathak
Department of Pharmacy, Invertis University, Bareilly, Uttar Pradesh, India

Humzah Postwala
Department of Pharmacology, L. M. College of Pharmacy, Opp. Gujarat University, Navrangpura, Ahmedabad, Gujarat, India

Bhupendra Prajapati
Professor, Department of Pharmaceutics and Pharmaceutical Technology, Shree S.K. Patel College of Pharmaceutical Education and Research, Ganpat University, Mehsana, Gujarat, India

Patel Priya
Department of Pharmaceutics and Pharmaceutical Technology, Ramanbhai Patel College of Pharmacy, Charotar University of Science and Technology (CHARUSAT), CHARUSAT Campus, Changa, Gujarat, India

Neetu Sachan
Maharana Pratap College of Pharmacy, Mandhana, Kanpur, Uttar Pradesh, India

Palak A. Shah
Department of Pharmacology and Pharmacy Practice, K. B. Institute of Pharmaceutical Education and Research, Gandhinagar, Gujarat, India

Yesha Shah
Department of Pharmacology, L. M. College of Pharmacy, Opp. Gujarat University, Navrangpura, Ahmedabad, Gujarat, India

Zahra Alidousti Shahraki
Department of Computer Engineering, University of Isfahan, Isfahan, Iran

Himanshu Sharma
Teerthanker Mahaveer College of Pharmacy, Teerthanker Mahaveer University, Moradabad, Uttar Pradesh, India

Rashmi A. Shukla
Department of Pharmacology, L. M. College of Pharmacy, Ahmedabad, Gujarat, India

Sonali Sundram
Department of Pharmacy, School of Medical and Allied Science, Galgotias University, Greater Noida, Uttar Pradesh, India

A. Chaudhary Sunita
Department of Pharmaceutics, Arihant School of Pharmacy and Bioresearch Institute, Adalaj, Gandhinagar, Gujarat, India

Alok Shiomurti Tripathi
Shree S.K. Patel Institute of Pharmaceutical Education and Research, Ganpat University, Kherva, Mehsana, Gujarat, India

Garima Varshney
Department of Pharmacy, School of Medical and Allied Science, Galgotias University, Greater Noida, Uttar Pradesh, India

Foreword

Cancer remains the second leading cause of death in developed and developing countries. This is despite the fact that some cancers are preventable and treatable if detected early. Surgery, chemotherapy, and radiation are still major approaches in cancer treatment. Developing effective immune protection through vaccination has proven to be technologically challenging due to various technical reasons. The book, which consists of 12 chapters, serves as a valuable resource in this field. The theme of the work is "Cancer Vaccination and Challenges," covering every aspect of cancer immunology and vaccine development. Each chapter is carefully curated to provide readers with a comprehensive understanding of the rapidly growing literature in this field. The compilation explores the most successful and promising approaches by incorporating the latest studies on preventive and therapeutic cancer vaccination. The information in this book serves as a comprehensive guide to cancer vaccine research.

The book offers up-to-date information on various topics, including tumor antigens for cancer vaccines, strategies, approaches, advancements, difficulties, and challenges in vaccine development for cancer treatment. Recent advances in DNA and RNA-based vaccines for cancer prevention and treatment are thoroughly discussed in this volume. This compendium enables researchers and practitioners to easily access the comprehensive concept of cancer vaccines and their implications in healthcare.

I commend the editors and contributors for their thorough and in-depth treatment of these complex topics in this book. In my opinion, this compendium will prove to be an essential resource in disseminating the vast and expanding literature on cancer immunization and vaccine development.

—Prof. Harish Padh
Former Vice-Chancellor,
Sardar Patel University, Vallabh Vidyanagar, Anand, Gujarat, India;
Former Director, B. V. Patel Pharmaceutical Education and
Research Development (PERD), Ahmedabad, Gujarat, India

About the Book

The development of potent therapeutic cancer vaccines that can activate both effector and memory T cells to respond specifically to tumor antigens is one of the main objectives of cancer immunotherapy. The customization of cancer vaccinations has received a lot of interest lately. Nevertheless, despite the numerous vaccine techniques targeting various tumors that have been studied recently, the performance of therapeutic cancer vaccines remains poor. Clinical trials have not yet produced satisfactory data despite positive findings typically seen in preclinical data. The main cause of such failures is the difficulty in finding certain target tumor antigens that should be distinct from or overexpressed only by tumor cells in comparison to normal cells. Most tumor antigens used in cancer vaccines are non-mutated, overexpressed self-antigens that primarily activate T cells with weak T cell receptors (TCR) that are unable to effectively mediate an anti-tumor response.

This book provides a summary of the most prevalent cancer vaccination, including the role of tumor antigens in cancer vaccine development, strategies, advancements, and challenges in the development of cancer vaccines, combined therapy of cancer vaccines and chemotherapeutic agents, delivery and prevention of cancer vaccines, cancer adjuvants in cancer management, and cancer vaccination for different diseases.

This book offers in-depth knowledge of cancer vaccination and its challenges in a concise and novel manner. It will assist graduates, researchers, and clinicians in advancing cancer vaccination.

—*Editors*

Preface

Planning to publish a book, whether as an author or editor, is a journey starting. It is a mission that involves raising, mining, diving, refining, and moving through. It takes time and needs passion. I am extremely happy to release the pages of our new book *Cancer Vaccination and Challenges.*

This book is a set of two volumes. The first volume deals with different strategies for cancer vaccine development, while the second volume presents an understanding of delivery strategies of cancer vaccines for the management of various forms of carcinoma. A lot of work has been done on developing cancer vaccines over the past 50 years, only with minimal success. Meanwhile, current research in genetics, molecular biology, biochemistry, and immunology has revived interest in these immunotherapies and enabled the development of potential cancer vaccine candidates. Since the US Food and Drug Administration approved the first therapeutic cancer vaccine, a number of therapeutic cancer vaccines have been evaluated in clinical studies with cancer patients who showed good antitumor immune responses.

The primary focus of this book is the techniques for the development of therapeutic cancer vaccines. We also emphasize on the recently developed targeted cancer vaccine as a cutting-edge therapeutic strategy for cancer patients.

Chapters of this volume (Volume I) make clear our understanding of the need for vaccines in cancer treatment and presents information on the role of tumor antigens, protein/peptides, microbial genes, and stem cells for the development of vaccines for cancer management. Also, this volume provides the latest insights on DNA- and RNA-based cancer vaccines. Finally, we outline the regulatory perspectives for cancer vaccine development.

This book will be a valuable resource for researchers of pharmaceutical background and professionals working in the fields of biomedical science, engineering, vaccine development, and cancer therapy, as well as advanced studies in these fields.

We feel highly grateful to all contributors for making this wonderful outcome. We enjoyed it a lot and also learned new things during our work on this book. We want to sincerely thank the esteemed Apple Academic Press for their biggest help in the publication process. Working with AAP was a beautiful and pleasant experience.

CHAPTER 1

Cancer Vaccines: Types, Mechanisms, and Recent Developments

LUCY MOHAPATRA,[1] DEEPAK MISHRA,[1] ALOK SHIOMURTI TRIPATHI,[2] and SAMBIT KUMAR PARIDA[3]

[1]*Amity Institute of Pharmacy, Amity University, Noida, Uttar Pradesh, India*

[2]*Shree S.K. Patel Institute of Pharmaceutical Education and Research, Ganpat University, Kherva, Mehsana, Gujarat, India*

[3]*Seth Vishambhar Nath Institute of Pharmacy, Barabanki, Uttar Pradesh, India*

ABSTRACT

Vaccines have been effective approaches in averting several infectious diseases for centuries. On the other hand, cancer vaccines (CVs) can be defined as a form of immunotherapy that instructs our immune system to differentiate between cancerous and healthy cells in our body and eliminate the cancerous ones. The main objective of CVs is to assist in treating certain types of cancer or reducing the recurrence of cancer after treatment. For example, Sipuleucel-T is used in the treatment and management of advanced forms of prostate cancer (PC). However, there are also vaccines that can prevent certain types of cancer. For instance, vaccinating girls approaching puberty and certain young adults against human papillomavirus (HPV) helps protect against cervical cancer and five other important types of cancer caused by HPV. However, the etiology of most cancers does not indicate the involvement of infectious agents, so the role of vaccines in their prevention is questionable. Nevertheless, several clinical trials are currently being conducted worldwide to find suitable answers to this question. "The Lynch

Cancer Vaccination and Challenges, Volume 1: Strategies for Therapeutic Cancer Vaccine Development.
Rishabha Malviya, Bhupendra Prajapati, and Sonali Sundram (Eds.)

vaccine trial" is one such trial, which aims to determine whether vaccines can be effectively used in the prevention of non-pathogenic forms of cancer. Additionally, researchers are also investigating several other important strategies. This chapter will discuss various types of CVs, their presumed mechanisms of action, the adverse drug reactions associated with their use, the important targets of CVs, and CVs in different phases of clinical trials.

1.1 INTRODUCTION

The concept of a cancer vaccine (CV) has evolved since the 1950s from an assumption into an authenticity that has fascinated peers of cancer investigators with its ever progressively gigantic potential over the existing approaches of combating and managing several forms of cancer, including surgery, chemotherapy, radiotherapy, and monoclonal antibodies (Abs). This is owing to the fact that, in concept, it can be administered just once or over a brief time with promoting immunity, much like current vaccination programs against bacterial and viral infections, with minimally disruptive capacity and the ability to intensify cancer prevention for an individual throughout their life [1, 2]. Two key applications of immunology in oncology are immunoprevention and immunotherapy. Immunoprevention uses immunologic techniques, primarily vaccines, to elicit an immunological reaction that prevents the host from developing cancerous conditions and tumor growth before they occur. Immunotherapy involves immunologic procedures given to cancer patients in an effort to eradicate already-existing tumor cells. The contrasts between the controlling principles, practical applications, and immunological processes of immunoprevention and immunotherapy have been emphasized in numerous cancer immunoprevention research studies [3, 4]. Vaccination is one such cancer prevention strategy. It is evident that vaccines used against cancer-causing viruses like hepatitis-B virus and human papillomavirus (HPV) are highly successful in averting the initial infection and, as a result, significantly lowering the chance of developing cancers caused by these viruses. However, the majority of malignancies lack an infectious origin, therefore alternative targets are to be considered as conceivable antigens for precautionary vaccines [5, 6]. Since the majority of tumor antigens are composed of autonomous molecules, creating a potent immune reaction towards them, they have been believed to violate the rule of tolerating own body cells and increase the likelihood of developing autoimmunity. Because of this, the majority of attempts to develop CVs have been concentrated on producing vaccines to counteract metastatic cancer, as for such types of cancers the risks associated with the target antigen selection have been adequately manageable. It has been consistently demonstrated that Abs are the cornerstone of traditional cancer immunotherapy rather than cytotoxic T-cells. This is because Abs are more important in mediating the long-lasting effectiveness of vaccines used to prevent carcinogenesis [7, 8]. Figure 1.1 depicts the different molecular targets that could be targeted by CVs to combat different forms of cancers. Prevention and treatment for the neoplastic condition is possible through the use of vaccines in today's world. There are now three categories of CVs, including preventive, therapeutic, and individualized vaccines. While vaccines target cancer cells, preventive vaccines can shield patients against oncoviruses, or viruses closely associated with cancer. Although therapeutic vaccinations are employed to treat people with persistent conditions of different forms of cancers, preventive vaccines including the HPV vaccine are predominantly designed to avoid the onset of cancer types and have also been found to be effective in preventing their spread [10]. The FDA has currently approved three therapeutic vaccinations and four preventive vaccines for use in the United States [9].

This chapter will provide a detailed description of the different categories of vaccines against several forms of cancer and will also focus on the target mechanisms of these vaccines. Additionally, the recently approved vaccines

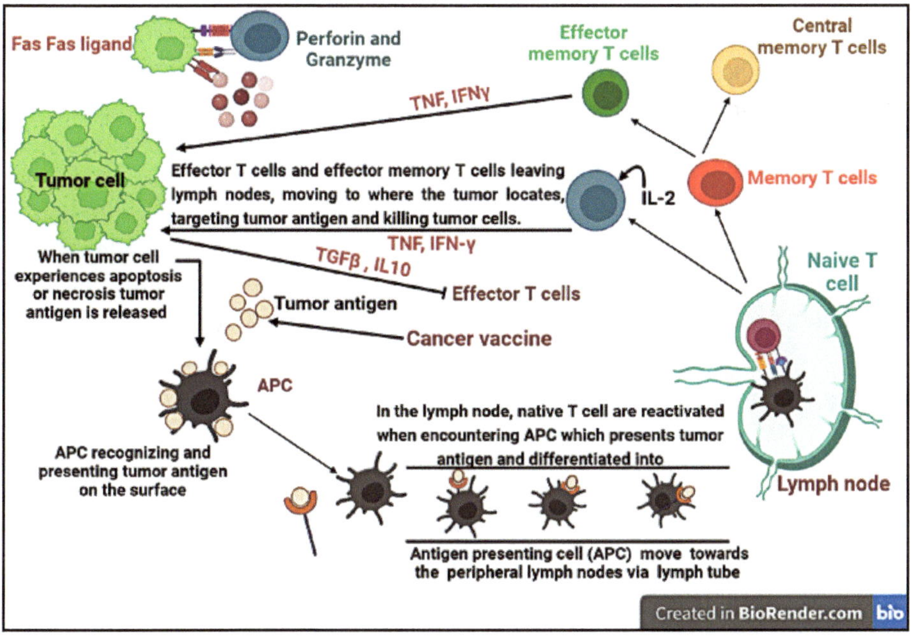

FIGURE 1.1 Different pathways to be targeted by the cancer vaccines primarily.

and the various preclinical and clinical trials focused on such vaccines will be discussed. Brief mechanistic approaches will also be explored in this chapter. Furthermore, the chapter will address the adverse drug reactions (ADRs) and challenges associated with these vaccines and their implementation.

Figure 1.1 illustrates the various pathways that can be considered for designing the CV. Antigen presenting cells (APC) play a crucial role in recognizing and presenting tumor antigens on their surface. Subsequently, they migrate to the lymph nodes where naïve T-cells are activated, initiating different signaling pathways. Activation of Fas-ligand leads to the release of perforin and granzyme, facilitating the movement of memory and effector T-cells from the lymph nodes to the tumor site. These cells target tumor antigens and eliminate tumor cells. Naïve T-cells in the lymph nodes release interleukin-2, TNF, and IFN-gamma, which exert their effects on tumor cells. This results in apoptosis or necrosis of the tumor cells, leading to the release of tumor antigens. APCs play a crucial role in discriminating and presenting tumor antigens on their cell surface. The mechanism of CVs primarily revolves around the recognition of tumor antigens. In the lymph nodes, naïve T-cells are reactivated upon encountering APCs. These APCs then migrate to peripheral lymph nodes via lymphatic vessels.

1.2 CANCER VACCINE TYPES

Preventive, therapeutic, and individualized vaccinations are the three categories into which anti-cancer vaccines can be classified. Personalized vaccinations consist of "target" vaccine antigens designed to achieve specific outcomes. This approach aims to enhance immune cell response while reducing the likelihood of vaccine noncompliance and adverse drug reactions, particularly in high-risk individuals with pre-existing serious illnesses or other conditions [10, 11]. Therapeutic vaccines are used to treat cancer patients, while preventive vaccines, such as the HPV vaccine, are used to prevent cancer from developing.

1.2.1 PREVENTIVE CANCER VACCINES

The primary category of anti-cancer vaccines is preventive in nature and works to stop cancer from further progressing in the body. All currently available licensed preventive anti-cancer vaccinations used in clinical trials focus on oncoviruses, which are known to cause cancer [12]. In 2006, Gardasil, one of the HPV immunization vaccines, was initially approved by the FDA. It was advised to be used in females aged 9 to 26 to prevent the occurrence of one of the most common forms of female cancer, cervical cancer. This vaccine was also proposed to efficiently prevent the onset of several other HPV-associated diseases, such as vaginal, vulvar, anal, and oral cancers [13]. Gardasil targets specific HPV subtypes (6, 11, 16, and 18), which contribute to roughly 70% of the total forms of cervical cancer incidence globally. In contrast to sipuleucel-T, a vaccine effective against many HPV-mediated cancers, the HPV vaccine primarily relies on inducing an immune response to prevent the first instance of HPV infection [14]. According to data from Phase 3 clinical research, females who have had 3 doses of the HPV vaccine are protected against HPV infections in over 90% of cases caused by HPV types 16 or 18 [15]. Cervical screening is still an essential means for the recognition and diagnosis of cervical cancer, despite the fact that the vaccine does not offer complete protection against all instances [16]. Furthermore, as the HPV vaccine was just recently made available in clinical settings, questions concerning its long-term effectiveness and safety, as well as disagreements over whether or not men should also take it, have not yet been entirely resolved [17]. The American Centers for Disease Control and Prevention (CDC) have recently recommended the addition of Gardasil to immunization regimens. Another precautionary vaccine is the Hepatitis B Virus (HBV) injection, which provides protection against chronic infections caused by the hepatitis-B virus. This virus significantly increases the risk of hepatocellular carcinoma (HCC), a common form of liver cancer. Similar to the HPV vaccine, the HBV vaccine works by triggering an antibody response to prevent the initial HBV infection [18]. Many countries worldwide currently include the HBV vaccine in their childhood or adolescent vaccination schedules. Research has shown that a 3-dose administration of the HBV vaccine can significantly reduce the incidence of chronic HBV infection, especially in individuals at higher risk [19].

1.2.2 THERAPEUTIC CANCER VACCINES

The rapid advancements in our understanding of the immune response and its involvement in oncology have created opportunities for therapeutic vaccinations that utilize the patient's immune system to train it to target, attack, and eliminate cancer cells [20]. This innovative vaccine approach involves the activation and proper functioning of T lymphocytes and antigen-presenting cells (APCs), including dendritic cells (DCs). Within the context of cell-mediated defense, CD4+ T-lymphocytes secrete cytokines that stimulate CD8+ cytotoxic T lymphocytes (CTLs), also known as CTLs, to attack cancer cells [21]. The immune response can target two different types of tumor antigens: tumor-specific antigens (TSAs) and tumor-associated antigens (TAAs). TSAs are antigens that are exclusively found on tumor cells, while TAAs are antigens that are broadly expressed on both tumor cells and normal cells within the host. Compared to TAAs, TSAs offer the advantage of being tailored to the specific tumor and often elicit a stronger immune response. However, TSAs are extremely rare, and their identification in a specific tumor is typically the main obstacle in the development of an anti-cancer vaccination that specifically targets TSAs [22]. TAAs, on the other hand, occur more frequently, despite typically causing a lesser immunological reaction than TSAs. Autoimmunity is a potential side effect of anti-cancer vaccinations that target TAAs [23]. When GM-CSF is administered as an adjuvant co-stimulant, DCs act on the PAP by taking it up with the help of PBMC and converting it into the major histocompatibility complex (MHC) on their cell surface. This allows the DCs to be recognized by specific CTLs. These specific CTLs can multiply to create a reservoir of CTLs that are resistant to PAP after becoming self-activated. The patient subsequently receives the sipuleucel-T vaccine, which is created using these CTLs. Each vaccine is autologous to the patient, providing the key benefit of preventing HLA formation that does not match and thus avoiding significant problems associated with tissue transplantation [24]. Despite no apparent changes in the timing of tumor growth between the two groups, a phase three randomized controlled trial (IMPACT) has successfully demonstrated higher survival rates for participants with m-CRPC who received sipuleucel-T compared to a placebo control group [25].

1.2.3 PERSONALIZED CANCER VACCINES

When it comes to vaccines, personalization refers to the targeting of vaccine antigens to achieve the best possible outcome in terms of increasing the response of immune cells and reducing the likelihood of adverse events (AEs) in patients who are at higher risk of experiencing certain complications. In the context of immunization goals, personalization aims to achieve improved outcomes. Personalization of vaccines can be related to the individual level, where it is understood that the presence of genotyping-x along with haplogroup-y may hinder the development of protective immunity or indicate a higher risk for a serious adverse reaction [26, 27]. The scientific study of how immunologic genetic polymorphisms influence the diversity of the innate immune system, cell-mediated responses, and antibody responses mediated by the innate system to vaccinations, both at the individual and population level, is known as vaccinomics (Figure 1.2).

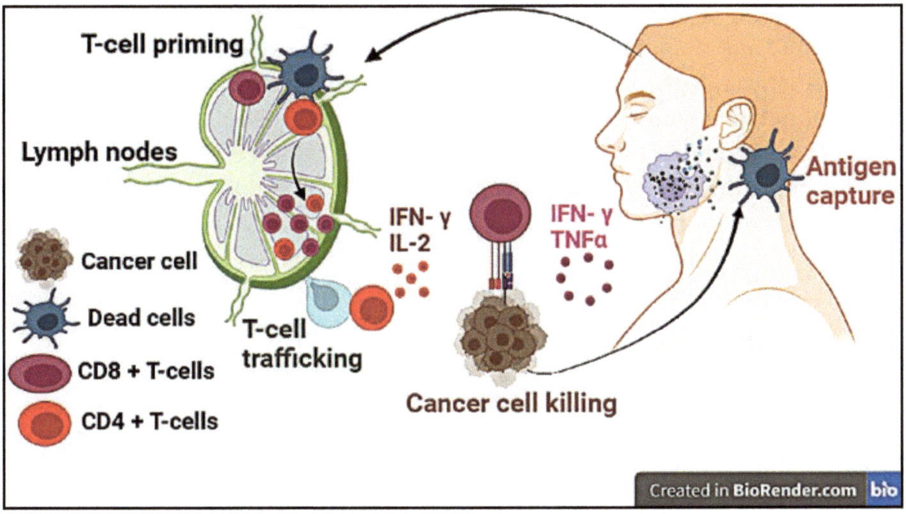

FIGURE 1.2 Important molecular targets of personalized anti-cancer vaccine.

As shown in Figure 1.2, A comprehensive study of T-cells and their ability to recognize pathogenic antigens (peptides) within the HLA system has sparked the concept of personalized vaccinations. These personalized vaccines are created by utilizing the patient's own immune cells, which are then exposed to the contents of cancer cells in a laboratory setting. Once prepared, the vaccine is administered through injection to stimulate a more robust immune response. The production of personalized anti-cancer vaccines becomes feasible once T-cell priming is achieved. Within the lymph nodes, cancer cells, dead cells, CD8+ T-cells, CD4+ T-cells, IFN-gamma, and IL-2 are present and released during T-cell trafficking. This binding of IFN-gamma and IL-2 to cancer cells leads to their destruction. The processes of T-cell priming and trafficking are observed within the lymph nodes. Personalized vaccines target the INF-gamma and TNF-alpha pathways by capturing antigens.

The initial stages of the Human Genome Project and the Global Hap-Map have been successfully completed, which have facilitated the advancement of vaccinomics and personalized vaccination studies. The understanding that immune responses to biologics, such as vaccines, can vary due to polymorphisms in crucial immune response genes has recently emerged [28–30].

As a result, we anticipate a time when both immune and adverse vaccine reactions can be predicted, measured, and used to guide clinical practice, leading to a more efficient and targeted vaccine development process. To achieve this, a validated database filled with phenotype-genotype data, affordable and accurate sequencing technologies, as well as bioinformatics and statistical tools for data evaluation and interpretation, are necessary. The field of personalized vaccines is primarily driven by the desire for improved patient-level outcomes (such as immunogenicity, reduced risk of side effects, and cost savings) at both the patient and payer levels, despite its reliance on laboratory methods.

A comprehensive understanding of how T-cells detect harmful peptides within the human leukocyte antigen (HLA) system has had a significant impact on the concept of personalized treatments. Figure 1.2 provides a brief overview of the working mechanism of personalized CVs. The human genome consists of approximately 1,000 human leukocyte antigen-A or B alleles, resulting in allelic variations that contribute to the complexity of the human leukocyte antigen complex, which is the most polymorphic gene group [31]. The HLA region is believed to have a significant influence on the range of immune responses to vaccinations and the genetic susceptibility to pathogens [32–34]. For example, the current (non-personalized) MMR vaccination is limited by various factors, including interference from maternal Abs, vaccine failure, and the inability to administer the vaccine to immunocompromised individuals. Extensive research is being conducted to support the development of novel vaccines to overcome these limitations. One potential solution is the use of tailored vaccines composed of proteins or peptides. In fact, researchers from different parts of the world have identified 13 immunogenic measles peptides that could be utilized in the development of a second-generation vaccine [35, 36]. These peptides were selected to overcome the challenges posed by HLA polymorphism. Furthermore, studies have demonstrated that measles peptides are immunogenic *in vitro* when administered to individuals [37].

1.3 MECHANISM OF ACTION OF CANCER VACCINES

Vaccines for carcinoma depict the anti-tumorigenesis effects of activating immune response signal transduction pathways, including targeting interferon-related signaling pathways, TLRs, NF-κB, and NLRP3.

1.3.1 TARGETING IFN-RELATED SIGNALING PATHWAYS

Literature supports that any cytokine with pronounced IFN-mediated immune regulation has anti-neoplastic properties [38]. The interferon (IFN) pathway plays a critical role in the human immune response. The body launches a sophisticated regulatory regime of natural and adaptive immune responses after a disease outbreak, aiming to combat the infection. The production of interferon (IFN), a pleiotropic cytokine, is one of the various reactions to viral invasion [39]. IFN gene induction increases cellular resistance to viral infection and may affect cell division. IFN is used to treat multiple sclerosis, several cancers, hepatitis type-B, type-C, and herpes simplex infections, among others [40–42]. In IFN research, we provide a wide range of products, including tyrosine kinase-2 (Tyk-2) and the IFNR-1 subunit, JAK1, and IFNR2 [43–45]. JAK1 and JAK2 are associated with the IFN-R1 and IFN-R2 subunits, respectively [46]. Following ligand binding, both Type 1 and Type 2 receptor activation cause the receptor subunits to reorganize and dimerize, which activates the associated JAKs by autophosphorylation and further activates the STAT proteins. After being phosphorylated, activated STAT assembles into homodimers and translocates into the nucleus, initiating the transcription of genes stimulated by IFN. The ISGF3 complex, composed of STAT1, STAT2, and IRF9, can also be formed by type-1 IFN. The ISGF-3 (ISG factor-3) Complex is another transcriptional complex that is significantly activated by Type-I IFN [47]. The ISREs are bound by the ISGF3 complexes, further stimulating the transcription of IFN-stimulated genes with ISREs in their promoters.

IFN-γ binds to the IFN-γ receptor, which causes STAT1 to be tyrosine phosphorylated at position Tyr-701. This phosphorylation of Tyr701 leads to the formation of STAT1 homodimers, which then translocate into the nucleus and bind to GAS elements [48, 49]. It has also been demonstrated that IFN signaling controls the MAPK pathway. Activated JAKs phosphorylate the guanine nucleotide exchange factor Vav, resulting in the activation of Rac1 downstream. Rac1 is further activated by MEKK1 (MAP3K1), which phosphorylates MAP2K3 and MAP2K6, ultimately leading to the phosphorylation of p38MAPK (MAPK14) [50]. Additionally, PI3K stimulates the translation process through the activation of the mammalian Target of Rapamycin, known as m-TOR [51, 52].

1.3.2 TARGETING TLR-RELATED SIGNALING PATHWAYS

This pattern recognition receptor (PRR) is also related to the typical strategy used by vaccines for anticancer therapy: TLR targeting. According to Mesa et al., VSSP activates toll-like receptor (TLR) type-4 on DCs, effectively eliciting a T1 cell-mediated immunological response [53]. In this study, VSSP also activates DCs in lipopolysaccharide (LPS)-hyporesponsive mice, indicating that other VSSP ingredients may also promote immunity. Additionally, VSSP performs a similar function to LPS while being less hazardous and having a better impact on people than MPL-A. Therefore, using VSSP as a DC activation agent may be practical [54]. The ability of VSSP to augment the immune response suggests that it could function as an additive in vaccinations. According to Torréns et al., the E-7 oncoprotein of HPV type-16 and the VSSP vaccine have been shown to protect mice from tumor formation, induce shrinkage of existing tumors, and activate CD8+ T lymphocytes [55]. The ERK pathway is believed to be the mechanism by which the cationic lipid DOTAP (1,2-dioleoyl-3-trimethylammonium-propane) acts as an enhancer and stimulates DCs [56]. When used as a vaccination adjuvant, the highest dose of DOTAP coupled with an HPV-16 E7-derived peptide blocks the development of TC-1 tumors [57]. Compared to S-DOTAP, R-DOTAP is considered a more effective excipient for stimulating CD8+ T-cells to produce interferon-γ (IFN-γ) in order to combat cancer [58]. Tumor regression is the ultimate outcome of the (R)-DOTAP adjuvant mechanism, which targets TLR7 and TLR9 to produce Myd88-dependent type I IFN [59]. Zhang et al. also identified a group of substances similar to cationic lipids. They found that a C-1 Lipid nanoparticle (LNP) with a 12-carbon tail efficiently delivered messenger ribonucleic acid (mRNA) encoding an antigen into DCs [60]. The C-1 LNP, stimulated by TLR-4 on DCs while delivering messenger RNA, induces an immune response.

1.3.3 TARGETING NF-KB (NUCLEAR FACTOR KAPPA-B) RELATED SIGNALING PATHWAYS

The upstream IκB kinase IKK, inhibitor IκB, and dimer transcription factor NF-κB/Rel constitute the NF-κB signaling pathway. IκB is phosphorylated by an activated IKK, degraded by IκB, and subsequently NF-κB translocates to the nucleus to modulate gene expression [60]. This signaling system is closely associated with malignancies and inflammation [62]. Furthermore, the additive function of specific nanoforms used in cancer vaccinations is linked to this signaling pathway. This pathway is connected to the activation of CCR7, promotion of DC maturation, and DC uptake of LDH. LDH nanomaterials, such as nanoparticles (NPs) or nanosheets, used as adjuvants in CVs, significantly impede tumor growth [63, 64]. The NPs were designed by Chattopadhyay et al. to deliver tumor lysates to macrophages [65, 66]. Subsequently,

TNF is released, activating the NF-κB signaling pathway and generating both *in vitro* and *in vivo* anti-tumor effects. The application of this adjuvant in other tumor models may be explored in future research. Su et al. developed a co-assembled hydrogel vaccination using supramolecular antigen Epitope Conjugated Peptides referred to as ECPs, which target CD8 or CD4 T-cell receptors (TCRs) [67]. The co-assembled peptide hydrogel vaccine exhibited a stronger anticancer impact in the E.G7-OVA tumor model as it effectively activated the MyD88-dependent NF-κB signaling pathway in DCs [68, 69].

1.3.4 MECHANISM OF ACTION FOR THERAPEUTIC VACCINES

The efficacy of CVs has been overlooked due to their lack of efficiency compared to adoptive T-cell transfer [70]. However, it has been observed that the therapeutic effectiveness of tumor infiltrating lymphocytes (TILs) is influenced by the epitope sensitivity of the transplanted T lymphocytes, and sometimes an increased number of TILs is found. Recent studies indicate that TILs often target neoantigens instead of the common antigens that are targeted by most therapeutic vaccines against nonviral malignancies. In fact, persistent CTL clones that target melanocyte differentiation antigens were insufficient in inducing significant regression of melanoma in patients [71]. Personalized immunizations that target immunogenic CD-4+ Th and CD-8+ CTL homologs reflecting neoantigens may have a strong potential to improve symptoms, as demonstrated by SLP vaccines in experimental animals [72].

The novel logistical and regulatory challenges posed by the production of customized vaccines can be addressed through innovative methods for rapid evaluation of cancer exome sequences. Demanding the effectiveness of cancer vaccinations without a significant reorganization of the cancer microenvironment is unreasonable [73]. It is necessary to establish a working mechanism for the vaccinations to effectively activate effector T-cell and memory T-cell populations in the bloodstream, and for these cells to migrate to cancerous areas and engage in tumor suppressive activity. Additional interventions are therefore required. For instance, targeting common chemotherapy drugs that have properties similar to thalidomide and other substances known to reduce immune-suppressive Treg and/or MDSC populations without affecting effector and memory T-cell count [74]. Notably, such depletion creates an ideal environment for immunotherapies, such as concurrent therapeutic vaccination, to achieve better results [75]. Another way in which chemoimmunotherapy enhances tumor eradication is through the reduction of apoptotic capacity of carcinomic or neoplastic cells by cisplatin, which is released by vaccine-induced intratumoral T-cells [76]. The inclusion of checkpoint-blocking drugs in therapeutic immunization is another viable option for combination therapy and has shown promising outcomes in preclinical models [77–79]. Figure 1.3 provides a brief overview of the mechanism by which therapeutic CVs enhance immunity by influencing the tumor microenvironment (TME).

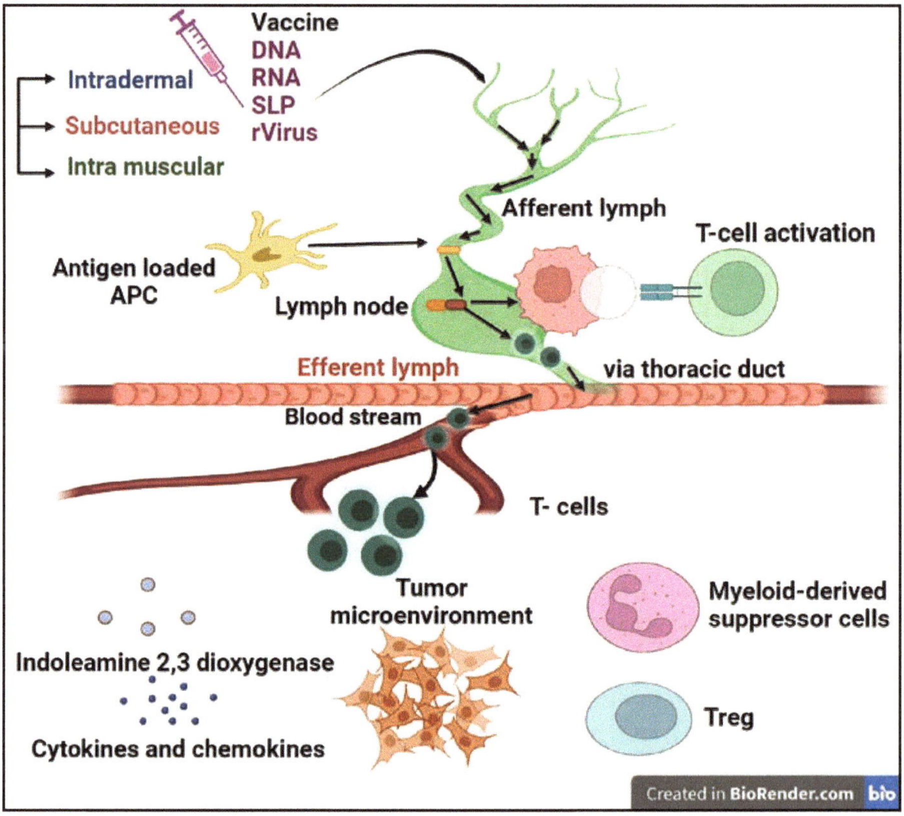

FIGURE 1.3 The primary mechanism of action generally demonstrated by therapeutic cancer vaccines.

Figure 1.3 shows routes for the administration of vaccines and the migration of immune cells, which can be achieved through subcutaneous (S.C.) administration or distribution into lymph nodes rich in DCs. These methods have proven to be highly effective. In addition to electroporation, intramuscular injection has also been successfully utilized for delivering DNA vaccinations. Therapeutic vaccines employ a sophisticated vaccine formulation that carries targeted antigens to DCs and stimulates their activity. This leads to the activation of CD4+ and CD8+ T-cells. To generate and maintain CD8+ memory cells, as well as to achieve the strongest and most durable immune response, effector CD8+ T-cell activity is required. Antigen-loaded DCs are transported to the lymph glands through the afferent lymph, where they activate lymphoid cells. Primed and activated T lymphocytes use routes such as the blood, thoracic duct, and efferent lymph to reach tumor tissue. In the context of melanoma, immunosuppressive cells like Tregs and MDSCs, as well as substances produced by tumor cells such as IDO, macrophages, and immunosuppressive chemokines, impede the movement and growth of T-cells. Overcoming these hostile factors is crucial for the effectiveness of T-cell immunization.

1.3.5 PERSONALIZED NEOANTIGEN PEPTIDE VACCINES

The functionality of CD8+ cytotoxic T-cells (Tc cells) in anti-tumor immunity has been extensively researched. However, recent studies have highlighted the deficiency of CD4+ Th cells, which hinders the responses of CTLs and underscores their crucial role [80]. Through CD40-CD40L interaction, CD4+ Th cells can activate DCs and enhance antigen presentation to CTLs, making it more effective [81]. In contrast to their typical role in the immune response through cytokine production, CD4+ T-cells exhibit cytotoxic characteristics and can directly eliminate tumors even in the absence of an MHC-I restricted CD8+ T-cell response [82–85]. Interestingly, it has been suggested that CD4+ T-cells detect a large number of immunostimulatory mutanome in mice [86], emphasizing the importance of incorporating CD4+ T-cell responses in anti-tumor immunity, particularly in the context of personalized neoantigen vaccines. Notably, recent clinical studies on neoantigen vaccination have also observed significant CD4+ T-cell responses, using MHC-1 binding prediction methods. Short peptides have been found to activate CD8+ T lymphocytes more rapidly than Synthetic Long Peptides (SLPs) as they can be directly presented on MHC-I molecules after vaccination, bypassing cellular uptake and internal processing stages [85, 86]. However, the higher signal transduction efficiency of short peptides compared to SLPs is only observed in the early stages after immunization, as the penetrating power of SLPs (explained in Figure 1.4) leads to a reversal of this trend over time [87–89]. This suggests that delivering dual markers on a single APC (Antigen-Presenting Cell) instead of two independent APCs provides a more optimal approach for developing anti-neoplastic immunity [90, 91].

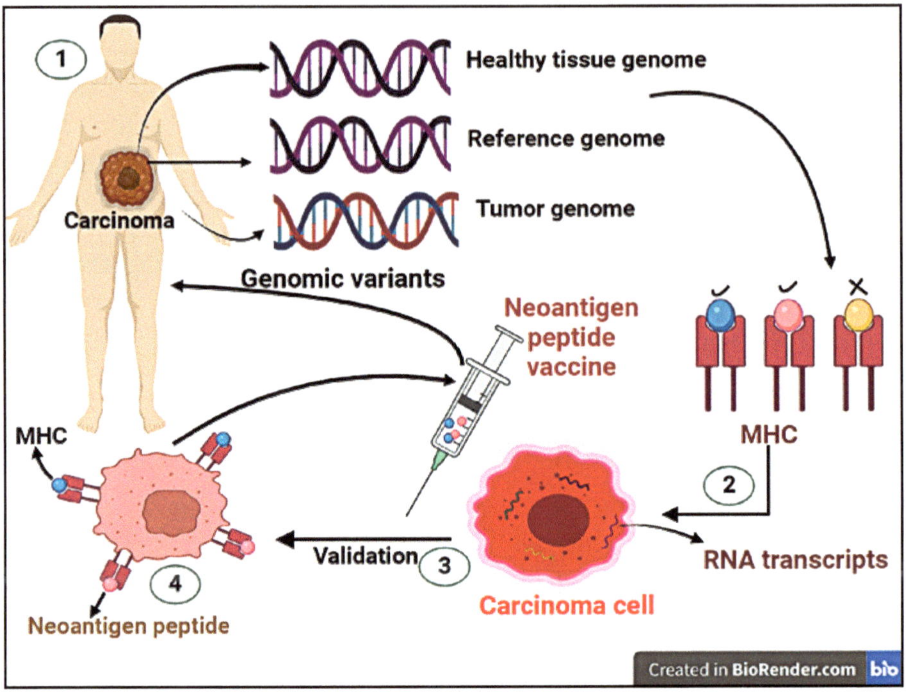

FIGURE 1.4 The most common workflow of personalized neoantigen peptide vaccine.

As shown in Figure 1.4, the classical antigen presentation pathway consists of MHC class I molecules, which present endogenous antigens to CD8+ lymphocytes, and MHC class II molecules, which present exogenous antigens to CD4+ T-cells. In this process, genomic variants such as the healthy tissue genome, reference genome, and tumor genome are obtained from the patient. Short peptides derived from these genomes are used to stimulate CD8+ T-cells that specifically recognize the antigens. These peptides bind directly to MHC class I molecules expressed on nearby DCs after injection, forming MHC class I-peptide complexes. The MHC molecules are loaded with neoantigen peptides, which are derived from RNA transcription in cancer cells. The neoantigen peptides bind to MHC class I molecules, and T-cells can be activated through cross-presentation, leading to various cellular and humoral responses. This personalized neoantigen peptide vaccine plays a crucial role in the anti-cancer immune response.

Cross-presentation enables the activation of CD8+ T-cells through DCs by binding foreign antigens to MHC class I molecules [92–94]. T-cells can be stimulated in various ways through cross-presentation, which orchestrates diverse cellular and humoral responses, facilitating the development of an anti-tumor immune response [95]. Upon injection, short peptides generate MHC class I peptide complexes that directly bind to MHC-I molecules expressed on local DCs, thereby stimulating CD8+ T lymphocytes specific to the antigen [96]. While MHC class I-peptide complexes are observed on the surface of mature DCs, they exhibit constant fluctuations. After 24 hours, the majority of MHC class I-peptide complexes become rarely detectable on the cell surface of mature DCs, whereas MHC class II-peptide complexes form. To promptly eliminate abnormal tissue cells and pathogens, MHC-I molecules continuously present ligands derived from proteins within the cytosol, including newly synthesized misfolded proteins or viral proteins [97]. Consequently, short peptide vaccines have a limited window for antigen presentation, which is believed to contribute to their remarkable efficacy [98]. However, SLP MHC-I molecules have been shown to sustain antigen presentation for at least 3 days and elicit stronger anti-tumor responses [99, 100]. Further research has demonstrated that antigens (such as SLPs) requiring intracellular internalization and processing to release MHC-I ligands can be stored intracellularly in DCs for several days. This ensures a continuous supply of antigens and allows for a significant number of extracellular MHC class I-peptide interactions, thereby prolonging the antigen bridge by DCs. These storage organelles have been described as lysosome-like compartments [100].

1.4 COMMON ADRS REPORTED WITH THE USE OF CANCER VACCINES

The roadblocks to developing any form of cancer treatment vaccines are both financial and scientific in nature. One major drawback associated with the use of CVs is the prevalence of adverse reactions. The ability of tumor cells to adapt to a changing microenvironment is a key characteristic of neoplastic cells [101]. The immune system must overcome these tumor cells in order to survive and spread throughout the body, and these are generally the most accepted approaches to immunotherapy against cancer. The TME is rich in immune system inhibitors, such as Tregs, which can hinder the immune response to anti-cancer vaccines [105].

Reported adverse effects have been observed following vaccination, specifically in relation to the disadvantages of the vaccines as outlined in Table 1.1. One major concern is the potential development of autoimmunity due to Tumor-Associated Antigens (TAAs). For example, individuals receiving a melanoma vaccine have reported experiencing vitiligo, a condition characterized by patches of skin color loss [41, 46]. This is believed to be a side effect resulting from the vaccine's lack of specificity in targeting TAAs. Another significant issue is the inclusion of excipients used to enhance the immunological action of the vaccine. Adjuvants are necessary for vaccine effectiveness, but it is crucial to strike a delicate balance between their desired impact and potential detrimental effects [102]. Currently, certain anti-cancer vaccine adjuvants are approved for medical use, including aluminum salts known as alum, and a mono-phosphoryl lipid type-A called MPL, which is prepared in an oil-in-water emulsion known as MF-59. When MPL and alum are combined, it is referred to as AS04 [103]. Both alum and oil-in-water emulsions deliver the vaccination antigen to the host immune system while controlling its release. However, they may also activate the innate immune system, leading to chronic inflammation that can exacerbate cell-mediated or systemic immunological responses [104]. MPL is produced by Toll-like receptor 4 (TLR-4), a type of PRR that identifies pathogen-associated molecular patterns (PAMPs) and LPSs [105, 106]. AS04 has been authorized for use as an additive in Cervarix and Fendrix, two variants of the HBV vaccination (an HPV vaccine) [107, 108].

1.5 IMPORTANT ANTIGEN TARGETS OF CANCER VACCINES

There are several levels of preventive strategies. In general, primary prevention focuses on preventing the onset of disease and may involve screening healthy individuals. It also focuses on preventing the recurrence of secondary cancers, such as contralateral breast cancer, or the early detection and treatment of initial cancers that are not malignant. Finally, prevention of tertiary type involves managing post-therapy for subjects to prevent the disease from returning [109]. Based on the stage of the vaccine's administration, the targeted antigens fluctuate. Therefore, numerous clinical trials for CPV's addressing all preventive levels are seen. Table 1.2 enlists the different target antigens for vaccines against different forms of cancers in different clinical testing.

1.5.1 TUMOR ASSOCIATED ANTIGENS (TAAS)

TAAs are a class of antigenic forms that are produced more or less frequently in cancerous cells or play a role in tissue differentiation, such as Muc-1

TABLE 1.1 Different Vaccines with Their Indication and Side Effects

Vaccine	FDA Approval	Use	Side Effect	References
Sipuleucel-T (Provenge)	April 29, 2010	• Used in the prevention of Prostate cancer.	• High body temperature • Chilling sensation • Tiredness • Back and joint discomfort • Nausea • Headache • Breathing • Higher blood pressure	[101]
Talimogene laherparepvec (T-VEC or Imlygic®)	Oct. 27, 2015	• Used in the prevention of advanced melanoma skin cancer.	• Fever • Chills • Fatigue • Symptoms resembling of flu for a short time is observed.	[102]
Other common anti-cancer immunization. Human papillomavirus bivalent of type 16 and 18 vaccines (Cervarix®). Human papillomavirus quadrivalent (types 6, 11, 16, and 18) vaccine (Gardasil®). Vaccine for human papillomavirus 9-valent (Gardasil-9®) Hepatitis B vaccine (HEPLISAV-B®)	16 Oct. 2009 May 1, 2017 Dec. 10, 2014 Nov. 9, 2017	• Used in preventing infection by the papilloma virus that cause major forms of cervical cancers. • Cervical as well as neck or head cancers can be prevented. • Genital warts caused by types 6 or 11 can be treated. • Immunization to protect against infection by the hepatitis-B virus.	• Inflammation, redness, and swelling may occur at the site where the vaccine is administered. This can also cause analgesia, warmth, itchiness, or a rash on the skin. • Anorexia • Pain at the back • Chilling sensation • Fatigue like condition • Elevation of body temperature or fever like condition. • Flu-like symptoms such as headache or neuralgia along with nausea and headache or joint discomfort can be seen.	[103, 104]

[114]. Cancer-testis antigens (CTAs), which are typically expressed in germ cells, including fetal cells, ovaries, testes, and trophoblasts, are also included in this category. They are minimally present in other healthy tissues. Since germ cells lack HLA-I molecules and are ectopically produced in various tumor forms, the immune system recognizes these antigens as exogenous substances [21, 116].

TABLE 1.2 Target Antigens for Vaccines Against Cancers

Target Antigen	Examples	Central Tolerance	Tumor Specificity	Preventive Settings	References
CTAs	Melanoma and carcinomas in the lung, esophageal, liver, ovarian, and bladder areas are collectively referred to as Ny/ESO.	Partial	Medium	People with a genetic predisposition to the ailment, risky behaviors (including tobacco smoking), precancerous lesions, and repeated therapy.	[110]
TAAs	An epithelial tumor, as well as colon cancer, can also affect the breast, bladder, or gastric cancers, which is referred to as MUC-1.	Yes	Medium	People with a genetic predisposition to the ailment, risky behaviors (including tobacco smoking), precancerous lesions, and repeated therapy.	[111]
Common neoantigens	Mutations in lynch syndrome are referred to as passenger/driver mutations, along with the mutated oncogenes.	No	High	Patients suffering from genetic mutations or engaging in risk behaviors (including smoking), those with pre-malignant lesions, and those seeking recurrence prevention.	[112]
Personalized neoantigens	Depending on the patient and each tumor.	No	High	None	[113]

1.5.2 CANCER-TESTIS ANTIGENS

One of the most heavily targeted families of melanoma proteins by vaccines is the CTA. A peptide called NY-ESO-one serves as an antigen for melanoma. PRAME, along with other malignancies, has been found to be highly expressed in ovaries, malignant melanoma, myelodysplastic syndrome, and various osteosarcoma [110, 118]. In patients with NSCLC [119] and metastatic melanoma, recombinant PRAME protein in combination with immunostimulant AS15, which produces related Abs and CD-4+ immunological action [120]. Due to its noteworthy demonstration of spontaneous humoral immune response, NY-ESO-one, a moiety with a limited expression profile to spermatids and placental cells and re-expression in a wide range of malignancies, is extensively used in various cancer immunotherapy strategies [121]. Several controlled trials have demonstrated that this vaccine generates specific CD-4+ and CD-8+ responses [122, 123].

1.5.3 TUMOR SPECIFIC ANTIGENS (TSAS) AND RELATED NEOANTIGENS

Personalized medicine is experiencing a rapid rise in interest due to the development of next-generation sequencing, known as NGS technology, along with general advancements in molecular biology. The standard approach for personalized neoantigen vaccination methods includes obtaining genotyping. The next step is to assess or predict neoantigens through a computer model specifically designed for each patient, taking into account their HLA haplotype. It is possible to develop a personalized vaccination that targets neoantigens or neoepitopes that are highly specific to malignancies [123]. The possibility of a preventive vaccination for early defensive mechanisms against carcinomic lesions that can develop into tumors is supported by the observation that neoantigens appear in preneoplastic lesions early, albeit in lower numbers than in malignancies [124, 125].

1.6 CANCER VACCINES IN DIFFERENT PHASES OF CLINICAL TRIALS

The primary challenge associated with any form of cancer vaccination is its high cost, rendering it unaffordable for the National Health Service [126]. Additionally, the clinical trials have shown only a marginal improvement in survival rates, with an average extension of approximately four months compared to a placebo [126]. This limited benefit may be attributed to unidentified predictive factors that were not accounted for in the human trials [115, 117]. The manufacturer of Provenge, Dendreon, was compelled to file for Chapter 11 bankruptcy, and the vaccine has since been acquired by valeant pharmaceuticals [127]. The STAND randomized Phase 2 trial evaluated various responses, including humoral and cellular responses, over a span of 24 months, demonstrating the efficacy of Provenge in stimulating a potent immune response in prostate cancer (PC) patients. However, there is currently insufficient data on the treatment's overall efficacy, making the sipuleucel-T odyssey an ongoing work in progress [61]. Table 1.3 provides an overview of CVs in different phases of clinical trials for various types of cancer, with most of them still in the initial recruiting phases.

TABLE 1.3 Preventive Cancer Vaccine under Clinical Trial with Their Current Status [61, 115, 117, 127]

Clinical Trial ID	Type of Vaccine	Patients Having	Current Status
NCT02134925 NCT00773097	Poly-ICLC with MUC1 100-mer	Colorectal adenoma in advanced stage.	Completed Recruiting
NCT00107211	HLA Class 2 promiscuous-binding peptides from Her-2 were pumped into autologous DC (DC1 vaccine).	Her2+ breast DCIS	Completed
NCT00639639	DC pumped due to autologous cytomegalovirus (CMV) form of mRNA carrying PP65-Lamp.	Glioblastoma multiforme.	Not recruiting but active.
NCT01431391	Pulsed autologous DC with combination of protein Pa-2024 called Sipuleucel-T.	Metastatic prostate cancer.	Completed
NCT00854789 NCT00841399	HLA Class 1 peptide (E75) from Her2 along the GM-CSF.	Breast cancer related to Her-2+.	Completed Not recruiting but active
NCT00524277	HLA Class 1 peptide (GP2) from Her2 along the GM-CSF.	Breast cancer related to Her-2.	Not recruiting but active.
NCT00077532	Class 1 peptide derived from Hla which resembles from gp100.	Melanoma	Completed
NCT02243371 NCT02004262 NCT01417000	CRS-207 and GM-CSF induced pancreatic cancer cells.	Metastatic adenocarcinoma of pancreas.	Not recruiting but active Recruiting Recruiting
NCT01510288	Allogeneic prostate cancer cells with GM-CSF transfection.	Metastatic prostate cancer.	Already terminated

Source: Huber, Haynes, Parker, Iversen [127].

1.7 CONCLUSION

The path to the establishment of preventative CVs is filled with obstacles. The main obstacles include the diminution of the immune system by the carcinomic cells. However, researchers are trying to exploit several adjuvants in order to address this issue. Additionally, as cancer cells develop from an individual's own body cells, they are often ignored by the immune cells. Furthermore, certain forms of cancer treatments can weaken an individual's immune system, limiting the effectiveness of vaccines. Despite these challenges, some investigators believe that treating and managing different types of cancers with vaccines may be more effective for minor tumors or cancers in their early stages. Despite the numerous disadvantages associated with the concept and strategy of CVs, current research suggests that the inoculation of this approach could be feasible for cancer prevention. Success in animal models provides a promising model for clinical development in this field. The FDA has recently approved three therapeutic vaccinations and four preventative vaccines for use in the U.S. Clinical trials are essential for further understanding of cancer immunotherapy, both in therapeutic and preventive contexts. Researchers are currently testing immunizations for a number of types of cancer, particularly immunizations for colon cancer (CC) based on NY-ESO-1, as well as MUC1, CEA, and guanylyl cyclase C. Doctors can administer the sipuleucel-T vaccine to patients with metastatic PC. Recent advancements have facilitated the development of various types of CVs. Many strategies discussed in this chapter appear to be feasible for future progress, and additional insights may emerge from recent and improved engineering solutions that combine different approaches. For instance, targeting and enhancing the host's adaptive immune system response to a preventive anti-cancer vaccine will be crucial for the successful implementation of effective prophylactic cancer vaccines. Therefore, there is significant potential and optimism that more types of such vaccines will be developed and widely implemented in the near future to prevent various forms of cancer.

KEYWORDS

- **cancer**
- **carcinomic cells**
- **clinical trials**
- **immune system**
- **immunity**
- **personalized neoantigens**
- **prophylactic cancer vaccines**
- **vaccine targets**
- **vaccines**

REFERENCES

1. Gilboa, E., (2004). The promise of cancer vaccines. *Nat. Rev. Cancer, 4*(5), 401–411.
2. Lollini, P. L., Cavallo, F., Nanni, P., & Forni, G., (2006). Vaccines for tumor prevention. *Nat. Rev. Cancer, 6*(3), 204–216.
3. Ryan, S. O., Gantt, K. R., & Finn, O. J., (2007). Tumor antigen-based immunotherapy and immunoprevention of cancer. *Int. Arch. Allergy Immunol., 142*, 179–189.
4. Finn, O. J., (2003). Cancer vaccines: Between the idea and the reality. *Nat. Rev. Immunol., 3*, 630–641.
5. Mariotto, A. B., Yabroff, K. R., Shao, Y., Feuer, E. J., & Brown, M. L., (2011). Projections of the cost of cancer care in the United States: 2010–2020. *J. Natl. Cancer Inst., 103*, 117–128.
6. Eniu, A. E., Martei, Y. M., Trimble, E. L., & Shulman, L. N., (2017). Cancer care and control as a human right: Recognizing global oncology as an academic field. *Am. Soc. Clin. Oncol. Educ. Book, 37*, 409–415.
7. Nanni, P., Landuzzi, L., Nicoletti, G., et al., (2004). Immunoprevention of mammary carcinoma in HER- 2/neutransgenic mice is IFN-g and B cell dependent. *J. Immunol., 173*, 2288–2296.
8. Curcio, C., Di Carlo, E., Clynes, R., et al., (2003). Nonredundant roles of antibody, cytokines, and perforin in the eradication of established Her-2/neucarcinomas. *J. Clin. Invest., 111*, 1161–1170.
9. Saslow, D., Castle, P. E., Cox, J. T., Davey, D. D., Einstein, M. H., Ferris, D. G., Goldie, S. J., et al., (2007). American Cancer Society Guideline for human papillomavirus (HPV) vaccine use to prevent cervical cancer and its precursors. *CA: A Cancer Journal for Clinicians, 57*(1), 7–28.
10. Finn, O. J., (2003). Cancer vaccines: Between the idea and reality. *Nat. Rev. Immunol., 3*(8), 630- 41.
11. Fang, J., Zhang, H., & Jin, S., (2014). Epigenetics and cervical cancer: From pathogenesis to therapy. *Tumor Biology, 35*(6), 5083–5093.
12. Schiller, J. T., & Lowy, D. R., (2010). Vaccines to prevent infections by oncoviruses. *Annual Review of Microbiology, 64*, 23.
13. Garland, S. M., & Smith, J. S., (2010). Human papillomavirus vaccines. *Drugs, 70*(9), 1079–1098.
14. Keam, S. L., & Harper, D. M., (2008). Human papillomavirus types 16 and 18 vaccine (recombinant, AS04 adjuvanted adsorbed) [Cervarix™]. *Drugs, 68*(3), 359–372.
15. Cutts, F. T., Franceschi, S., Goldie, S., Castellsague, X. D., De Sanjose, S., Garnett, G., Edmunds, W. J., et al., (2007). Human papillomavirus and HPV vaccines: A review. *Bulletin of the World Health Organization, 85*, 719–726.
16. Kitchener, H. C., Denton, K., Soldan, K., & Crosbie, E. J., (2013). Developing role of HPV in cervical cancer prevention. *BMJ, 7*, 347.
17. Vanderpool, R. C., Crosby, R. A., & Stradtman, L. R., (2014). Protecting a new generation against HPV: Are we willing to be bold? *Human Vaccines & Immunotherapeutics, 10*(9), 2559–2561.
18. Shepard, C. W., Simard, E. P., Finelli, L., Fiore, A. E., & Bell, B. P., (2006). Hepatitis B virus infection: Epidemiology and vaccination. *Epidemiologic Reviews, 28*(1), 112–125.

19. Aspinall, E. J., Hawkins, G., Fraser, A., Hutchinson, S. J., & Goldberg, D., (2011). Hepatitis B prevention, diagnosis, treatment, and care: A review. *Occupational Medicine, 61*(8), 531–540.
20. Guo, C., Manjili, M. H., Subjeck, J. R., Sarkar, D., Fisher, P. B., & Wang, X. Y., (2013). Therapeutic cancer vaccines: Past, present, and future. *Advances in Cancer Research, 119*, 421–475.
21. Coulie, P. G., Van, D. E. B. J., Van, D. B. P., & Boon, T., (2014). Tumor antigens recognized by T lymphocytes: At the core of cancer immunotherapy. *Nature Reviews Cancer, 14*(2), 135–146.
22. Buonaguro, L., Petrizzo, A., Tornesello, M. L., & Buonaguro, F. M., (2011). Translating tumor antigens into cancer vaccines. *Clinical and Vaccine Immunology, 18*(1), 23–34.
23. Amos, S. M., Duong, C. P., Westwood, J. A., Ritchie, D. S., Junghans, R. P., Darcy, P. K., & Kershaw, M. H., (2011). Autoimmunity associated with immunotherapy of cancer. *Blood, The Journal of the American Society of Hematology, 118*(3), 499–509.
24. Cheever, M. A., & Higano, C. S., (2011). PROVENGE (Sipuleucel-T) in prostate cancer: The first FDA-approved therapeutic cancer vaccine therapeutic prostate cancer vaccine. *Clinical Cancer Research, 17*(11), 3520–3526.
25. Kantoff, P. W., Higano, C. S., Shore, N. D., Berger, E. R., Small, E. J., Penson, D. F., Redfern, C. H., et al., (2010). Sipuleucel-T immunotherapy for castration-resistant prostate cancer. *New England Journal of Medicine, 363*(5), 411–422.
26. Black, F. L., Schiffman, G., & Pandey, J. P., (1995). HLA, Gm, and Km polymorphisms and immunity to infectious diseases in South Amerinds. In: *Immunogenetic Risk Assessment in Human Disease* (pp. 206–216). Karger Publishers.
27. Granoff, D. M., Pandey, J. P., Boies, E., Squires, J., Munson, R. S., & Suarez, B., (1984). Response to immunization with haemophilus influenzae type b polysaccharide-pertussis vaccine and risk of haemophilus meningitis in children with the Km (1) immunoglobulin allotype. *The Journal of Clinical Investigation, 74*(5), 1708–1714.
28. Poland, G. A., & Jacobson, R. M., (1998). The genetic basis for variation in antibody response to vaccines. *Current Opinion in Pediatrics, 10*(2), 208–215.
29. Poland, G. A., (1998). Variability in immune response to pathogens: Using measles vaccine to probe immunogenetic determinants of response. *The American Journal of Human Genetics, 62*(2), 215–220.
30. Poland, G. A., Ovsyannikova, I. G., & Jacobson, R. M., (2008). Personalized vaccines: The emerging field of vaccinomics. *Expert Opinion on Biological Therapy, 8*(11), 1659–1667.
31. Liesmaa, I., Paakkanen, R., Järvinen, A., Valtonen, V., & Lokki, M. L., (2018). Clinical features of patients with homozygous complement C4A or C4B deficiency. *PLoS One, 13*(6), e0199305.
32. Klein, J., & Sato, A., (2000). Advances in immunology: The HLA system. Second of two parts. *The New England Journal of Medicine, 343*(11), 782–786.
33. Jin, P., & Wang, E., (2003). Polymorphism in clinical immunology–from HLA typing to immunogenetic profiling. *Journal of Translational Medicine, 1*(1), 1–1.
34. Burgner, D., Jamieson, S. E., & Blackwell, J. M., (2006). Genetic susceptibility to infectious diseases: Big is beautiful, but will bigger be even better? *The Lancet Infectious Diseases, 6*(10), 653–663.
35. Johnson, K. L., Ovsyannikova, I. G., Poland, G. A., & Muddiman, D. C., (2005). Identification of class II HLA-DRB1* 03-bound measles virus peptides by 2D-liquid

chromatography tandem mass spectrometry. *Journal of Proteome Research, 4*(6), 2243–2249.

36. Ovsyannikova, I. G., Johnson, K. L., Muddiman, D. C., Vierkant, R. A., & Poland, G. A., (2004). Identification and characterization of novel, naturally processed measles virus class II HLA-DRB1 peptides. *Journal of Virology, 78*(1), 42–51.
37. Ota, M. O., Ndhlovu, Z., Oh, S., Piyasirisilp, S., Berzofsky, J. A., Moss, W. J., & Griffin, D. E., (2007). Hemagglutinin protein is a primary target of the measles virus—Specific HLA-A2—Restricted CD8+ T cell response during measles and after vaccination. *The Journal of Infectious Diseases, 195*(12), 1799–1807.
38. Saleiro, D., & Platanias, L. C., (2019). Interferon signaling in cancer. Non-canonical pathways and control of intracellular immune checkpoints. In: *Seminars in Immunology* (Vol. 43, p. 101299). Academic Press.
39. Barber, G. N., (2001). Host defense, viruses, and apoptosis. *Cell Death & Differentiation, 8*(2), 113–126.
40. Aaronson, D. S., & Horvath, C. M., (2002). A road map for those who don't know JAK-STAT. *Science, 296*(5573), 1653–1655.
41. Platanias, L. C., (2005). Mechanisms of type-I-and type-II-interferon-mediated signaling. N*ature Reviews Immunology, 5*(5), 375–386.
42. David, M., (2002). Signal transduction by type I interferons. *Biotechniques, 33*, S58–65.
43. Sethuraman, G., Dev, T., & Ramesh, V., (2019). Mycobacterial skin infections. Harper's *Textbook of Pediatric Dermatology, 20*, 485–502.
44. Nguyen, K. B., Cousens, L. P., Doughty, L. A., Pien, G. C., Durbin, J. E., & Biron, C. A., (2000). Interferon α/β-mediated inhibition and promotion of interferon γ: STAT1 resolves a paradox. *Nature Immunology, 1*(1), 70–76.
45. Chen, J., Baig, E., & Fish, E. N., (2004). Diversity and relatedness among the type I interferons. *Journal of Interferon & Cytokine Research, 24*(12), 687–698.
46. Katsoulidis, E., Li, Y., Mears, H., & Platanias, L. C., (2005). The p38 mitogen-activated protein kinase pathway in interferon signal transduction. *Journal of Interferon & Cytokine Research, 25*(12), 749–756.
47. Paolicelli, D., Direnzo, V., & Trojano, M., (2009). Review of interferon beta-1b in the treatment of early and relapsing multiple sclerosis. *Biologics: Targets & Therapy, 3*, 369.
48. Goldstein, D., & Laszlo, J., (1988). The role of interferon in cancer therapy: A current perspective. *CA: A Cancer Journal for Clinicians, 38*(5), 258–277.
49. Cooksley, W. G., (2004). Gastroenterology expert column: The role of interferon therapy in hepatitis B. *Medscape General Medicine, 6*(1), 16.
50. Wilhelmus, K. R., (2015). Antiviral treatment and other therapeutic interventions for herpes simplex virus epithelial keratitis. *Cochrane Database of Systematic Reviews, 2015*(1).
51. McNab, F., Mayer-Barber, K., Sher, A., Wack, A., & O'garra, A., (2015). Type I interferons in infectious disease. *Nature Reviews Immunology, 15*(2), 87–103.
52. González-Navajas, J. M., Lee, J., David, M., & Raz, E., (2012). Immunomodulatory functions of type I interferons. *Nature Reviews Immunology, 12*(2), 125–135.
53. Mesa, C., De León, J., Rigley, K., & Fernández, L. E., (2004). Very small size proteoliposomes derived from *Neisseria meningitidis*: An effective adjuvant for Th1 induction and dendritic cell activation. *Vaccine, 22*(23, 24), 3045–3052.

54. Mesa, C., De León, J., & Fernández, L. E., (2006). Very small size proteoliposomes derived from *Neisseria meningitidis*: An effective adjuvant for generation of CTL responses to peptide and protein antigens. *Vaccine, 24*(14), 2692–2699.
55. Torréns, I., Mendoza, O., Batte, A., Reyes, O., Fernández, L. E., Mesa, C., & Guillén, G., (2005). Immunotherapy with CTL peptide and VSSP eradicated established human papillomavirus (HPV) type 16 E7-expressing tumors. *Vaccine, 23*(50), 5768–5774.
56. Yan, W., Chen, W., & Huang, L., (2007). Mechanism of adjuvant activity of cationic liposome: Phosphorylation of a MAP kinase, ERK and induction of chemokines. *Molecular Immunology, 44*(15), 3672–3681.
57. Chen, W., Yan, W., & Huang, L., (2008). A simple but effective cancer vaccine consisting of an antigen and a cationic lipid. *Cancer Immunology, Immunotherapy, 57*(4), 517–530.
58. Vasievich, E. A., Chen, W., & Huang, L., (2011). Enantiospecific adjuvant activity of cationic lipid DOTAP in cancer vaccine. *Cancer Immunology, Immunotherapy, 60*(5), 629–638.
59. Gandhapudi, S. K., Ward, M., Bush, J. P., Bedu-Addo, F., Conn, G., & Woodward, J. G., (2019). Antigen priming with enantiospecific cationic lipid nanoparticles induces potent antitumor CTL responses through novel induction of a type I IFN response. *The Journal of Immunology, 202*(12), 3524–3536.
60. Zhang, H., You, X., Wang, X., Cui, L., Wang, Z., Xu, F., Li, M., et al., (2021). Delivery of mRNA vaccine with a lipid-like material potentiates antitumor efficacy through toll-like receptor 4 signaling. *Proceedings of the National Academy of Sciences, 118*(6), e2005191118.
61. Antonarakis, E. S., Kibel, A. S., Adams, G. W., Karsh, L. I., Elfiky, A., Shore, N. D., Vogelzang, N. J., et al. (2015). Antigen-Specific Immune Responses Through 24 Months in the STAND Trial: A Randomized Phase 2 Study Evaluating Optimal Sequencing of Sipuleucel-T (sip-T) and Androgen Deprivation Therapy (ADT) in Biochemically Recurrent Prostate Cancer (BRPC). *Journal of Clinical Oncology. 33,* 171–171.
62. DiDonato, J. A., Mercurio, F., & Karin, M., (2012). NF-κB and the link between inflammation and cancer. *Immunological Reviews, 246*(1), 379–400.
63. Xu, Y., Wang, Y., Yang, Q., Liu, Z., Xiao, Z., Le, Z., Yang, Z., & Yang, C., (2019). A versatile supramolecular nanoadjuvant that activates NF-κB for cancer immunotherapy. *Theranostics, 9*(11), 3388.
64. Li, A., Qin, L., Zhu, D., Zhu, R., Sun, J., & Wang, S., (2010). Signaling pathways involved in the activation of dendritic cells by layered double hydroxide nanoparticles. *Biomaterials, 31*(4), 748–756.
65. Li, A., Qin, L., Wang, W., Zhu, R., Yu, Y., Liu, H., & Wang, S., (2011). The use of layered double hydroxides as DNA vaccine delivery vector for enhancement of anti-melanoma immune response. *Biomaterials, 32*(2), 469–477.
66. Yan, S., Gu, W., Zhang, B., Rolfe, B. E., & Xu, Z. P., (2018). High adjuvant activity of layered double hydroxide nanoparticles and nanosheets in anti-tumor vaccine formulations. *Dalton Transactions, 47*(9), 2956–2964.
67. Chattopadhyay, S., Dash, S. K., Ghosh, T., Das, S., Tripathy, S., Mandal, D., Das, D., et al., (2013). Anticancer and immunostimulatory role of encapsulated tumor antigen containing cobalt oxide nanoparticles. *JBIC Journal of Biological Inorganic Chemistry, 18*(8), 957–973.

68. Chattopadhyay, S., Dash, S. K., Mandal, D., Das, B., Tripathy, S., Dey, A., Pramanik, P., & Roy, S., (2016). Metal based nanoparticles as cancer antigen delivery vehicles for macrophage based antitumor vaccine. *Vaccine, 34*(7), 957–967.
69. Su, Q., Song, H., Huang, P., Zhang, C., Yang, J., Kong, D., et al., (2021). Supramolecular co-assembly of self-adjuvanting nanofibrious peptide hydrogel enhances cancer vaccination by activating myD88-dependent NF-κB signaling pathway without inflammation. *Bioact Mater., 6*, 3924–3934.
70. Rosenberg, S. A., Yang, J. C., & Restifo, N. P., (2004). Cancer immunotherapy: Moving beyond current vaccines. *Nature Medicine, 10*(9), 909–915.
71. Chandran, S. S., Paria, B. C., Srivastava, A. K., Rothermel, L. D., Stephens, D. J., Dudley, M. E., Somerville, R., et al., (2015). Persistence of CTL clones targeting melanocyte differentiation antigens was insufficient to mediate significant melanoma regression in humans persistence of MDA Clones insufficient for tumor response. *clinical Cancer Research, 21*(3), 534–543.
72. Kreiter, S., Diken, M., Selmi, A., Türeci, Ö., & Sahin, U., (2011). Tumor vaccination using messenger RNA: Prospects of a future therapy. *Current Opinion in Immunology, 23*(3), 399–406.
73. Hinrichs, C. S., & Rosenberg, S. A., (2014). Exploiting the curative potential of adoptive T-cell therapy for cancer. *Immunological Reviews, 257*(1), 56–71.
74. Verdegaal, E. M., Visser, M., Ramwadhdoebé, T. H., Van, D. M. C. E., Van, S. J. A., Kapiteijn, E., Haanen, J. B., et al., (2011). Successful treatment of metastatic melanoma by adoptive transfer of blood-derived polyclonal tumor-specific CD4+ and CD8+ T cells in combination with low-dose interferon-alpha. *Cancer Immunology, Immunotherapy, 60*(7), 953–963.
75. Alizadeh, D., & Larmonier, N., (2014). Chemotherapeutic targeting of cancer-induced immunosuppressive cells immunosuppressive cell targeting with anticancer agents. Cancer *Research, 74*(10), 2663–2668.
76. Van, D. S. T. C., Van, D. S., Huppelschoten, S., Jordanova, E. S., Beyranvand, N. E., Sloots, A., Boon, L., et al., (2015). Vaccine-induced tumor necrosis factor–producing T cells synergize with cisplatin to promote tumor cell death synergy between T cell–produced cytokines and cisplatin. *Clinical Cancer Research, 21*(4), 781–794.
77. Fu, J., Malm, I. J., Kadayakkara, D. K., Levitsky, H., Pardoll, D., & Kim, Y. J., (2014). Preclinical evidence that PD1 blockade cooperates with cancer vaccine TEGVAX to elicit regression of established tumors DC activation and anti-PD1 can induce cancer regression. *Cancer Research, 74*(15), 4042–4052.
78. Mkrtichyan, M., Chong, N., Abu, E. R., Wallecha, A., Singh, R., Rothman, J., & Khleif, S. N., (2013). Anti-PD-1 antibody significantly increases therapeutic efficacy of *Listeria monocytogenes* (Lm)-LLO immunotherapy. *Journal for Immunotherapy of Cancer, 1*(1), 1–9.
79. Wada, S., Jackson, C. M., Yoshimura, K., Yen, H. R., Getnet, D., Harris, T. J., Goldberg, M. V., et al., (2013). Sequencing CTLA-4 blockade with cell-based immunotherapy for prostate cancer. *Journal of Translational Medicine, 11*(1), 1–4.
80. Borst, J., Ahrends, T., Bąbała, N., Melief, C. J., & Kastenmüller, W., (2018). CD4+ T cell help in cancer immunology and immunotherapy. *Nature Reviews Immunology, 18*(10), 635–647.

81. Hunder, N. N., Wallen, H., Cao, J., Hendricks, D. W., Reilly, J. Z., Rodmyre, R., Jungbluth, A., et al., (2008). Treatment of metastatic melanoma with autologous CD4+ T cells against NY-ESO-1. *New England Journal of Medicine, 358*(25), 2698–2703.
82. Quezada, S. A., Simpson, T. R., Peggs, K. S., Merghoub, T., Vider, J., Fan, X., Blasberg, R., et al., (2010). Tumor-reactive CD4+ T cells develop cytotoxic activity and eradicate large established melanoma after transfer into lymphopenic hosts. *Journal of Experimental Medicine, 207*(3), 637–650.
83. Perez, S. A., Von, H. E., Kallinteris, N. L., Gritzapis, A. D., Peoples, G. E., Papamichail, M., & Baxevanis, C. N., (2010). A new era in anticancer peptide vaccines. *Cancer, 116*(9), 2071–2080.
84. Bos, R., & Sherman, L. A., (2010). CD4+ T-cell help in the tumor milieu is required for recruitment and cytolytic function of CD8+ T lymphocytes mechanisms of tumor-specific CD4 help. *Cancer Research, 70*(21), 8368–8377.
85. Schumacher, T., Bunse, L., Pusch, S., Sahm, F., Wiestler, B., Quandt, J., Menn, O., et al., (2014). A vaccine targeting mutant IDH1 induces anti-tumor immunity. *Nature, 512*(7514), 324–327.
86. Kreiter, S., Vormehr, M., Van, De. R. N., Diken, M., Löwer, M., Diekmann, J., Boegel, S., et al., (2015). Mutant MHC class II epitopes drive therapeutic immune responses to cancer. *Nature, 520*(7549), 692–696.
87. Faure, F., Mantegazza, A., Sadaka, C., Sedlik, C., Jotereau, F., & Amigorena, S., (2009). Long-lasting cross-presentation of tumor antigen in human DC. *European Journal of Immunology*, (2), 380–390.
88. Li, F., Chen, C., Ju, T., Gao, J., Yan, J., Wang, P., Xu, Q., Hwu, P., Du, X., & Lizée, G., (2016). Rapid tumor regression in an Asian lung cancer patient following personalized neo-epitope peptide vaccination. *Oncoimmunology, 5*(12), e1238539.
89. Tobias, J., Jasinska, J., Baier, K., Kundi, M., Ede, N., Zielinski, C., & Wiedermann, U., (2017). Enhanced and long term immunogenicity of a Her-2/neu multi-epitope vaccine conjugated to the carrier CRM197 in conjunction with the adjuvant Montanide. *BMC Cancer, 17*(1), 1–3.
90. Riccione, K. A., He, L. Z., Fecci, P. E., Norberg, P. K., Suryadevara, C. M., Swartz, A., Healy, P., et al., (2018). CD27 stimulation unveils the efficacy of linked class I/II peptide vaccines in poorly immunogenic tumors by orchestrating a coordinated CD4/CD8 T cell response. *Oncoimmunology, 7*(12), e1502904.
91. Shirai, M., Pendleton, C. D., Ahlers, J., Takeshita, T., Newman, M., & Berzofsky, J. A., (1994). Helper-cytotoxic T lymphocyte (CTL) determinant linkage required for priming of anti-HIV CD8+ CTL *in vivo* with peptide vaccine constructs. *The Journal of Immunology, 152*(2), 549–556.
92. Urban, J. L., & Schreiber, H., (1992). Tumor antigens. *Annual Review of Immunology, 10*(1), 617–644.
93. Mandelboim, O., Berke, G., Fridkin, M., Feldman, M., Eisenstein, M., & Eisenbach, L., (1994). CTL induction by a tumor-associated antigen octapeptide derived from a murine lung carcinoma. *Nature, 369*(6475), 67–71.
94. Wölfel, T., Hauer, M., Schneider, J., Serrano, M., Wölfel, C., Klehmann-Hieb, E., De Plaen, E., et al., (1995). A p16INK4a-insensitive CDK4 mutant targeted by cytolytic T lymphocytes in a human melanoma. *Science, 269*(5228), 1281–1284.
95. Gjertsen, M. K., Buanes, T., Rosseland, A. R., Bakka, A., Gladhaug, I., Søreide, O., Eriksen, J. A., et al., (2001). Intradermal ras peptide vaccination with granulocyte-macrophage

colony-stimulating factor as adjuvant: Clinical and immunological responses in patients with pancreatic adenocarcinoma. *International Journal of Cancer, 92*(3), 441–450.
96. Bezu, L., Kepp, O., Cerrato, G., Pol, J., Fucikova, J., Spisek, R., Zitvogel, L., Kroemer, G., & Galluzzi, L., (2018). Trial watch: Peptide-based vaccines in anticancer therapy. *Oncoimmunology, 7*(12), e1511506.
97. Yewdell, J. W., (2007). Plumbing the sources of endogenous MHC class I peptide ligands. *Current Opinion in Immunology, 19*(1), 79–86.
98. Bijker, M. S., Van, D. E. S. J., Franken, K. L., Melief, C. J., Van, D. B. S. H., & Offringa, R., (2008). Superior induction of anti-tumor CTL immunity by extended peptide vaccines involves prolonged, DC-focused antigen presentation. *European Journal of Immunology, 38*(4), 1033–1042.
99. Van, M. N., Camps, M. G., Khan, S., Filippov, D. V., Weterings, J. J., Griffith, J. M., Geuze, H. J., et al., (2009). Antigen storage compartments in mature dendritic cells facilitate prolonged cytotoxic T lymphocyte cross-priming capacity. *Proceedings of the National Academy of Sciences, 106*(16), 6730–6735.
100. Appay, V., Douek, D. C., & Price, D. A., (2008). CD8+ T cell efficacy in vaccination and disease. *Nature Medicine, 14*(6), 623–628.
101. World Health Organization, (2017). Human papillomavirus vaccines: WHO position paper, May 2017–recommendations. *Vaccine, 35*(43), 5753–5755.
102. Schiller, J. T., Castellsagué, X., & Garland, S. M., (2012). A review of clinical trials of human papillomavirus prophylactic vaccines. *Vaccine, 30*, F123–138.
103. Kuter, B. J., Garland, S. M., Giuliano, A. R., & Stanley, M. A., (2021). Current and future vaccine clinical research with the licensed 2-, 4-, and 9-valent VLP HPV vaccines: What's ongoing, what's needed? *Preventive Medicine, 144*, 106321.
104. Meacham, C. E., & Morrison, S. J., (2013). Tumor heterogeneity and cancer cell plasticity. *Nature, 501*(7467), 328–337.
105. Hanahan, D., & Weinberg, R. A., (2011). Hallmarks of cancer: The next generation. *Cell, 144*(5), 646–674.
106. Banchereau, J., Palucka, A. K., Dhodapkar, M., Burkeholder, S., Taquet, N., Rolland, A., Taquet, S., et al., (2001). Immune and clinical responses in patients with metastatic melanoma to CD34+ progenitor-derived dendritic cell vaccine. *Cancer Research, 61*(17), 6451–6458.
107. Luiten, R. M., Kueter, E. W., Mooi, W., Gallee, M. P., Rankin, E. M., Gerritsen, W. R., Clift, S. M., et al., (2005). Immunogenicity, including vitiligo, and feasibility of vaccination with autologous GM-CSF–transduced tumor cells in metastatic melanoma patients. *Journal of Clinical Oncology, 23*(35), 8978–8991.
108. Reed, S. G., Orr, M. T., & Fox, C. B., (2013). Key roles of adjuvants in modern vaccines. *Nature Medicine, 19*(12), 1597–1608.
109. Wojtowicz, M. E., Dunn, B. K., & Umar, A., (2016). Immunologic approaches to cancer prevention—Current status, challenges, and future perspectives. In: *Seminars in Oncology* (Vol. 43, No. 1, pp. 161–172). WB Saunders.
110. Roszik, J., Wang, W. L., Livingston, J. A., Roland, C. L., Ravi, V., Yee, C., Hwu, P., et al., (2017). Overexpressed PRAME is a potential immunotherapy target in sarcoma subtypes. *Clinical Sarcoma Research, 7*(1), 1–7.
111. Beatty, P. L., Narayanan, S., Gariépy, J., Ranganathan, S., & Finn, O. J., (2010). Vaccine against MUC1 antigen expressed in inflammatory bowel disease and cancer lessens colonic inflammation and prevents progression to colitis-associated colon cancerMUC1

vaccine controls IBD and prevents colon cancer. *Cancer Prevention Research, 3*(4), 438–446.

112. Finn, O. J., (2018). The dawn of vaccines for cancer prevention. *Nature Reviews Immunology, 18*(3), 183–194.
113. Hodi, F. S., O'day, S. J., McDermott, D. F., Weber, R. W., Sosman, J. A., Haanen, J. B., Gonzalez, R., et al., (2010). Improved survival with ipilimumab in patients with metastatic melanoma. *New England Journal of Medicine, 363*(8), 711–723.
114. Nath, S., & Mukherjee, P., (2014). MUC1: A multifaceted oncoprotein with a key role in cancer progression. *Trends in Molecular Medicine, 20*(6), 332–342.
115. Drake, C. G., (2012). Re: Interdisciplinary critique of sipuleucel-T as immunotherapy in castration-resistant prostate cancer. *Journal of the National Cancer Institute, 104*(18), 1422.
116. Gjerstorff, M. F., Andersen, M. H., & Ditzel, H. J., (2015). Oncogenic cancer/testis antigens: Prime candidates for immunotherapy. *Oncotarget, 6*(18), 15772.
117. Kantoff, P. W., Higano, C. S., Small, E. J., Whitmore, J. B., Frohlich, M. W., & Schellhammer, P. F., (2012). Re: Interdisciplinary critique of sipuleucel-T as immunotherapy in castration-resistant prostate cancer. *Journal of the National Cancer Institute, 104*(14), 1107–1109.
118. Xu, Y., Zou, R., Wang, J., Wang, Z. W., & Zhu, X., (2020). The role of the cancer testis antigen PRAME in tumorigenesis and immunotherapy in human cancer. *Cell Proliferation, 53*(3), e12770.
119. Pujol, J. L., De Pas, T., Rittmeyer, A., Vallières, E., Kubisa, B., Levchenko, E., Wiesemann, S., et al., (2016). Safety and immunogenicity of the PRAME cancer immunotherapeutic in patients with resected non–small cell lung cancer: A phase I dose escalation study. *Journal of Thoracic Oncology, 11*(12), 2208–2217.
120. Gutzmer, R., Rivoltini, L., Levchenko, E., Testori, A., Utikal, J., Ascierto, P. A., Demidov, L., et al., (2016). Safety and immunogenicity of the PRAME cancer immunotherapeutic in metastatic melanoma: Results of a phase I dose escalation study. *ESMO Open, 1*(4), e000068.
121. Thomas, R., Al-Khadairi, G., Roelands, J., Hendrickx, W., Dermime, S., Bedognetti, D., & Decock, J., (2018). NY-ESO-1 based immunotherapy of cancer: Current perspectives. *Frontiers in Immunology, 9*, 947.
122. Kakimi, K., Isobe, M., Uenaka, A., Wada, H., Sato, E., Doki, Y., Nakajima, J., et al., (2011). A phase I study of vaccination with NY-ESO-1f peptide mixed with Picibanil OK-432 and montanide ISA-51 in patients with cancers expressing the NY-ESO-1 antigen. *International Journal of Cancer, 129*(12), 2836–2846.
123. Blass, E., & Ott, P. A., (2021). Advances in the development of personalized neoantigen-based therapeutic cancer vaccines. *Nature Reviews Clinical Oncology, 18*(4), 215–229.
124. Pećina-Šlaus, N., Kafka, A., Salamon, I., & Bukovac, A., (2020). Mismatch repair pathway, genome stability and cancer. *Frontiers in Molecular Biosciences*, 122.
125. Mardis, E. R., (2019). Neoantigens and genome instability: Impact on immunogenomic phenotypes and immunotherapy response. *Genome Medicine, 11*(1), 1–2.
126. Ledford, H., (2015). Therapeutic cancer vaccine survives biotech bust: Pharmaceutical company rescues landmark prostate-cancer treatment, Provenge. *Nature, 519*(7541), 17–19.

127. Huber, M. L., Haynes, L., Parker, C., & Iversen, P., (2012). Interdisciplinary critique of sipuleucel-T as immunotherapy in castration-resistant prostate cancer. *Journal of the National Cancer Institute, 104*(4), 273–279.

CHAPTER 2

Strategies to Develop Cancer Vaccines

J. PATEL DHAVAL,[1] A. CHAUDHARY SUNITA,[2] B. CHAUDHARY ANKIT,[1] and BHUPENDRA G. PRAJAPATI[3]

[1]*Department of Pharmaceutics, Saraswati Institute of Pharmaceutical Sciences, Dhanap, Gandhinagar, Gujarat, India*

[2]*Department of Pharmaceutics, Arihant School of Pharmacy and Bioresearch Institute, Adalaj, Gandhinagar, Gujarat, India*

[3]*Department of Pharmaceutics and Pharmaceutical Technology, Shree S.K. Patel College of Pharmaceutical Education and Research, Ganpat University, Gujarat, India*

ABSTRACT

An efficient vaccine-derived immune response to tumors could potentially provide a lifelong solution for cancer, in contrast to conventional procedures such as chemotherapy, surgery, immunotherapy, and radiation. Cancer vaccine (CV) research has made significant advancements alongside developments in basic tumor immunology knowledge. Promising results from recent phase II/III clinical trials, such as Provenge®, a dendritic cell (DC)-based vaccine, indicate the emergence of a new generation of immunotherapy vaccines. Over the past 20 years, several tumor-associated antigens (TAAs) have been identified and clinically tested, showing promising results as immunotherapy against various types of cancer. Various therapeutic CV techniques, including cell-based, DNA or RNA-based, protein or peptide-based, and vector-based vaccines, have been developed to address the weakened immune response to TAAs. This review focuses on examining the most promising phase II/ III clinical trials, with a particular emphasis on vaccines designed to prevent

Cancer Vaccination and Challenges, Volume 1: Strategies for Therapeutic Cancer Vaccine Development. Rishabha Malviya, Bhupendra Prajapati, and Sonali Sundram (Eds.)

carcinomas. Additionally, this review provides an overview of potential combination strategies that can unlock the unique potential of CVs.

2.1 INTRODUCTION

The second largest cause of mortality worldwide, cancer was responsible for 10 million fatalities in 2020. In that same year, over 70% of the 10 million cancer-related deaths occurred in low- and middle-income countries [1]. Breast cancer is the most prevalent type, with an estimated 2,90,560 new cases expected in the United States in 2022. Prostate cancer (PC) and lung cancer (LC) follow as the next most common malignancies [2]. Men are more susceptible to developing prostate, lung, and colorectal cancers (CRCs), while women are more prone to breast, lung, and CRCs [3]. Overall, both men's and women's cancer incidence and mortality rates have consistently decreased due to early screenings and advancements in treatment, including chemotherapy, surgery, radiotherapy, and targeted therapy [4].

Despite these positive advancements, cancer remains a significant global public health issue that requires innovative approaches and therapeutic interventions to enhance the treatment experience. Immunotherapy has long been considered a promising and viable option for cancer patients. Vaccination theoretically induces specific cellular and humoral immune responses to inhibit tumor growth and ultimately eliminate malignant cells. The discovery of Melanoma-associated antigen 1 (MAGE-1), the first human tumor antigen in the early 1990s, opened a new chapter in the application of tumor antigens in cancer vaccines (CVs) [5]. The successful use of Sipuleucel-T (Dcs vaccine) to combat PC in 2010 demonstrated the potential of CVs and sparked great interest in the field [6]. Currently, almost all CVs are in the preclinical and clinical development stages [7].

It is necessary to establish platforms for the development of vaccines and more specific antigens. CVs can be classified as either therapeutic or preventive vaccines. Preventive vaccinations can hinder oncogenesis by limiting infection from carcinogenic microorganisms, without providing direct therapeutic benefits for malignancies. The most well-known examples of these are the hepatitis B virus (HBV) and human papillomavirus (HPV) vaccines, which have been widely utilized and FDA-approved, resulting in significantly reduced incidences of liver and cervical cancer, respectively [8]. Therapeutic vaccines function by specifically triggering autoimmune reactions to eliminate tumor cells. MAGE-1 is a human tumor antigen found in melanoma tissues but not in healthy tissues or cells, as indicated by research conducted by Van der Bruggen [9]. This discovery has ushered in a new era in vaccine research.

This has given rise to the theory that utilizing tumor antigens in a "vaccine" format, due to their preferential targeting of cancer cells, offers a safe and effective approach to cancer treatment, capable of generating lifelong immunity. Over the past 40 years, extensive preclinical and clinical research on CVs has been conducted for various types of malignancies. In the United States and the European Union, only two therapeutic CVs, talimogene laherparepvec (T-VEC) and sipuleucel-T, have obtained regulatory approval [10].

CVs are categorized into four groups based on various production techniques: (i) cell-based vaccines; (ii) virus-based vaccines; (iii) peptide-based vaccines; and (iv) nucleic acid-based vaccines. Cell-based CVs are the primary type of vaccinations that utilize whole cells as antigen carriers, while DCs vaccines have shown promise in clinical studies. Viruses are primarily used as vectors in virus-based CVs for the treatment and prevention of cancers. Peptide-based vaccines contain epitopes of known or anticipated tumor antigens. Adjuvants must be added to peptide-based vaccines to enhance their immunogenicity, as they are often less immunogenic. DNA/RNA vaccines consist of coding genes with microorganism antigens as carriers, representing examples of nucleic acid vaccinations [11]. DNA-based vaccines utilize DNA plasmids that encode TAAs compounds to induce specific tumor responses. *In vitro*-produced mRNA vaccines may contain antigens and, upon internalization, generate proteins to elicit an immune response. The combination of CVs with different immunotherapies or treatments has successfully overcome tumor resistance and improved clinical outcomes [12]. This chapter summarizes the four approaches to producing CVs, along with the introduction of CV antigens. The development of CV research is also addressed, with a focus on its clinical application and therapeutic efficacy, which should be valuable for future design.

2.2 CANCER VACCINE ANTIGENS

Antigen selection is a critical step in the development of a CV. The efficacy of cancer vaccinations relies heavily on the T-lymphocytes' ability to recognize tumor antigens. A CV antigen should be highly immunogenic, exclusively present in cancer cells (not expressed in normal cells), and vital for the survival of cancerous cells [13]. There are primarily two categories of tumor immunotherapy: (i) passive (or adaptive) tumor immunotherapy involving the administration of cells or antibodies (Abs); and (ii) active tumor immunotherapy, represented by vaccines, which aims to induce a specific immune response against tumor-specific antigens (TSAs) and tumor-associated antigens (TAAs) (Figure 2.1) [14].

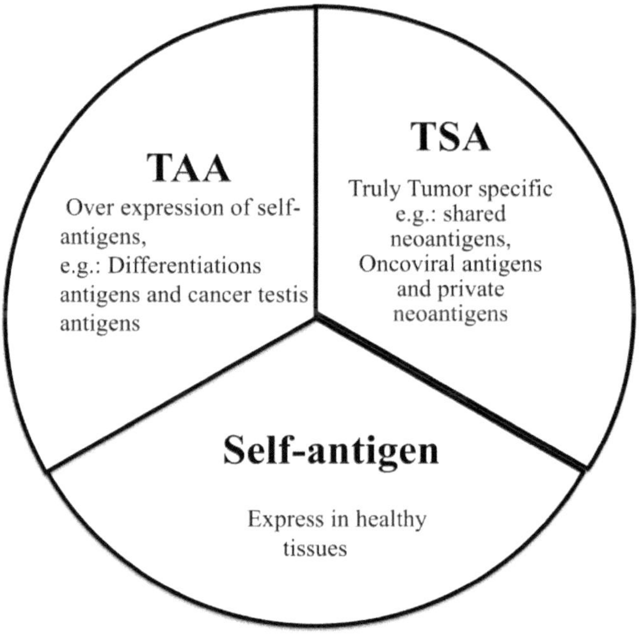

FIGURE 2.1 Tumor antigen classification.

TAAs, also known as tumor-shared antigens, encompass "self-antigens" that are overexpressed, differentiated, cancer-testicular, and "non-self" antigens resulting from viral infections. Notable tumor antigens with overexpression include human telomerase reverse transcriptase (hTERT) and human epidermal growth factor receptor 2 (HER2). Prostate-specific antigen (PSA) and melanoma antigen tyrosinase are expressed in PC and melanoma cells, respectively. These antigens are referred to as tissue differentiation antigens [15]. TAAs that are common to most tumors of a specific histological type have received significant attention in the history of CV development. For example, it is well-known that the surface antigen HER2 is also an oncogene product that drives cell division. Monoclonal antibody therapy against

HER2 is a key component of treatment for breast cancer and other malignancies indicating HER2+; however, there is currently no vaccination that can stimulate the production of these Abs in humans [16].

TAAs are available in various forms and can be applied to a range of patients. TAAs were the primary targets of initial CVs. The formation of high-affinity T lymphocytes for TAAs may not be adequate to generate an immunological response due to tolerance mechanisms at central and peripheral levels. Therefore, TAA-based vaccines need to be effective and sufficient to "break the tolerance." Despite years of attention to TAAs, clinical studies for CVs based on TAA have yielded inconsistent results. Additionally, the expression of TAAs in nonmalignant tissues raises concerns about vaccine-induced autoimmune damage [17].

Therapeutic CVs aim to "train" the host defense system to identify and precisely target tumor cells for elimination. Tumor-Specific Antigens (TSAs), which are antigens found exclusively on tumor cells, are considered ideal targets for cancer immunotherapy due to their specificity. TSAs are sometimes referred to as neoantigens, which are abnormal peptides that arise in cancer cells as a result of genetic and epigenetic changes such as point or frameshift mutations, and insertion/deletion events [18]. The advantage of targeting TSAs lies in the fact that several proteins play a crucial role in tumorigenesis and cancer development. However, targeting TSAs also has a significant disadvantage in that most of the alterations found are unique to each tumor, potentially requiring the development of individualized immunotherapy for specific individuals. The production of neoantigens is more common in tumors with a higher tumor mutational burden, presenting a unique target opportunity. Tumor mutational burden refers to the cumulative frequency of simultaneous mutational events occurring within malignant tumor cells.

Neoantigens differ from TAAs in several ways that make them more desirable [19]:

- They have greater affinities for binding to human leukocyte antigen (HLA) and T-cell receptors;
- Their manifestation is only found on tumor cells; and
- Their combination of specificity and binding affinities allows them to avoid central tolerance and problems with autoimmunity.

Recently, the primary focus of tumor vaccines has been to target neoantigens. Neoantigen vaccines have shown promise in several recent clinical trials, leading to improved patient survival. The formulation of a CV follows a personalized approach, as it specifically targets a patient's neoantigens. In this process, the tumor genome is sequenced, mutations are identified, and neoantigens are predicted using computer systems (validated through experimentation to produce and bind major histocompatibility complex (MHC) proteins). Subsequently, a vaccine containing the predicted neoantigens is produced and administered to the patient.

Evidence-based studies conducted on mice have examined the viability and effectiveness of this strategy [13]. For example, it has been observed that vaccination with synthetic long peptide (SLP) neoantigens resulted in significant anticancer activity in three different mouse models [20]. Another experiment involving neoantigen-containing SLP immunization in mice demonstrated potent T-cell responses but did not extend lifespan. Interestingly, both studies found that most neoantigens elicited CD4 T-cell responses instead of CD8 T-cell responses [21].

A notable example is the mRNA neoantigen-based melanoma vaccine, which induced T-cell infiltration and specific elimination of tumor cells expressing the neoantigens. Subsequent immunization significantly reduced the occurrence of metastasis, leading to sustained progression-free survival (PFS) [5, 22]. The efficacy of vaccines containing individualized neoantigens in melanoma patients was recently demonstrated in two phase-I clinical studies. In both cases, all patients showed immunological responses, including activation and expansion of CD4 and CD8 T-cells that recognized multiple neoantigens [13]. The antigen pool for vaccination has been expanded by scientists, utilizing a combination of shared antigens and neoantigens to enhance the efficiency of the vaccine. The APVAC1/2 vaccines can effectively stimulate the immune (T-cell) response in glioblastoma, as they contain both shared tumor antigens and personalized neoantigens [23]. Furthermore, preliminary therapeutic trials using immunosuppressive receptor inhibitors programmed death-1 (PD-1) or programmed cell death ligand-1 (PD-L1), along with personalized neoantigen vaccines, have demonstrated anti-tumor efficacy [24].

2.3 CANCER VACCINES PLATFORMS

Several types of CVs have been developed, each relying on a different type of antigen. These include – (i) cell-based vaccines; (ii) peptide-based vaccines; (iii) viral-based vaccines; and (iv) nucleic acid-based vaccines.

2.3.1 CELL-BASED CANCER VACCINES

The first type of tumor vaccine is cell-based vaccines, which are frequently made of entire cells or cell fragments that mimic tumor antigens to elicit a broader immune reaction to the antigen. DCs vaccine is an essential subset of cell-based vaccines. Clinical trials of individualized neoantigen vaccines based on DCs have shown positive anti-tumor outcomes. DCs, being the powerful Antigen-Presenting Cells (APCs) within the human immune system, are crucial for eliciting an immunological response. DC-prepared vaccines are more effective at eliciting anti-tumor immune responses [25]. DCs vaccines are typically made by collecting patient monocytes that are activated by cytokines like Granulocyte-Macrophage Colony Stimulating Factor (GM-CSF) and interleukin-4 (IL-4) and allowing them to differentiate into DCs. The patient's tumor antigens are loaded into the DCs to create an autologous antigen-pulsed DCs vaccine. The DCs vaccines are composed of various antigens, DCs fused with tumor cells or lysate, and combinations of DNA, RNA, peptides, and proteins, as well as other types of antigens. Therefore, DCs vaccines could enhance their effectiveness by improving antigen performance, priming immature CD8+ T-cells, promoting T-cells to differentiate into memory T-cells, etc. [26]. The FDA-approved representative DCs vaccine, named Sipuleucel-T, has demonstrated positive effects against castration-resistant PC (Table 2.1). It has been supported by the findings of the phase-III IMPACT trial, which showed that sipuleucel-T therapy increased overall survival (OS) by approximately 26 months compared to 22 months with placebo (hazard ratio: 0.8; $p = 0.03$) [27]. The immunogenicity of whole tumor cell vaccines needs to be improved. Live tumor cells release soluble chemicals that suppress immune cells, thereby reducing the immunogenicity of the tumor cells. For example, dead cells stimulate the immune system more effectively than living ones. Tumor cell death can trigger an adaptive immune response. There are several ways to induce immunogenic cell death:

- Interferon genes, stimulator of interferon genes (STING)-activating NPs induces apoptosis in neuroblastoma cells [28];
- RAS selective lethal 3-induced ferroptosis tumor cells;
- Photosensitizer-induced dead tumor cells may both exhibit significant immunogenicity [29, 30].

Additionally, modifying tumor cells may potentially enhance the efficacy of whole tumor cell vaccines. The use of IL-21 and IL-7 genetic modifications in whole tumor cell vaccines has shown remarkable effectiveness [31]. Various techniques have been employed to modify the vaccine, such as coating dead tumor cells with CpG-loaded NPs to enhance antigen presentation [32]. Furthermore, combining immune checkpoint inhibitors (ICI) with whole tumor cell vaccines has become more common, aiming to block pathways that inhibit the action of self-reactive T-cells [33].

The majority of DCs vaccines are derived from monocytes, which exhibit good tolerance and effectiveness. In contrast, naturally occurring DCs express a higher number of MHC molecules, making them more efficient in antigen delivery. It has been established that the type of DCs used can impact the effectiveness of DCs vaccines [34]. Myeloid/plasmacytoid DCs and conventional DCs are two popular alternatives to monocyte-derived DCs. Two clinical trials have treated patients with stage III melanoma and metastatic castration-resistant PC using natural DCs, including Myeloid DCs, conventional DCs, or a combination of both [35].

Another widely used optimization technique is radiotherapy. DCs vaccine and radiotherapy can be used in combination to enhance the immune response. Radiation therapy has the potential to attract tumor-associated neutrophils, resulting in increased oxygen radicals and the oxidative destruction of tumor cells. The DCs vaccine could also be combined with adjuvants, cytokines, ICI, or chemotherapy to enhance its effectiveness [36].

2.3.2 VIRUS-BASED CANCER VACCINES

The primary advantage of vaccines derived from viruses is their ability to stimulate the inherent and innate immune systems to collaborate, resulting in an efficient and long-term immune response. Viral-origin malignancies, such as HBV and HPV, offer significant potential for inhibiting primary cancer through viral-directed vaccines, despite the inherent challenges of cancer vaccination techniques compared to those targeting viruses [37]. FDA-approved vaccines have successfully prevented viral infections like HBV and HPV, which are the main causes of hepatocellular and cervical cancer, respectively [38]. According to a study by Ahmed et al. [39], the Epstein-Barr virus protein (BZLF1) may be a suitable candidate for a DCs-based vaccine in immune-suppressed solid organ transplants. There are three types of virus-based vaccines: (i) oncolytic virus-related vaccines; (ii) virus-related vector vaccines; and (iii) live attenuated vaccines against tumor-causing viruses. It is estimated that approximately 12% of cancer cases are caused by viral infections [40]. An innovative treatment called oncolytic virus specifically targets and kills tumor cells while promoting an anti-tumor immune response. Oncolytic virus-infected tumor cells produce oxygen radicals and cytokines that stimulate immune cells. This leads to oncolysis, and TAAs are also released [41]. Several clinical studies have already demonstrated the effectiveness of oncolytic viruses (OVs) against tumors. The classes of OVs include adenovirus, measles, herpes simplex virus, reovirus, vaccinia, vesicular stomatitis virus, and others [42]. T-VEC was the first FDA-approved oncolytic virus for cancer treatment (Table 2.1). T-VEC utilizes direct intratumoral injection to enhance anticancer immune responses, avoiding dilution and neutralization in the blood and inducing cell lysis in peripheral tumors [43]. Adenovirus is a common oncolytic virus used as a gene delivery vector due to its ease of handling, well-defined gene structure, and broad host cell tropism. It can be easily produced in large quantities and is widely employed for gene transfer and tumor antigen production [44]. Viruses have also been engineered to express specific antigens and immunomodulatory molecules simultaneously to disrupt the tumor microenvironment (TME). TG4010, for example, consists of a modified vaccinia virus (MVA) that produces the tumor antigen human mucin-1 (MUC-1) and the immunostimulatory cytokine interleukin 2 (IL2) [45]. A heterologous prime-boost strategy is employed to train T-cells and generate a robust immune response. This approach involves delivering a tumor antigen using a single viral vector, followed by administration of the same tumor antigen using a different viral vector or vector type. PROSTVAC-VF/Tricom has been shown to improve OS in PC patients by priming with a vaccinia virus encoding PSA and then administering booster doses of fowlpox virus encoding PSA [46].

2.3.3 PEPTIDE-BASED CANCER VACCINES

Cancer research is currently primarily focused on the development of subunit-based vaccines rather than the study of whole, inactivated, or attenuated viruses. Peptide-based vaccines are composed of polypeptides containing anticipated tumor antigen epitopes. This protein (or epitope) loaded vaccines offer two key advantages over tumor cells/lysates: (i) faster manufacturing, storage, and supply; and (ii) isolation and supply of tumor-specific antigens (TSAs) since most tumor cell compositions contain self-proteins that lack medicinal qualities and may even trigger autoimmune reactions. However,

TABLE 2.1 Ongoing Phase III Clinical Studies of Cancer Vaccines

Clinical Trial Identifier	Product Name	Cancer Type	Category	Phases
NCT00769704 (OPTiM)	Talimogene laher-parepvec (T-VEC)	Unresected stage IIIB to IV melanoma.	Cell-based (oncolytic virus)	III
NCT00065442 (IMPACT)	Sipuleucel-T	Metastatic castration-resistant prostate cancer.	Cell-based	III
NCT03548571	–	Glioblastoma	Cell-based (trivalent DC)	II/III
NCT04206254	GP96 heat shock protein-peptide complex.	Liver cancer	Peptide	II/III
NCT01436968	ProstAtak® (AdV-tk) + valacyclovir	Localized prostate cancer	Viral	III
NCT03091660	BCG Tokyo-172 strain	Urothelial carcinoma	Cell-based (bacterial)	III
NCT01983748	DCs plus autologous tumor RNA	Melanoma	Cell-based (DCs)	III
NCT02448173	OncoVAX®	Stage II colon cancer	Cell-based (tumor cell)	III
NCT02948543	OncoTICE®	Non-muscle-invasive bladder cancer	Cell-based (bacterial)	III
NCT04165317	BCG	Non-muscle-invasive transitional cell carci-noma of the urothelium.	Cell-based (bacterial)	III
NCT02993315	nDC	Stage III cutaneous melanoma	Cell-based (DCs)	III
NCT03495921	Vigil	Ewing sarcoma family of tumors.	Cell-based (tumor cell)	III
NCT02654587	OSE2101/Tedopi/ EP-2101	Advanced non-small-cell lung cancer.	TAA vaccine	III
NCT04229979	Galinpepimut-S	Acute myelogenous leukemia.	Analog peptide vaccine	III

this strategy has some drawbacks. Firstly, a single protein or epitope has low immunogenicity. Secondly, tumors can quickly evade immune identification through antigen mutation or inactivation. Thirdly, their usage is HLA-limited, mostly for epitope-based vaccines, and restricted to specific patient populations, typically HLA-A2+. Finally, they have limited capacity to trigger stable stimulation of CD4 and CD8 cells, which is crucial for efficient and long-term antitumor immunotherapy. Many of the aforementioned issues can be addressed by using longer peptides or combining multiple epitopes in a single vaccination. However, the relatively low immunogenicity of peptides may require the use of dendritic cells (DCs) or adjuvants. Peptide-based vaccinations are often paired with adjuvants to enhance the overall immune response. Protein antigens exhibit different immunogenic properties for B cells and T-cells. Peptide-based vaccines elicit a more effective immune response, particularly against essential neutralizing epitopes, compared to deactivated tumor cell vaccines. This immunological advantage is known as Immunodominance [5, 47].

The effectiveness of the peptide vaccine is greatly influenced by the length of the peptide chain. The small epitope of CD8+ T-cells is typically a short peptide consisting of 15 amino acids, which has a shortened half-life *in vivo*. Consequently, short peptides often only temporarily activate cytotoxic T lymphocytes (CTLs) and may even lead to CTL tolerance. Furthermore, HLA types frequently impose restrictions on shorter peptides. In comparison, longer peptides can cover a wider range of HLA types, presenting various epitopes and supporting recognition and binding to enhance immunogenicity [48]. Long peptides cannot be directly loaded onto MHC molecules; they must first be processed by APCs. Long peptides undergo substantial digestion, breaking down through the endocytic process, and then get deposited on MHC-II molecules, where they are recognized by CD4+ T helper cells. On the other hand, to activate CD8+ T lymphocytes, the other components follow the cytoplasmic pathway and merge with MHC-I molecules [49]. Consequently, long peptide vaccines have a greater potential to induce long-lasting and robust anti-tumor activity reactions. Long peptides are generally produced using protein expression systems, while short peptides are typically synthesized chemically. Various expression platforms, such as Escherichia coli, plants, insect cells, mammalian cells, and yeasts, have been employed in the development of CVs. Mammalian cells are capable of expressing proteins that closely resemble natural tumor antigens [50].

Small antigens can fuse with carrier proteins to increase their immunogenicity. For example, the MUC1 CV combines tetanus toxins, resulting in a significant slowdown of breast tumor growth in mouse models [51]. Another example is the fusion of β-cell epitopes in the HER2 extracellular domain with the polypeptide IMU-131 and the diphtheria toxin. IMU-131 induces the production of HER2-specific Abs and cellular reactions, as demonstrated in a published phase I clinical investigation [52]. NeuVax, a CV targeting HER2, is currently in phase III clinical trials. Additionally, Trastuzumab is being tested in two ongoing Phase II clinical trials for HER2+ breast cancer [53].

Adjuvants are frequently used in conjunction with peptide-based vaccinations. The type and magnitude of the T-cell response following immunization are significantly influenced by adjuvants. Several new adjuvants, such as nanomaterials, synthetic Toll-like receptor (TLR) ligands, Nod-like receptors (NLRs), CD40 agonists, retinoic acid-inducible gene (RIG) receptors, STING, and cytokines, show promising anti-tumor benefits [54]. These adjuvants are extensively utilized in the development of various CVs, including nucleic acid and cell-based vaccines. Theoretically, these new adjuvants enhance T-cell proliferation and activation, facilitate their migration to lymph nodes, and assist in recruiting leukocytes to the site of immunization [5]. Recent studies have demonstrated the essential role of T helper cells in generating potent and persistent immune-mediated anti-tumor responses. Targeting T helper type 1 (Th1) cells has significant effects on initiating and maintaining anti-tumor immunity, while T helper type 2 (Th2) cells may promote neoplastic transformation [55].

Adjuvants made of aluminum increase Th2-type responses, which makes them less appropriate for use in CVs. On the other hand, AS04, an aluminum salt, when combined with MPL (Monophosphoryl lipid A-purified lipopolysaccharide (LPS)), can strongly elicit a Th1 reaction and result in the production of Interferon-gamma (INF-γ) [56]. Nanomaterials are also extensively researched as interesting antigen-delivery devices as they preserve proteins from immediate protease breakdown, extending the half-life of antigens. They can be designed to target specific cells and organ types. Polyethylene glycol (PEG) and other biodegradable polymers are typically treated with liposomes to increase the half-life of antigens [57]. TLR-targeting adjuvants have demonstrated considerable promise as adjuvants in cancer vaccinations. For instance, a TLR3 agonist, polyinosinic-polycytidylic acid (poly I:C), is the most effective inducer of Th1-type reactions and when paired with the vaccine, produces a powerful CTL response. Clinical trials for melanoma and glioma have examined poly(I:C) [51]. Additionally, cytokines including GM-CSF, IL-2, and IFN have shown promise in clinical trials as adjuvants for CVs. Lung, breast, prostate, and skin cancer clinical trials have been conducted using GM-CSF as a vaccine adjuvant for anti-tumor immunotherapy [58].

2.3.4 NUCLEIC ACID (DNA/MRNA) VACCINES

One of the prominent areas in biomedical advancements in the 1990s was the use of nucleic acids to develop gene treatments for generating specific antibody (T-cell) responses against pathogens (prophylactic vaccinations) or tumor antigens (CVs). The concept of utilizing non-live vaccinations presents an opportunity to enhance cellular and humoral immunity [16]. Cancer DNA vaccines are constructed using bacterial plasmids, incorporating multiple oncology antigens that activate innate immunity and trigger immune responses. DNA vaccines stimulate cellular and humoral immune responses. To be translated into cytoplasmic antigens through transcription, DNA vaccines must penetrate the nucleus. The antigen is processed by MHC I and MHC II molecules and presented to CD8+ T and CD4+ T-cells to elicit specific immunological responses. DNA vaccines work in three different ways: (i) DNA is directly transfected into a somatic cell, such as a muscle cell; (ii) Apoptotic bodies (or secreting bodies) release the antigen encoded in somatic cells' DNA; and (iii) DNA is directly transfected into APCs. The most important approach for DNA CVs is the direct transfer of DNA plasmids into APCs through intradermal (ID) administration [59].

DNA vaccines have elicited immunological responses in humans, although their effectiveness is still quite low. Different types of antigen-presenting cells (APCs), such as myocytes, monocytes, and DCs, can take up naked DNA and RNA. However, cellular absorption of naked nucleic acids is inefficient, so research has focused on formulating and improving uptake methods. Transfection can be significantly enhanced by delivery methods such as microneedle arrays, NPs, *in situ* electroporation, and gene guns [60]. Over the past decade, both preclinical and clinical research have been conducted on numerous DNA CVs. However, only a few clinical trials have published their findings. Most of the research on DNA vaccination has been centered around cervical cancer. Currently, two Phase III clinical trials are examining the protection and effectiveness of the first DNA vaccine, VGX-3100, against HPV, alongside the "ZycoV-D" COVID-19 vaccine. INOVIO, a pharmaceutical company, recently released promising outcomes from the phase III trial of VGX-3100 for cervical precancerous lesions [5, 61].

Previously, DNA-based nucleic acid vaccines (NAVs) were developed because DNA is more stable and remains in the body for a longer period compared to mRNA. However, modern mRNA CVs are showing encouraging clinical results in the treatment of various solid cancers. RNA CVs may have some advantages over DNA vaccinations. Unlike DNA molecules, mRNA moves into the cytoplasm to translate and directly express antigens, without the need to enter the cell nucleus for transcription. As a result, the generation of mRNA antigens is rapid and efficient. In comparison to mRNA vaccines, DNA vaccines elicit a lesser immune response due to the requirement of an additional step for entry into the cell nucleus. However, once plasmid DNA has entered the nucleus, it can produce multiple copies of mRNA, resulting in the production of a greater number of antigens compared to a single mRNA molecule. Additionally, mRNA vaccines do not involve integration into the genome, whereas DNA vaccines may carry a risk of insertion mutations [62]. Recent advancements have improved the stability and translation efficiency of mRNA through various modifications to the backbone and untranslated regions (UTRs). By employing new purification techniques such as fast protein liquid chromatography to eliminate double strands from mRNA products, non-specific immune activation is reduced. Moreover, the use of lipid nanoparticles (LNPs) and innovative techniques has significantly enhanced the *in vivo* distribution of mRNA [5]. The majority of LNPs consist of cholesterol, phospholipids, and lipid-anchored PEG (Ionizable amino lipid-like complexes). Cholesterol and phospholipids play a crucial role in membrane fusion, *in vivo* escape, and stabilization of the liposome structure. Lipid-anchored PEG helps in reducing macrophage-mediated clearance, preventing particle aggregation, and improving storage stability. Optimization of ionizable lipids is essential for enhancing the delivery and functionality of LNPs [63]. Recent modifications have enabled targeted delivery of LNPs. For example, mannose-modified LNPs can potentially target DCs by binding to the mannose receptor. Furthermore, the stability and reproducibility of LNPs produced by chip-based microfluidic devices are advantageous for the GMP manufacturing of LNPs [63].

Clinical trials conducted on *in vitro* transcribed (IVT) mRNA-based tumor vaccines have shown promising results. mRNA vaccines have been used in the treatment of aggressive and metastatic malignant solid tumors, including CRC, non-small cell lung cancer (NSCLC), and melanoma. In a recent phase I study involving patients with metastatic melanoma, mRNA encoding four different TAAs was tested in a liposomal formulation. One example of such an mRNA vaccine is BNT111, which contains four TAAs (NY-ESO-1, MAGE-A3, tyrosinase, and TPTE). BNT111 demonstrates the potential of targeting common non-mutated tumor antigens and providing effective treatment for melanoma patients receiving ICIs [64]. Clinical trials of patient-specific mRNA vaccines developed by Moderna and BioNTech have also shown promising results in terms of anticancer benefits. For instance, Moderna has developed mRNA-4157, a customized neoantigen vaccine capable of encoding 34 neoantigens in LNPs (LNPs). Phase I studies have demonstrated satisfactory anticancer activity in patients with unresectable solid tumors who received mRNA-4157 alone or in combination with Pembrolizumab [22].

2.4 CANCER VACCINE AND COMBINATION STRATEGIES

Despite tremendous breakthroughs, cancer vaccinations still have limited clinical promise when used as a single therapy for patients with progressive malignancies. Tumors spread and develop a variety of escape mechanisms from the immune system, such as angiogenesis of the tumor, recruitment of immune-suppressive cells, and overexpression of inhibitory compounds [65]. The T-cells produced by immunization are not successful in combating the tumor. Therefore, it is crucial to combine these immunological escape mechanisms with other approaches. Therapeutic vaccination strategies have played a significant role as cancer-fighting tools. Encouraging outcomes have been observed when radiotherapy or chemotherapy is combined in numerous published trials [66]. Combinations of therapeutic vaccines with antiangiogenic treatments (ATs) or ICIs are being explored, with an emphasis on their potential synergy and an update of preclinical and clinical findings (Figure 2.2) [67].

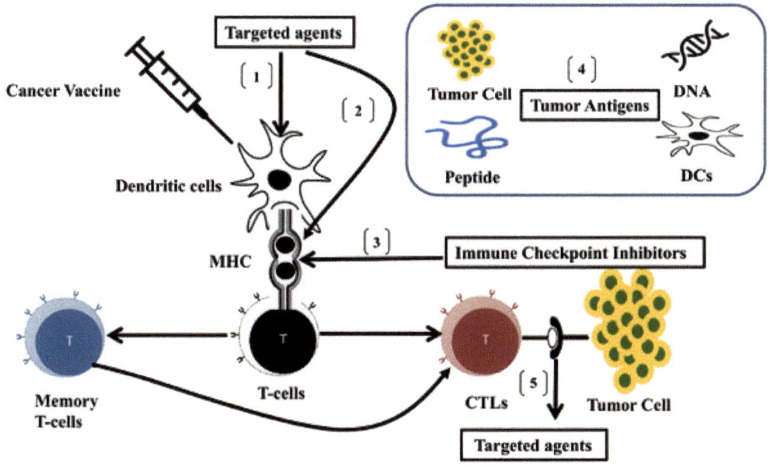

FIGURE 2.2 Types of a cancer vaccine with combination strategies. (1) Targeted agents cause the DC to mature; (2) targeted agents encourage CD8+ T-cell development and antigen presentation; (3) the tumor immune escape effect is prevented by ICIs; (4) different vaccine antigens; and (5) targeted agents increase the CLTs' ability to kill.

2.4.1 CANCER VACCINES AND ANTIANGIOGENIC TREATMENTS COMBINATIONS

Tumors express an increased amount of pro-angiogenic molecules during carcinogenesis, including vascular endothelial growth factor (VEGF), which plays a role in tumor angiogenesis. Despite the abnormal formation of blood vessels, this process facilitates the tumor's access to nutrients and oxygen, promoting growth and metastatic spread. Tumor angiogenesis also aids in immune escape due to its immunosuppressive functions [68]. Several antiangiogenic compounds have been developed to target tumor angiogenesis within this context. Adoptive therapies (ATs) work by inhibiting suppressor cells produced by the tumor, promoting DC maturation and differentiation, enhancing antigen-presenting reactions to tumor cells, inducing T-cell formation, activating CD8+ T-cells, and enhancing their ability to kill tumor cells. They also target various components of the TME. The majority of antiangiogenic drugs currently used in clinical settings target the VEGF/VEGF receptor (VEGFR) pathway. Examples include bevacizumab (anti-VEGF antibody), sorafenib, sunitinib, or axitinib (tyrosine kinase inhibitors) [69]. The combination of ATs and therapeutic vaccines is now being evaluated in human patients based on promising results from preclinical trials. In a phase II clinical study, sunitinib was combined with AGS-003, a personalized DC vaccine derived from monocyte-derived DCs treated with allogeneic tumor RNAs and CD40L RNA, in intermediate and low-risk metastatic renal cell cancer (mRCC) patients. The second cycle of sunitinib began with the introduction of the vaccine. Combination therapy has been shown to increase immune responses and extend the survival period [70]. Breast malignancies often exhibit high levels of HER2 expression. In a mouse model, the combination of trastuzumab and a whole-cell vaccine demonstrated the ability to enhance the impact of CD8+ T-cells and prolong survival time [71]. However, certain targeted medicines may not increase vaccine efficiency and can even decrease immune response due to their interaction with other tumor signaling pathways. For example, Hipp et al. discovered that Sorafenib, by inhibiting DCs activity, could diminish the immune reaction induced by the vaccine [72]. Despite ongoing clinical trials, the combination of vaccines and targeted therapies has received less attention compared to other combinations. Therefore, targeted medicines and vaccines are still in the early stages of cancer treatment, with further positive outcomes anticipated.

2.4.2 CANCER VACCINES AND CHECKPOINT INHIBITORS COMBINATIONS

Immunological checkpoint inhibitor (ICI) Abs have been recognized as an effective approach in cancer treatment when used in conjunction with other immunotherapies because tumor cells restrict T-cell function through a selection of immune checkpoints [73].

The FDA has approved many immune checkpoint-blocking Abs (Table 2.2). For example, ipilimumab blocks CTLA-4 (Cytotoxic T lymphocyte antigen-4) in melanoma, Pembrolizumab, and Nivolumab block PD-1 in melanoma, NSCLC, Atezolizumab blocks PD-L1 in NSCLC, and other specific malignancies [74].

In a phase Ib research, individuals with pancreatic adenocarcinoma received ipilimumab in addition to the GM-CSF gene (GVAX) vaccine. Compared to ipilimumab alone, the vaccine and anti-CTLA-4 therapy enhanced OS [75].

In a similar study, a synthetic cancer therapeutic vaccine and CTLA-4 blockading antibody were used to administer in triple-negative breast cancer (TNBC) [76]. The study discovered that TCR and CD28 on the T-cell surface, which are involved in vaccine-induced T-cell activation, would promote the expression of CTLA-4, which ultimately diminishes and abolishes T-cell responses. The therapeutic properties of the vaccine on TNBC were significantly improved when combined with anti-CTLA-4 Abs. Additionally, it was found that T-cell penetration in tumor tissue had improved, which is another factor in the successful inhibition of the CTLA-4 pathway [77].

A current phase II trial assessing the use of the PD-1 antagonist, nivolumab, to an HPV E6 and E7 multi-antigen vaccine produced an average OS in patients with incurable HPV-positive oropharyngeal cancer that was nearly twice as long as ICIs alone [78].

Above all, there was a synergistic impact when immune checkpoint blockade (ICB) therapy was combined with therapeutic CVs. However, the combined impact might be influenced by a significant number of nonresponsive cases that have appeared in clinical applications. Therefore, it is necessary to provide additional input for assessing the patient's response to ICI and determining a reasonable combination of multiple immune checkpoint-blocking techniques.

2.4.3 RADIOTHERAPY AND CHEMOTHERAPY COMBINATION WITH CANCER VACCINE

Finally, several recent papers have described the utilization of CVs in combination with chemotherapy and radiotherapy. Preclinical investigations have demonstrated that radiotherapy can induce immunogenic stress and cell death in tumor cells, leading to enhanced T-cell reactivity and synergy with CVs [79]. In the case of *in-situ* vaccination, bacterial membrane-coated nanoparticles (BNP) were combined with radiotherapy to enhance immune response activation. BNP could capture neoantigens generated by radiation therapy through intratumoral injection, which would then be taken up by DCs and stimulate the innate immune response. The combined effect of BNP and radiation therapy resulted in significant tumor regression and the development of tumor-specific immunological memory [80]. Chemotherapeutic drugs induce cytotoxicity in tumor cells, leading to tumor regression. Cancer immunotherapy using chemotherapy was designed to induce immunogenic cell death in targeted tumor cells, which in turn generate a diverse array of tumor antigens *in vivo* and release various immune-stimulating molecules. Furthermore, studies have shown that low doses of certain chemotherapeutic drugs, such as Gemcitabine, can modulate immunosuppressive cells and reshape immunological tolerance in the TME [81].

Conventional cancer treatment, such as cyclophosphamide (CP), damages DNA and enhances the body's ability to combat tumors. Studies have shown that a single low-dose CP treatment prior to administering DCs vaccine to tumor-bearing mice also improves the anti-tumor immune response by reducing regulatory T-cells (TReg) and increasing the release of immunostimulatory cytokines, such as IFN-γ [82]. The order and dosage of paclitaxel administration significantly impact vaccine-induced tumor regression, even when used in combination with GM-CSF-modified CV and other drugs. It has been confirmed that low-dose paclitaxel primarily enhances the anti-tumor immune response when administered after the vaccine, rather than before. The appropriate timing following the administration of chemotherapeutics must also be considered when using cancer therapeutic vaccines [83].

2.5 CONCLUSION AND PERSPECTIVE

Beyond surgery, chemotherapy, and radiation, cancer immunotherapy has established itself as the fourth element of anti-cancer therapy. Based on effective preclinical models, various vaccination platforms have been

TABLE 2.2 Examples of Combining CVs with Antiangiogenic Therapies (ATs) or ICIs (ICIs) (Continuing Clinical Trials)

Clinical Trials Identifier	Cancer Type	Vaccine	Combination with	Other Therapy/ Adjuvants	Trial Phase
NCT02754362	Glioblastoma	Peptide vaccine clinical, EphA2, HER2/neu peptide)	Bevacizumab	–	Phase II
NCT02432846	mRCC	Intuvax (allogenic cell-based therapy)	Sunitinib	Poly-ICLC+/– KLH	Phase II
NCT02616185	Prostate cancer	PF-06755990 (vaccine)	Sunitinib	–	Phase I
NCT02287528	Glioblastoma	NeoVax (neoantigen peptide)	Pembrolizumab	Radiation therapy	Phase I
NCT03029403	Advanced ovarian cancer	DPX-Survivac (encapsulated peptide)	Pembrolizumab	Cyclophosphamide	Phase II
NCT02451982	Pancreatic cancer	GVAX (GM-CSF-secreting tumor cells)	Nivolumab	–	Phase I/II
NCT02506114	Pancreatic cancer	PROSTVAC (poxviral vector expressing PSA)	Ipilimumab	–	Phase II
NCT03199040	Triple-negative breast cancer	Neoantigen DNA vaccine	Durvalumab	–	Phase I
NCT03206047	Recurrent ovarian cancer	CDX-1401 vaccine	Atezolizumab	Guadecitabine	Phase II

applied in clinical trials. Major barriers to the generation of tumor antigens include insufficient immunogenicity and tumor heterogeneity. Tumor cell lysates, RNA, peptides, viral and microbial vectors, as well as tumor cells, can serve as sources of tumor antigens. There may not be enough information from clinical trials to determine which antigen or collection of antigens is the better option. No single antigen or set of antigens will be used for every type of cancer instance. The efficacy of vaccines containing the same antigen varies depending on the malignancy being treated, as is evident from the limited clinical trial data available. Additionally, CV adjuvants have encountered numerous challenges. The FDA has only approved a small number of adjuvants, and there are restrictions on CVs. Furthermore, most FDA-approved adjuvants have uncertain mechanisms, which further complicates the development of CVs.

Therapeutic CVs are playing an increasingly prominent role in cancer immunotherapy, as evidenced by encouraging findings in both basic science and clinical studies. However, there is a pressing need for improved vaccine technologies, which will likely involve combining them with checkpoint inhibitors or other techniques to counteract immune suppression by the tumor. Despite numerous vaccines demonstrating efficacy in generating a tumor-specific immune response, they have not been successful in eliciting a clinical response. It is crucial to consider various constraints at each step, including tumor antigens, vaccination adjuvants, delivery methods, and combination strategies, as these unresolved issues limit the effectiveness of therapeutic vaccines.

CVs hold great promise as a potential cornerstone of long-term cancer therapy regimens. They offer a platform that can be easily integrated with existing medicines, exhibit low toxicity, and possess desirable safety characteristics observed in vaccines that have been extensively researched thus far.

KEYWORDS

- **antiangiogenic treatments**
- **cancer**
- **cancer vaccines**
- **chemotherapy**
- **combination strategies**
- **dendritic cells**
- **immunological checkpoint inhibitor**
- **metastatic renal cell cancer**
- **radiotherapy**
- **tumor antigens**
- **vaccine platforms**

REFERENCES

1. American Cancer Society, (2020). *The Global Cancer Burden Why Global Cancer Rates are Rising [Internet]*. Available from: https://www.cancer.org/about-us/our-global-health-work/global-cancer-burden.html (accessed on 11 March 2024).
2. National Institute of Health, (2021). *Common Cancer Types [Internet]*. National Cancer Institute. Available from: https://www.cancer.gov/types/common-cancers (accessed on 11 March 2024).
3. Schlom, J., Vergati, M., Intrivici, C., Huen, N. Y., & Tsang, K. Y., (2010). Strategies for cancer vaccine development. In: *Journal of Biomedicine and Biotechnology*, 1–13.

4. Sung, H., Ferlay, J., Siegel, R. L., Laversanne, M., Soerjomataram, I., Jemal, A., et al., (2021). Global cancer statistics 2020: GLOBOCAN estimates of incidence and mortality worldwide for 36 cancers in 185 countries. *CA Cancer J. Clin., 71*(3), 209–249.
5. Liu, J., Fu, M., Wang, M., Wan, D., Wei, Y., & Wei, X., (2022). Cancer vaccines as promising immuno-therapeutics: Platforms and current progress. *J. Hematol. Oncol., 15*(1).
6. Gardner, T. A., Elzey, B. D., & Hahn, N. M., (2012). Sipuleucel-T (provenge) autologous vaccine approved for treatment of men with asymptomatic or minimally symptomatic castrate-resistant metastatic prostate cancer. *Hum. Vaccines Immunother., 8*(4), 534–539.
7. Saxena, M., Van, D. B. S. H., & Melief, C. J. M., (2021). BNT cancer vaccines. NRC 2021. therapeutic cancer vaccines. *Nat. Rev. Cancer., 21*, 360–378.
8. Qin, H., Sheng, J., Zhang, D., Zhang, X., Liu, L., Li, B., et al., (2018). New strategies for therapeutic cancer vaccines. *Anticancer Agents Med. Chem., 19*(2), 213–221.
9. Van, D. B. P., Traversari, C., Chomez, P., Lurquin, C., De Plaen, E., Van, D. E. B., Knuth, A., & Boon, T., (1991). A Gene encoding an antigen recognized by cytolytic T lymphocytes on a human melanoma. *Science (80-), 254*, 1643–1647.
10. Morse, M. A., Gwin, W. R., & Mitchell, D. A., (2021). Vaccine therapies for cancer: Then and now. *Target Oncol., 16*(2), 121–152.
11. Tiptiri-Kourpeti, A., Spyridopoulou, K., Pappa, A., & Chlichlia, K., (2016). DNA vaccines to attack cancer: Strategies for improving immunogenicity and efficacy. *Pharmacol. Ther., 165*, 32–49.
12. Jou, J., Harrington, K. J., Zocca, M. B., Ehrnrooth, E., & Cohen, E. E. W., (2021). The changing landscape of therapeutic cancer vaccines-novel platforms and neoantigen identification. *Clin. Cancer Res., 27*(3), 689–703.
13. Hollingsworth, R. E., & Jansen, K., (2019). Turning the corner on therapeutic cancer vaccines. *NPJ Vaccines, 4*(1).
14. Miao, L., Zhang, Y., & Huang, L., (2021). mRNA vaccine for cancer immunotherapy. *Mol. Cancer, 20*(1).
15. Buonaguro, L., Petrizzo, A., Tornesello, M. L., & Buonaguro, F. M., (2011). Translating tumor antigens into cancer vaccines. *Clin. Vaccine Immunol., 18*(1), 23–34.
16. Maeng, H. M., & Berzofsky, J. A., (2019). Strategies for developing and optimizing cancer vaccines. *F1000Research, 8*, 654.
17. Melero, I., Gaudernack, G., Gerritsen, W., Huber, C., Parmiani, G., Scholl, S., et al., (2014). Therapeutic vaccines for cancer: An overview of clinical trials. *Nat. Rev. Clin. Oncol., 11*(9), 509–524.
18. Olweus, J., & Lund-Johansen, F., (2019). Finding neo (antigens, that is). *Blood, 134*(2), 108–109.
19. Yarchoan, M., Johnson, B. A., Lutz, E. R., Laheru, D. A., & Jaffee, E. M., (2017). Targeting neoantigens to augment anti-tumor immunity. *Nat. Rev. Cancer, 17*(4), 209–222.
20. Kreiter, S., Vormehr, M., Van, De. R. N., Diken, M., Löwer, M., Diekmann, J., et al., (2015). Mutant MHC class II epitopes drive therapeutic immune responses to cancer. *Nature, 520*(7549), 692–696.
21. Martin, S. D., Brown, S. D., Wick, D. A., Nielsen, J. S., Kroeger, D. R., Twumasi-Boateng, K., et al., (2016). Low mutation burden in ovarian cancer may limit the utility of neoantigen-targeted vaccines. *PLoS One, 11*(5), 5.

22. Sahin, U., Derhovanessian, E., Miller, M., Kloke, B. P., Simon, P., Löwer, M., et al., (2017). Personalized RNA mutanome vaccines mobilize poly-specific therapeutic immunity against cancer. *Nature, 547*(7662), 222–226.
23. Hilf, N., Kuttruff-Coqui, S., Frenzel, K., Bukur, V., Stevanović, S., Gouttefangeas, C., et al., (2018). Actively personalized vaccination trial for newly diagnosed glioblastoma. *Nature, 565*, 240–245.
24. Gillison, M. L., Awad, M. M., Twardowski, P., Sukari, A., Johnson, M. L., Stein, M. N., et al., (2021). Long term results from a phase 1 trial of GEN-009, a personalized neoantigen vaccine, combined with PD-1 inhibition in advanced solid tumors. *J. Clin. Oncol., 39*(15_suppl), 2613–2613.
25. Santos, P., & Butterfield, L., (2018). Dendritic cell-based cancer vaccines. *J. Immunol., 200*, 443–449.
26. Hardin, M. O., Vreeland, T. J., Clifton, G. T., Hale, D. F., Herbert, G. S., Greene, J. M., et al., (2018). Tumor lysate particle loaded dendritic cell vaccine: Preclinical testing of a novel personalized cancer vaccine. *Immunotherapy, 10*(5), 373–382.
27. Kantoff, P., Higano, C., Shore, N., Berger, E., Small, E., Penson, D., et al., (2010). Sipuleucel-T immunotherapy for castration-resistant prostate cancer. In: *N Engl. J. Med.* (Vol. 363. pp. 411–422).
28. Wang-Bishop, L., Wehbe, M., Shae, D., James, J., Hacker, B. C., Garland, K., et al., (2020). Potent STING activation stimulates immunogenic cell death to enhance antitumor immunity in neuroblastoma. *J. Immunother. Cancer, 8*(1).
29. Turubanova, V. D., Balalaeva, I. V., Mishchenko, T. A., Catanzaro, E., Alzeibak, R., Peskova, N. N., et al., (2019). Immunogenic cell death induced by a new photodynamic therapy based on photosens and photodithazine. *J. Immunother. Cancer, 7*(1).
30. Efimova, I., Catanzaro, E., Van, D. M. L., Turubanova, V. D., Hammad, H., Mishchenko, T. A., et al., (2020). Vaccination with early ferroptotic cancer cells induces efficient antitumor immunity. *J. Immunother. Cancer, 8*(2).
31. Gu, Y. Z., Fan, C. W., Lu, R., Shao, B., Sang, Y. X., Huang, Q. R., et al., (2016). Forced co-expression of IL-21 and IL-7 in whole-cell cancer vaccines promotes antitumor immunity. *Sci. Rep., 6*.
32. Fan, Y., Kuai, R., Xu, Y., Ochyl, L. J., Irvine, D. J., & Moon, J. J., (2017). Immunogenic cell death amplified by co-localized adjuvant delivery for cancer immunotherapy. *Nano Lett., 17*(12), 7387–7393.
33. Srinivasan, P., Wu, X., Basu, M., Rossi, C., & Sandler, A. D., (2018). PD-L1 checkpoint inhibition and anti-CTLA-4 whole tumor cell vaccination counter adaptive immune resistance: A mouse neuroblastoma model that mimics human disease. *PLoS Med., 15*(1).
34. Wculek, S. K., Cueto, F. J., Mujal, A. M., Melero, I., Krummel, M. F., & Sancho, D., (2020). Dendritic cells in cancer immunology and immunotherapy. *Nat. Rev. Immunol., 20*(1), 7–24.
35. Bol, K. F., Schreibelt, G., Rabold, K., Wculek, S. K., Schwarze, J. K., Dzionek, A., et al., (2019). The clinical application of cancer immunotherapy based on naturally circulating dendritic cells. *J. Immunother. Cancer, 7*(1).
36. Goto, T., (2019). Radiation as an in-situ auto-vaccination: Current perspectives and challenges. *Vaccines, 7*(3), 100.
37. Donninger, H., Li, C., Eaton, J. W., & Yaddanapudi, K., (2021). Cancer vaccines: Promising therapeutics or an unattainable dream. *Vaccines, 9*(6), 668.

38. Chudasama, R., Phung, Q., Hsu, A., & Almhanna, K., (2021). Vaccines in gastrointestinal malignancies: From prevention to treatment. *Vaccines, 9*(6), 647.
39. Ahmed, E. H., Brooks, E., Sloan, S., Schlotter, S., Jeney, F., Hale, C., Mao, C., et al., (2021). Targeted delivery of BZLF1 To DEC205 drives EBV-protective immunity in a spontaneous model of EBV-driven lymphoproliferative disease. *Vaccines, 9*, 555.
40. Tashiro, H., & Brenner, M. K., (2017). Immunotherapy against cancer-related viruses. *Cell Res., 27*(1), 59–73.
41. Kaufman, H. L., Kohlhapp, F. J., & Zloza, A. (2015). Oncolytic viruses: a new class of immunotherapy drugs [published correction appears in Nat Rev Drug Discov. 2016 Aug 30;15(9):660]. *Nat Rev Drug Discov.14*(9), 642–662. doi:10.1038/nrd4663.
42. Raja, J., Ludwig, J. M., Gettinger, S. N., Schalper, K. A., & Kim, H. S. (2018). Oncolytic virus immunotherapy: future prospects for oncology. *J Immunother Cancer.6*(1), 140. Published on 4 December 2018. doi:10.1186/s40425-018-0458-z.
43. Conry, R. M., Westbrook, B., McKee, S., & Norwood, T. G., (2018). Talimogene laherparepvec: First in class oncolytic virotherapy. *Hum. Vaccines Immunother., 14*(4), 839–846.
44. Chang, J., (2021). Adenovirus vectors: Excellent tools for vaccine development. *Immune. Netw., 21*(1), 1–11.
45. Tosch, C., Bastien, B., Barraud, L., Grellier, B., Nourtier, V., Gantzer, M., et al., (2017). Viral based vaccine TG4010 induces broadening of specific immune response and improves outcome in advanced NSCLC. *J. Immunother. Cancer, 5*(1).
46. Stenzl, A., (2020). Re: Phase III trial of PROSTVAC in asymptomatic or minimally symptomatic metastatic castration-resistant prostate cancer. *Eur. Urol., 77*(1), 131–132.
47. Graham, D. B., Luo, C., O'Connell, D. J., Lefkovith, A., Brown, E. M., Yassour, M., et al., (2018). Antigen discovery and specification of immunodominance hierarchies for MHCII-restricted epitopes. *Nat. Med., 24*(11), 1762–1772.
48. Tay, R. E., Richardson, E. K., & Toh, H. C., (2021). Revisiting the role of CD4+ T cells in cancer immunotherapy—New insights into old paradigms. *Cancer Gene Ther., 28*(1, 2), 5–17.
49. Jhunjhunwala, S., Hammer, C., & Delamarre, L., (2021). Antigen presentation in cancer: Insights into tumor immunogenicity and immune evasion. *Nat. Rev. Cancer, 21*(5), 298–312.
50. Pollet, J., Chen, W. H., & Strych, U., (2021). Recombinant protein vaccines, a proven approach against coronavirus pandemics. *Adv. Drug Deliv. Rev., 170*(170), 71–82.
51. Steinhagen, F., Kinjo, T., Bode, C., & Klinman, D. M., (2011). TLR-based immune adjuvants. *Vaccine, 29*(17), 3341–3355.
52. Wiedermann, U., Garner-Spitzer, E., Chao, Y., Maglakelidze, M., Bulat, I., Dechaphunkul, A., et al., (2021). Clinical and immunologic responses to a B-Cell epitope vaccine in patients with HER2/neu-overexpressing advanced gastric cancer—Results from phase Ib trial IMU.ACS.001. *Clin. Cancer Res., 27*(13), 3649–3660.
53. Mittendorf, E. A., Lu, B., Melisko, M., Hiller, J. P., Bondarenko, I., Brunt, A. M., et al., (2019). Efficacy and safety analysis of nelipepimut-S vaccine to prevent breast cancer recurrence: A randomized, multicenter, phase III clinical trial. *Clin. Cancer Res., 25*(14), 4248–4254.
54. Bowen, W. S., Svrivastava, A. K., Batra, L., Barsoumian, H., & Shirwan, H., (2018). Current challenges for cancer vaccine adjuvant development. *Expert Rev. Vaccines, 17*(3), 207–215.

55. Kumai, T., Lee, S., Cho, H. I., Sultan, H., Kobayashi, H., Harabuchi, Y., et al., (2017). Optimization of peptide vaccines to induce robust antitumor CD4 T-cell responses. *Cancer Immunol. Res., 5*(1), 72–83.
56. Khong, H., & Overwijk, W., (2016). Adjuvants for peptide-based cancer vaccines. *J. Immunother Cancer., 4*, 1.
57. Milla, P., Dosio, F., & Cattel, L., (2011). PEGylation of proteins and liposomes: A powerful and flexible strategy to improve the drug delivery. *Curr. Drug Metab., 13*(1), 105–119.
58. Petrina, M., Martin, J., & Basta, S., (2021). Granulocyte macrophage colony-stimulating factor has come of age: From a vaccine adjuvant to antiviral immunotherapy. *Cytokine Growth Factor Rev., 59*, 101–110.
59. Rezaei, T., Davoudian, E., Khalili, S., Amini, M., Hejazi, M., De La Guardia, M., et al., (2021). Strategies in DNA vaccine for melanoma cancer. *Pigment Cell Melanoma Res., 34*(5), 869–891.
60. Jorritsma, S. H. T., Gowans, E. J., Grubor-Bauk, B., & Wijesundara, D. K., (2016). Delivery methods to increase cellular uptake and immunogenicity of DNA vaccines. *Vaccine, 34*(46), 5488–5494.
61. Trimble, C. L., Morrow, M. P., Kraynyak, K. A., Shen, X., Dallas, M., Yan, J., et al., (2015). Safety, efficacy, and immunogenicity of VGX-3100, a therapeutic synthetic DNA vaccine targeting human papillomavirus 16 and 18 E6 and E7 proteins for cervical intraepithelial neoplasia 2/3: A randomized, double-blind, placebo-controlled phase 2b trial. *Lancet, 386*(10008), 2078–2088.
62. Papachristofilou, A., Hipp, M. M., Klinkhardt, U., Früh, M., Sebastian, M., Weiss, C., et al., (2019). Phase Ib evaluation of a self-adjuvanted protamine formulated mRNA-based active cancer immunotherapy, BI1361849 (CV9202), combined with local radiation treatment in patients with stage IV non-small cell lung cancer. *J. Immunother. Cancer, 7*(1), 1.
63. Ramishetti, S., Hazan-Halevy, I., Palakuri, R., Chatterjee, S., Naidu, G. S., Dammes, N., et al., (2020). A combinatorial library of lipid nanoparticles for RNA Delivery to leukocytes. *Adv. Mater., 32*(12), 12.
64. Sahin, U., Oehm, P., Derhovanessian, E., Jabulowsky, R. A., Vormehr, M., Gold, M., et al., (2020). An RNA vaccine drives immunity in checkpoint-inhibitor-treated melanoma. *Nature, 585*(7823), 107–112.
65. Schaaf, M. B., Garg, A. D., & Agostinis, P., (2018). Defining the role of the tumor vasculature in antitumor immunity and immunotherapy article. *Cell Death Dis., 9*(2), 2.
66. Van, G. M., Dammeijer, F., Aerts, J. G. J. V., & Vroman, H., (2018). Combination strategies to optimize efficacy of dendritic cell-based immunotherapy. *Front Immunol.*, 9.
67. Mougel, A., Terme, M., & Tanchot, C., (2019). Therapeutic cancer vaccine and combinations with antiangiogenic therapies and immune checkpoint blockade. *Front Immunol., 10*.
68. Lapeyre-Prost, A., Terme, M., Pernot, S., Pointet, A. L., Voron, T., Tartour, E., et al., (2017). Immunomodulatory activity of VEGF in cancer. *Int. Rev. Cell Mol. Biol., 330*, 295–342.
69. Van, B. J. R., Nowak-Sliwinska, P., Huijbers, E. J. M., Thijssen, V. L., & Griffioen, A. W., (2015). The great escape; the hallmarks of resistance to antiangiogenic therapy. *Pharmacol. Rev., 67*(2), 441–461.

70. Amin, A., Dudek, A. Z., Logan, T. F., Lance, R. S., Holzbeierlein, J. M., Knox, J. J., Master, V. A., et al., (2015). Survival with AGS-003, an autologous dendritic cell–based immunotherapy, in combination with sunitinib in unfavorable risk patients with advanced renal cell carcinoma (RCC): Phase 2 study results. *J. Immunother. Cancer, 3*(1), 1–3.
71. Nagini, S., (2017). Breast cancer: Current molecular therapeutic targets and new players. In: *Anti-Cancer Agents in Medicinal Chemistry, 17*(2), 152–163.
72. Hipp, M. M., Hilf, N., Walter, S., Werth, D., Brauer, K. M., Radsak, M. P., et al., (2008). Sorafenib, but not sunitinib, affects function of dendritic cells and induction of primary immune responses. *Blood, 111*(12), 5610–5620.
73. Zhou, Z., Pang, J., Wu, X., Wu, W., Chen, X., & Kong, M., (2020). Reverse immune suppressive microenvironment in tumor draining lymph nodes to enhance anti-PD1 immunotherapy via nanovaccine complexed microneedle. *Nano Res., 13*(6), 1509–1518.
74. Golay, J., & Andrea, A. E., (2020). Combined anti-cancer strategies based on anti-checkpoint inhibitor antibodies. *Antibodies, 9*(2), 17.
75. Le, D. T., Lutz, E., Uram, J. N., Sugar, E. A., Onners, B., Solt, S., et al., (2013). Evaluation of ipilimumab in combination with allogeneic pancreatic tumor cells transfected with a GM-CSF gene in previously treated pancreatic cancer. *J. Immunother., 36*(7), 382–389.
76. Liu, L., Wang, Y., Miao, L., Liu, Q., Musetti, S., Li, J., et al., (2018). Combination immunotherapy of MUC1 mRNA nano-vaccine and CTLA-4 blockade effectively inhibits growth of triple negative breast cancer. *Mol. Ther., 26*(1), 45–55.
77. Xue, S., Hu, M., Iyer, V., & Yu, J., (2017). Blocking the PD-1/PD-L1 pathway in glioma: A potential new treatment strategy. *J. Hematol. Oncol., 10*(1), 1.
78. Massarelli, E., William, W., Johnson, F., Kies, M., Ferrarotto, R., Guo, M., et al., (2019). Combining immune checkpoint blockade and tumor-specific vaccine for patients with incurable human papillomavirus 16-related cancer: A phase 2 clinical trial. *JAMA Oncol., 5*(1), 67–73.
79. Cadena, A., Cushman, T. R., Anderson, C., Barsoumian, H. B., Welsh, J. W., & Cortez, M. A., (2018). Radiation and anti-cancer vaccines: A winning combination. *Vaccines, 6*(1), 9.
80. Patel, R. B., Ye, M., Carlson, P. M., Jaquish, A., Zangl, L., Ma, B., et al., (2019). Development of an *in-situ* cancer vaccine via combinational radiation and bacterial-membrane-coated nanoparticles. *Adv. Mater., 31*(43), 31.
81. Zhang, X., Wang, D., Li, Z., Jiao, D., Jin, L., Cong, J., et al., (2020). Low-dose gemcitabine treatment enhances immunogenicity and natural killer cell-driven tumor immunity in lung cancer. *Front Immunol.*, 11.
82. Hughes, E., Scurr, M., Campbell, E., Jones, E., Godkin, A., & Gallimore, A., (2018). T-cell modulation by cyclophosphamide for tumor therapy. *Immunology, 154*(1), 62–68.
83. He, Q., Li, J., Yin, W., Song, Z., Zhang, Z., Yi, T., et al., (2011). Low-dose paclitaxel enhances the anti-tumor efficacy of GM-CSF surface-modified whole-tumor-cell vaccine in mouse model of prostate cancer. *Cancer Immunol. Immunother., 60*(5), 715–730.

CHAPTER 3

Role of Tumor Antigens for Cancer Vaccine Development

HIMANSHU SHARMA,[1] RASHMI PATHAK,[2] NEETU SACHAN,[3] and PHOOL CHANDRA[1]

[1]*Teerthanker Mahaveer College of Pharmacy, Teerthanker Mahaveer University, Moradabad, Uttar Pradesh, India*

[2]*Department of Pharmacy, Invertis University, Bareilly, Uttar Pradesh, India*

[3]*Maharana Pratap College of Pharmacy, Mandhana, Kanpur, Uttar Pradesh, India*

ABSTRACT

The customization of cancer vaccinations has received a lot of interest lately. Nevertheless, despite the numerous vaccine techniques targeting various tumors that have been studied recently, the performance of therapeutic cancer vaccines (CVs) remains poor. Clinical trials have not yet produced data that is satisfactory, despite the fact that preclinical data typically demonstrates positive findings. The key purpose for such disappointments is the challenge of recognizing tumor-specific antigens that should be overexpressed simply by cells of the tumor in comparison to healthy cells. The majority of tumor antigens employed in malignancy immunotherapy are non-mutated and overexpressed self-antigens that primarily activate T-cells with suboptimal T-cell binding sites, rendering them unable to efficiently facilitate an anti-tumor reaction. The creation of potent therapeutic CVs that can activate both effector and memory T-cells to respond specifically to tumor antigens is one of the main objectives of CVs. The objective tumor regression following the

Cancer Vaccination and Challenges, Volume 1: Strategies for Therapeutic Cancer Vaccine Development. Rishabha Malviya, Bhupendra Prajapati, and Sonali Sundram (Eds.)

adoptive transmission of interleukin (IL) and tumor-infiltrating lymphocytes (TILs) into autologous cancer individuals suggests that T-cells are critical for tumor regression. The tumor antigens that have been discovered so far can be categorized into three groups: tumor-specific differentiation antigens, tumor-specific shared antigens, and tissue-specific unique antigens. In organizing the host immune reaction against cancer, viral illnesses, and autoimmune diseases, CD4+ T-cells are crucial. Telomerase is a prototype for a universal cancer antigen since it is expressed in the majority of tumors and plays an inherent practical role in oncogenic alteration. These desirable characteristics have prompted research into the possibility of using the cancer antigen, the catalytic subclass of telomerase, human telomerase reverse transcriptase (hTERT), as a generally applicable immunological target. Early clinical tests of different immunotherapy preparations have confirmed that it is safe to elicit hTERT-specific immune responses in individuals, which can impact clinical outcomes. Fundamental immunological tests have demonstrated that hTERT is immunogenic. The tumor antigens targeted in recent years in the development of cancer immunotherapy vaccination approaches are described in this chapter, along with novel categories of tumor antigens that may exist, such as unconventional antigens, HERVs, and heteroclitic peptides.

3.1 INTRODUCTION

Vaccines are successful in preventing bacterial and viral illnesses. Since the creation of the first vaccination over 200 years ago, they have saved the lives of hundreds of millions of people worldwide by preventing some of the worst illnesses of the 20th century. Vaccines work by exposing individuals to a weakened version of the threat posed by illnesses caused by bacteria (e.g., diphtheria, tetanus, and TB) and viruses (e.g., measles, polio, and smallpox). This enables their immune system to identify these threats based on their unique markers, or "antigens," and to mount a defense against them. When administered to a person before they become ill with the bacterium or virus, these vaccines often work most effectively in a preventive capacity. Cancer is the leading cause of death globally. However, the situation is more complex in the case of cancer for several reasons, making it more challenging to develop immunotherapies to treat or prevent cancer. Specifically, cancer cells more closely resemble our normal, healthy cells than bacteria and viruses, which our immune system recognizes as foreign. Additionally, each person's tumor is unique in some way and contains distinct antigens. Therefore, more advanced approaches are needed to develop effective CVs. Presently, more than 8 million patients die from malignancy every year, and it's predicted that the number will rise to 22 million cases over the next 20 years from the current 14 million. Colorectal, lung, prostate, and breast malignancies are the most frequent types of malignancy in the Western world, making it one of the leading reasons for death. Pancreatic, liver, and thyroid malignancies are expected to rise sharply as populations and medical practices alter [1]. Over a century has been spent researching CVs but compared to the enormous effects of vaccinations against infectious diseases, they have provided more hope than actual benefit. Given the current influence of tumors on the immune system at poorly recognizable levels, cancer immunotherapies have focused mostly on illness therapy, a significantly higher hurdle to overcome than infectious disease vaccines, which place a greater emphasis on disease prevention [2].

3.2 PREVALENCE OF CANCER

The primary cause of mortality globally has been cancer. Currently, over 8 million patients have succumbed to malignancy annually, with a projected increase to 22 million cases within the next 20 years from the existing 14 million. Colorectal, lung, prostate, and breast malignancies represent the most prevalent forms of cancer in the Western world [3]. Lung, prostate, breast, pancreatic, and colorectal cancers (CRCs) exhibit the highest fatality rates in the United States and are the most frequently diagnosed malignancies. The molecular characterization of human malignancies has revealed the complexity of individual tumors and the diversity among tumors in different individuals. Consequently, a more personalized treatment approach could enhance the outcome. Treatments are gradually shifting from being based on tissue to focusing on the mutational profile of the specific tumor. For example, trastuzumab (anti-Her2 antibody), previously used exclusively for treating breast cancer, is now being applied to treat other cancers that have an elevated Her2 cell surface.

3.3 TUMOR ANTIGENS

The development of cancer immunotherapy focuses on identifying tumor antigens that are produced in the presence of Major Histocompatibility Complex (MHC) molecules and have the potential to trigger a CTL response specific to the tumor. This response may impact the destruction of tumor cells and the recurrence of malignancy. Early cancer immunotherapies targeted a broad range of antigens to prevent the emergence of escape variants. These approaches were evaluated in numerous clinical trials involving different tumor types, such as irradiated autologous tumor cells, autologous tumor lysates, whole tumor-derived mRNA, and allogeneic tumor cell lines [4, 5]. However, these methods face challenges, including the requirement for a substantial amount of tumor tissue, complexities in collection and processing, and issues in meeting regulatory standards like Good Manufacturing Practice (GMP) guidelines. One of the key obstacles is the abundance of non-tumor self-antigens in whole tumor cells, which reduces the number of tumor-specific antigens and promotes immunological tolerance [6, 7].

To address these challenges, therapeutic CVs have been developed, often focusing on one or a few cancer antigens. Tumor antigens fall into two categories: tumor-specific antigens (TSAs) and tumor-associated antigens (TAAs). TSAs are considered more effective in CVs as they are unique to cancer cells and arise from tumor-specific genetic mutations or viral antigens [8, 9].

3.4 HISTORY OF IDENTIFYING CANCER ANTIGENS

Since the 1890s, there have been reports of the first immunotherapy attempts involving the use of bacterial extracts to promote anti-tumor immunity. According to a study, thermal-killed Serratia marcescens and Streptococcus pyogenes were administered to individuals with incurable soft tissue sarcomas. At the time of therapy, the tumor had completely regressed in more than 50% of cases, leading to over five-year survival rates. A favorable correlation between the temperature level and treatment effectiveness was also discovered. It is widely recognized that the anticancer immune response is responsible for the effectiveness of Coley's toxin. Despite the unmatched reaction rates reported by contemporary immunotherapy, other studies did not show similar effectiveness, especially in carcinoma patients who had received Coley's toxin treatment. Coley's toxin may have failed in cancer patients because its effectiveness relied on non-specific immune activation [10].

There has been a long-standing controversy over whether host immune cells could distinguish tumor cells from healthy cells. In 1953, Foley suggested that sarcomas induced by methylcholanthrene could be prevented from growing by implanting an immunization against them. The concept of "immune tolerance" was met with skepticism [11].

After the 1970s, multiple studies showed that chemical carcinogenesis produced a variety of cancer antigens for each mouse that could not be transferred into another mouse to produce immunity, even though that mouse had the same genetic background and used the same chemical carcinogenesis procedure. Since Ikeda et al. discovered that chemically produced sarcomas evoked cytotoxic T-cell responses that were substantially greater and more durable than those created naturally [12]. Their findings could indicate that using cancer immunotherapy in clinical settings is challenging. After that, "cancer antigens" identified by host T-cells in mouse models since the 1980s were based on unique acquired genetic changes, or "neo-antigens." Also recognized as "cancer antigens" are the cancer-testis antigens P1A (mouse) and MAGE (human) [13].

Except for the testis, they are stated in the variability of malignancies but not in standard mature tissues. Consequently, cancer-testis antigen developed as a popular immunotherapy target. Then, methods for recognizing cancer antigens have steadily improved, particularly gene sequencing technology, which has advanced quickly. In addition to the three peptides from tyrosinase that have already been recognized, a discovery of five novel antigens resulting from somatic mutations established a leading role for T-cells specific to neoantigen in the regulation of melanoma. Since subsequent generation sequencing has been commercially adopted for medical therapy and research, neo-antigens have now taken Centre stage in immunotherapy [14, 15].

3.5 HUMAN TUMOR ANTIGENS

3.5.1 TUMOR ANTIGENS IDENTIFICATION

Cancer patient blood or tumors have been used to isolate antitumor CTL clones [16]. Expression cloning, which involves separating the peptide-encoding gene by transducing a library of tumoral cDNA and assessing the transduced cells for their capability to stimulate the CTL clone, is a common technique used to identify the peptides recognized by these CTL. The region of the discovered gene that encodes the antigenic peptide can then be defined by transfecting fragments of it, and candidate peptides with suitable HLA-binding patterns are evaluated to determine if they can sensitize target cells to CTL lysis. Numerous antigenic peptides have been successfully identified using this method [17, 18].

To identify TAAic peptides, the "reverse immunology" technique involves selecting peptides with sufficient motifs bound to HLA within a protein, such as a protein produced by a mutated oncogene that is either preferentially overexpressed or expressed by tumors, is now widely used. The development and *in vitro* testing of potential peptides for HLA binding [19]. The best binders are loaded onto antigen cells and used to stimulate T lymphocytes *in vitro* to generate clones or CTL lines that recognize peptide-pulsed target cells. This method has the limitation that tumors may not efficiently process the identified peptides. Therefore, it is essential to confirm that tumor cells expressing the gene recognized by the CTL code for the peptide. The regular transfectants of the cells or genes in which the gene expression has been silenced by siRNA or shRNA should also be examined [20].

The immunopurification of MHC class I molecules from tumor cell surfaces and liberating antigenic peptides is the third technique for finding antigens. This technique has been effective in determining or verifying the significance of peptides that have undergone posttranslational alterations such as glycosylation-dependent asparagine deamidation, serine/threonine phosphorylation, and peptide splicing, even though the series of the eluted peptides must be directly recognized by mass spectrometry [21, 22]. Using these diverse techniques, a substantial antigenic peptide recognized by antitumor CTL has been discovered. According to the parent gene's expression pattern, these antigens can be easily categorized [23] (Figure 3.1). A consistently updated list of the antigenic peptides that tumor cells use to promote themselves [24].

3.6 CLASSES OF TUMOR ANTIBODIES THAT T-CELLS CAN RECOGNIZE

T-cells sometimes recognize "tumor antigens" that aren't necessarily made in a tumor specific manner. Antigens that are overexpressed in cancerous cells and those that are only expressed in cancerous cells and not in healthy cells are among them as depicted in Figure 3.1, and some features of tumor antigens are specified in Table 3.1.

3.6.1 HIGH-TUMOR SPECIFICITY ANTIGENS

Because of their tumor-specific expression pattern, antigens from three different categories, such as those made by point mutations, viral proteins, and cancer-germline genes, can induce T-cell reactions specific to tumors [25, 26] (Figure 3.2).

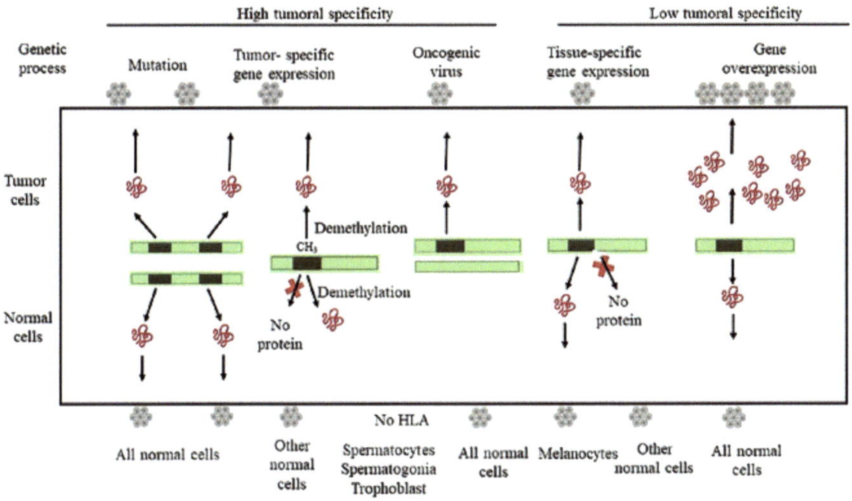

FIGURE 3.1 Major tumor antigen categories that CD8 T-cells can recognize.

TABLE 3.1 The Features of Various Tumor Antigen Classes

Characteristics	Types of Tumor Antigens				
	Differentiation Antigens	**Overexpressed Antigens**	**Cancer-Testis Antigens**	**Viral Antigens**	**Neo-Antigens**
Antigenicity	+ or +/–	+ or +/–	+/++	++	++
Affinity towards T-cell	+ or +/–	+ or +/–	+/++	++	++
Tolerability towards Immune	+/–	+/–	– or +/–	–	–

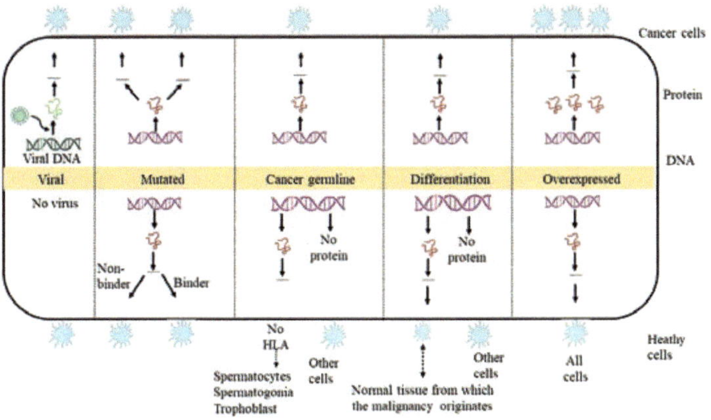

FIGURE 3.2 Tumor antigens that cytotoxic T-cells can detect. According to how the parental gene expresses, tumor antigens are categorized.

Healthy cells (bottom panel) and cancer cells (upper panel) are shown to produce antigenic peptides. Only virally infected cells express viral antigens. A changed peptide that can bind to the HLA class I molecules can result from mutated genes. Whereas the wild peptide is incapable (left). The alteration may potentially change a peptide that binds to the HLA class I molecule, causing circulating CTL to identify the changed peptide as nonself. Whole genome demethylation leads to cancer germline gene expression in malignancies. Antigens of MAGE type produced by cancer germline genes are neither expressed on the surfaces of germline nor normal healthy cells because germline cells do not contain molecules of HLA class I. Genes that are differentially represented in various tissues are responsible for encoding differentiation antigens. Some tumor types and the matching healthy tissue, therefore, express them. Due to enhanced transcription or gene amplification, some genes are overexpressed in malignancies. The resultant peptides have low levels of expression in some or all healthy organs but are abundantly expressed in certain malignancies.

3.6.1.1 VIRAL ANTIGENS

It is estimated that viruses are responsible for 20% of cancer cases, and genetic variants generated from the virus may serve as tumor antigen peptides. Some viruses are thought to be involved in carcinogenesis. The immune system recognizes the Epstein-Barr virus and human T-cell leukemia virus linked to high-risk human papillomavirus (HPV) and head and neck malignancy as cancer rejection antigens linked to oncogenesis. Among these, cervical cancer displays elevated expression of the HPV viral antigens E6 and E7, which can serve as targets for cytotoxic T-cells. To date, very few clinical trials have examined the effectiveness and viability of virus-related antigen vaccination treatment [27].

3.6.1.2 ANTIGENS PRODUCED BY MUTATED GENES

A significant number of CTL obtained from malignancy individuals' blood, or tumors have been used to identify antigens resulting from mutations in widely expressed genes. One amino acid is frequently added to the peptide sequence to allow it to link to the molecule of MHC class I or to produce a novel antigenic factor that the CTL can recognize (Figure 3.2). A frameshift brought on by the mutation can occasionally create a novel antigenic peptide [28]. Oncogenic mutations cause some mutant antigenic peptides. It has been demonstrated that a CDK4 gene mutation can alter how CDK4 binds to its p16/INK4a inhibitor, encouraging the unrestrained division of the cell [29]. A change in the CTNNB1 gene stabilizes beta-catenin, enhances its interaction with the Lef-Tcf transcription factor, and generates an antigenic peptide. Genes implicated in the development of melanoma may become persistently transactivated as a result. It has been demonstrated that a CTL-detected mutation in the gene CASP8 reduces the susceptibility of cells to apoptosis. However, the antigenic peptides that go along with these changes are typically passenger mutations specific to the malignancies in which they were found. Microsatellite instability (MSI)+ colorectal carcinoma, melanoma, lung carcinoma, and other tumors with a high mutation rate are anticipated to include more mutant antigens and be more immunogenic. Notably, the antitumor CTL response in some patients is primarily focused on mutant epitopes [30, 31].

Surprisingly, only a small amount of CD8 T-cells were isolated that could identify peptides originating from genes like P53, KRAS, or NRAS that are frequently mutated in cancer. Probably, the associated peptides are not extremely immunogenic because the utmost of these CTL was discovered utilizing reverse immunology rather than tumor culture mixed with lymphocyte. Peptides from chromosomal translocations like ETV6-AML1 and BCR-ABL were also discovered [32].

3.6.1.3 CANCER GERMLINE GENES

The cancer germline genes represent a significant additional source of tumor-specific antigens (TSA) as illustrated in Figure 3.2. Among these genes are the MAGE genes, which consist of functional genes [25] located in three regions of the X chromosome known as MAGEA, MAGEC, and MAGEB. Other cancer-germline gene groups found on the X chromosome include BAGE, SSX, LAGE/NY-ESO1, and GAGE [33, 34].

Cancer germline genes are present across a wide range of cancer types, with the exception of trophoblastic and germline cells. These genes exhibit tumor-specific expression patterns due to demethylation of their promoter regions, a process observed genome-wide in progressive malignancies and male germ cells. Trophoblastic and male germline cells lack MHC class I molecules, preventing the presentation of antigens to T-cells. The MAGE type antigen, derived from antigenic peptides produced by cancer germline genes, appears to be specific to tumors, making them potential targets for immunotherapy without adverse effects on patients. However, it is important to note that recent findings indicate low expression of MAGE-A12 in brain cells.

Certain aberrant transcripts, either expressed at low levels or silent in healthy tissues but upregulated in tumors, have been identified to generate antigenic peptides. These transcripts are distinct from cancer-associated genes. Additionally, peptides derived from cyclin-A1, a protein expressed in the testis and acute myeloid leukemia with anti-apoptotic and pro-proliferative properties, have been discovered [35].

3.6.2 LOW TUMOR SPECIFICITY ANTIGENS

This class of antigens includes antigens for differentiation and ones produced from genes that are overexpressed in malignancies [36] (Figure 3.2).

3.6.2.1 TISSUE DIFFERENTIATION ANTIGENS

Differentiation antigens are exclusively expressed in tumor cells and cells generated from normal tissues due to the special trait of tissue differentiation. One research on these antigens showed that "tumor antigens" enable cytotoxic T lymphocytes (CTLs) in melanoma individuals to differentiate between normal melanocytes and cells of tumor [37]. Gp100, MART-1/Melan A, tyrosinase, and other tissue differentiation antigens were expressed by tumor cells and melanocytes in melanoma patients. They were enzymes that were exclusively expressed in melanosomes of healthy melanocytes and overexpressed in cells of malignant melanoma to allow for the differentiation of expression levels [37]. In an integrated review of the clinical trials, immunization treatment with melanocyte differentiation antigens such as Melan-A/MART-1, gp100 and tyrosinase exhibited a 2.5% reaction rate in individuals with metastatic melanoma [38]. In addition, side effects from the use of melanocyte differentiation antigens that were connected to systemic and ocular autoimmunity were also documented. As an immunotherapy target, tissue differentiation antigens have thus lost some of their relevance [37].

3.6.2.2 OVEREXPRESSED ANTIGENS

The detection of tumor antigens, which only need a few HLA-peptide complexes to be detectable on the cell surface, should be the first step in the antitumor immune reaction. The threshold should help T lymphocytes recognize peptides produced from proteins that are overexpressed in cells of tumor compared to healthy cells, resulting in tumor specificity. These tumor antigens are referred to as "overexpressed antigens" because they are abundantly expressed in cancerous cells and primarily contribute to the development and existence of tumor cells. For instance, some cancer types have overexpressed CEA, WT-1, and Her-2/neu. An important issue is that overexpressed antigens lack adequate antigenicity and specificity for clinical application. To be particularly identified by T-cells, there has to be a significant differential in expression levels between healthy tissue and malignant tissues. Many epithelial cancers, such as ovarian and breast malignancy, have overexpressed Her-2/neu, the oncogene that codes for the growth factor receptor. Several clinical studies looking at overexpressed antigens have shown minimal effectiveness and some negative consequences [13].

3.6.3 CANCER-TESTIS ANTIGENS

Only normal testicles, a normal placenta, and cancerous tissue express cancer-testis antigens, which are shared by individuals with all cancer types. HLA class I is not expressed in gonadal tissue, which prevents it from being identified as a target by cytotoxic T-cells; therefore, it cannot trigger a tumor-specific immune response. In the 1990s, MAGEA1, the initial cancer-testis antigen, was discovered. Numerous cancer-testis antigens, such as MAGE, XAGE1, and NY-ESO-1, have been discovered and may be used as potential immunotherapeutic targets due to their high tumor antigenicity [39]. NY-ESO-1 and MAGE, two testis-cancer antigens that have been the focus of clinical trials, were found to be only marginally effective as CV therapeutics. It is important to note that few testis-cancer antigens are expressed in normal or healthy tissues during the embryonic period, some adult tissues, and also, certain molecules of the homologous family are present in healthy tissue, although most of them have higher antigenicity and may be considered promising targets for CVs [40].

3.6.4 NEO-ANTIGENS

Somatic genes accumulate during the multi-step process of carcinogenesis, involving significant mutations of genes associated with cell proliferation, development, and survival (referred to as "driver gene mutations"), as well as numerous concurrent mutations caused by genomic instability related to melanoma risk factors such as UV rays and tobacco use (known as "passenger mutations") [32]. Neo-antigens are non-self-proteins unique to each individual, arising from cancer-specific gene changes and not present in healthy tissues. T-cells can mount a robust immune response against neoantigens due to the lack of immunological tolerance developed in the thymus during their maturation. This response is specific to tumors and does not occur in healthy cells, making novel antigens promising targets for cancer immunotherapy in various cancers [38].

3.6.5 ONCOFETAL ANTIGENS

Typical fetal antigens that are switched off throughout maturation but turn back on in modified cells include fetoprotein and oncofetal antigens. *In vitro*, in mouse models, and clinical studies, these antigens have shown immunogenicity and some utility as targets [41]. The National Cancer Institute held a workshop in 2008 to rate and categorize the antigens discovered at the time, publishing the antigens with the greatest priority and the most experimental evidence of tumor rejection. The majority of the antigens were non-mutated antigens, conventional, and they were grouped following previously established criteria, such as their tumor specificity, immunogenicity, oncogenicity, the incidence of overexpression, and expression in malignancy stem cells. The aforementioned categories of tumor-specific antigens are not tumor-specific [42].

The P53 tumor suppressor gene and RAS oncogene both have widespread mutations. It was thought that these mutations were unlikely to result in an immunogenic, processed, and presented MHC-restricted epitope in adequately mutual MHC class I molecules or MHC class II molecules and that it would be problematic to estimate and pretty much difficult to identify by immunotherapies in a clinical study. Highly beneficial T-cells that aim for these person-specific alterations have started to be recognized. However, sequencing technology and algorithms that can predict how peptide epitopes will be translated, presented, and bound have improved. Targetable, patient-specific mutations are difficult to find, which delays therapy. However, many researchers are concentrating on strategies to find such tumor-specific targets more rapidly [42].

3.6.6 EPITOPE SPREADING

Limiting autoimmunity and establishing a very precise memory and effector cell reaction are results of targeting particular tumor antigens. Such a technique might have the opposite impact biologically. This could increase the number of tumor cells resistant to vaccination therapy because they lack the targeted tumor antigen. However, the ability of resistant tumors to express additional antigens and subsequently increase the immune reaction in contrast to the tumor, i.e., create an immune reaction for the antigens not present in the vaccination, may be a benefit. Epitope spreading is the term for the phenomenon. Epitope spreading, also recognized as determinant spreading, is the process by which an immune reaction is transferred from one antigen to another produced in the same tissue. Research on autoimmune illnesses and studies on cancer immunotherapy have both observed this spontaneous phenomenon [43, 44].

A significant event for tumor rejection can be the phenomenon of tumor antigens being liberated from tumor cells as a result of T-cell attack and breaking them to help succeeding waves of T-cells directed against various antigens [45, 46]. To offer episodes of immunization against unidentified, patient-specific mutant antigens that may cause tumor rejection, immunotherapies that aim for common antigens can be useful. Many clinical trials have depicted no connection between the clinical outcome and immune reactions to the vaccine-targeted antigen. This might be connected to the distribution of an antigen and the emergence of immune responses against unmeasured tumor antigens. TAA-containing vaccines' key role may be to transfer epitopes to TSA, which in turn stimulate T lymphocytes more fervently and effectively to induce tumor rejection [47, 48].

3.6.7 GLYCOLIPIDS AND GLYCOPROTEINS AS ANTIGENS

On the membrane of malignant cells are carbohydrates that have been altered by glycosyltransferases and glycosidases to differ from the typical molecules seen on healthy cells. Tumor cell carbohydrate structure mainly changes quantitatively rather than qualitatively. This phrase indicates that, in comparison to normal cells, tumor cells show less carbohydrate structure. Despite this, tumor cells' cell surface carbohydrates may have a variety of structures that allow immunological methods to identify them as TAAs. Antigens containing carbohydrates may be crucial in identifying cells throughout development and differentiation. Two main methods have been used to determine the carbohydrate antigens linked to tumors (TACAs). The first approach involves systematically analyzing carbohydrates' chemical makeup and structure; if such a structure is discovered, it leads to the formation of antibodies (Abs) against an aberrant component. Another approach uses monoclonal Abs that specifically target tumor-specific carbohydrate antigens. Several antigens of glycolipids have been discovered in a diversity of human malignancies using this method [49, 50].

3.7 BIOMARKERS UTILIZING TUMOR ANTIGENS

In addition to serving as potential therapeutic targets, tumor-specific antigens can also be used for the identification and classification of malignancies. Tumor antigens are often detected in the blood as they are commonly secreted into the bloodstream by the tumor. Moreover, following the destruction of tumor cells by chemotherapy or radiation, a higher quantity of tumor antigens, which are typically located within intracellular structures or organelles, are released. Consequently, they can aid in evaluating the response to treatment and the risk of relapse after therapy [51]. Protein-based biomarkers, such as tumor antigens, can now be easily obtained due to recent technological advancements. For instance, tissue differentiation antigen and prostate-specific antigen (PSA) are frequently employed as prognostic and diagnostic markers in prostate cancer (PC). Apart from PSA, other tumor antigens like survivin for breast cancer, MAGE for hepatocellular carcinoma (HCC), and TP53 and NY-ESO-1 for ovarian cancer can also serve as diagnostic markers. However, except for PSA, these markers are not commonly used and are not widely available in clinical practice due to their limited specificity and sensitivity [51].

3.8 PROCESSING OF HUMAN TUMOR ANTIGENS

When the proteasome breaks down cellular proteins, peptides are created that can be detected by CTL. There are four different types of proteasomes, including the regular proteasome and the immunoproteasome. The TAP transporter then moves the peptides remaining from proteasome breakdown into the ER lumen. The peptide-loaded complex, consisting of TAP, the chaperone calreticulin, the oxidoreductase ERp57, and tapasin (Tpn), can further trim peptides with an extended N-terminus before loading them onto HLA class I molecules (CRT). The secretory route then transports peptide/HLA complexes to the cell membrane.

3.8.1 THE MHC CLASS I PRESENTATION PATHWAY

The proteasome, a large cytosolic complex that degrades cellular proteins to yield antigenic peptides identified by CTL, is primarily responsible for this (Figure 3.3). The transporter related to antigen processing (TAP) carries liberated peptides from proteasomal degradation into the endoplasmic reticulum lumen (ER) (Figure 3.3). Once inside the aminopeptidases, ER like ERAP1 and ERAP2 can further trim peptides. The complex, which is made up of TAP, Tpn, ERp57, and calreticulin, helps peptides with the right sequence and size link to MHC class I molecules. The ER will be left behind as stable MHC/peptide class I complexes move to the cell membrane (Figure 3.3). Most of the MHC/peptide complexes shown on the membrane of normal cells are not identified by T-cells as a consequence of tolerance. In these situations, though, tumor cells or cells that a virus has attacked have a novel repertoire of peptides that are formed from viral-associated proteins and for which an immunological reaction can be developed. The immune system can recognize minute modifications in cellular protein content because CTL can identify cells with as few as 10 MHC/peptide complexes [52, 53].

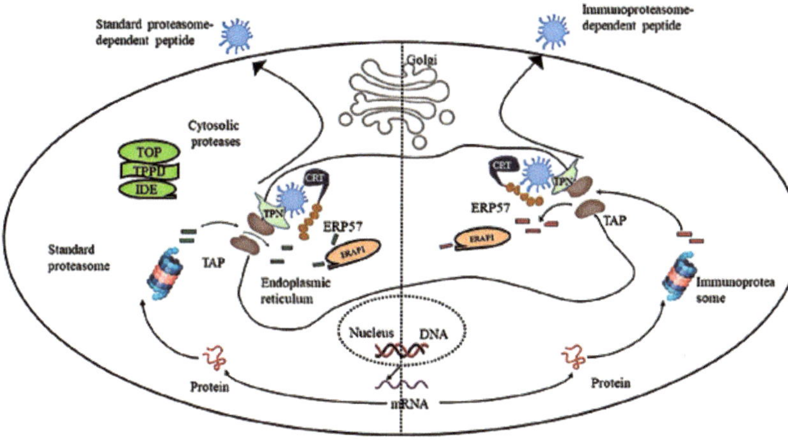

FIGURE 3.3 Processing by T-lymphocytes $CD8^+$ of tumor antigens.

3.8.2 THE PROTEASOME

The proteasome 20S is sizable, having a structure of barrel-shaped consisting of 4 stacked heptameric rings that define an enzyme-degrading chamber. While the two inner rings of the 20S proteasome comprise subclass ($\alpha 1$–$\alpha 7$), the two outer rings are made of subclass ($\beta 1$–$\beta 7$), of which three ($\beta 1$, $\beta 2$, and $\beta 5$) are catalytic in the proteasome of vertebrates. The proteasome 26S, which is formed when the proteasome 20S joins forces with the 19S regulatory compound, is in charge of breaking down polyubiquitinated proteins [54]. One of the initial manifestations that the proteasome was necessary for the production of antigenic peptides was the finding that ubiquitination was essential for displaying the antigen ovalbumin (OVA) model [55]. Later, the use of specific cellular-permeable proteasome blockers that link up to the hydroxyl group of the catalytic subtypes of the proteasome either irreversibly or reversibly confirmed the involvement of the proteasome in the outcome of antigenic peptides and the general peptide delivery to molecules of MHC class I [17, 56]. The hydroxyl group of each catalytic subunit's N-terminal threonine site is thought to carry out the nucleophilic approach that starts the breakage of the peptide bond, according to studies of mutagenesis and a study of the crystal structures of inhibitor-bound proteasomes [57, 58]. This nucleophilic hit on the glycosidic bonds leads to developing a peptide fragment that is still linked to the proteasome via an ester connection. After being immediately broken down by molecules of water in the catalytic zone, the liberated peptide is subsequently returned to the cytosol [58].

Three important catalytic activity types connected to the proteasome feature are the peptidyl-glutamyl peptide bond hydrolyzing (PGPH), chymotrypsin-like action, and the trypsin-like action, which break down after acidic, hydrophobic, and basic residues, respectively. Examination of mutants of yeast proteasome revealed that activity like caspase was related to the subclass $\beta 1$, whereas chymotrypsin and trypsin-like actions were linked to subclass $\beta 2$ and subclass $\beta 5$, respectively [59, 60]. However, additional factors are also at play because some proteasome subunits exhibits overlap in their specificities and because the sequences close to the cleavage site also affect the intensity of the cleavage [61, 62].

Depending on how effectively the proteasome destroys peptides, molecules of MHC class I will display different types of peptides on the cell membrane. It has been discovered that the proteasome can potentially combine distant peptide fragments from the parent protein to form antigenic peptides. Peptide splicing occurs as a result of protein degradation within the proteasome's catalytic chamber. Peptides are spliced through transpeptidation, where the N-terminus of the peptide fragment inside the chamber attacks the acyl-enzyme with a nucleophilic attack [57]. This process results in the formation of a new peptide bond between the N-terminus of the nucleophilic peptide and the C-terminus of the peptide released from the acyl-enzyme intermediate. Five spliced antigenic peptides have been identified [63, 64], three of which consist of fragments arranged in reverse order compared to their sequence in the parent protein. By examining a spliced peptide made from gp100 that contained the RSYVPLAH fragment connected to a single arginine (R) at its C-terminus, we could show that the defending nucleophilic peptide needed a minimum size of three amino acids to accomplish the splicing process (RSYVPLAH R). In the spliced RSYVPLAH R peptide, the proteasome must further trim the prolonged C-terminus to create the final antigenic peptide [63]. Following a new study, the nucleophilic peptide would be accommodated in a separate space within the catalytic region from the substrate receptor site. This allows for the C-terminal region of the peptide substrate from the point of proteolysis cleavage [65]. The splicing of peptide procedure can be facilitated and accelerated by the nucleophilic peptide's accommodation inside a designated pocket [65].

3.8.3 STANDARD AND IMMUNOPROTEASOME

To generate immunoproteasomes, LMP-7(5i), MECL-1(2i), and LMP-2(1i), three additional catalytic subunits, are incorporated into proteasomes in place of subunits 1, 2, and 5 in immune cells. The action of cytokines [66] enhances the immunoproteasome's tendency to break after hydrophobic and basic residues, improving the cell's capacity to create peptides for deposition on molecules of MHC class I, which commonly require these residues at their C-terminus [67, 68]. Research on mice deficient in 2 (5i or 1i) or 3 immuno-subclass, unable to display several epitopes of MHC class I, supported this theory [69]. The evolution of cancer therapy strategies may be significantly impacted by the fact that cells producing immunoproteasomes have a distinct peptide repertoire compared to cells producing regular proteasomes [70, 71]. In the first case, a kidney cell carcinoma provided CTL with a peptide from the common protein RU1, but not autologous EBV-transformed B cells [72]. This was clarified by the observation that EBV-transformed B cells have a significant number of immunoproteasomes compared to tumor cells, which are incapable of synthesizing the antigenic peptide. Subsequently, it was discovered that several additional peptides, including the Melan-A/MART-126–35, HLA-A2-restricted peptides, tyrosinase 369–377, and gp100209-217, are digested by the regular proteasome but not the immunoproteasome [71, 73]. Long peptide precursors digested *in vitro* reveal that the failure of one type of proteasome to process antigenic peptides is often due to an internal breakage leading to peptide loss [73, 74]. Interestingly, analysis of the spliced peptides' processing by the proteasome showed that one peptide was predominantly generated by the immunoproteasome, while three spliced peptides were mainly produced by the standard proteasome. Successful peptide splicing required each proteasome type to be capable of generating the implicated peptide fragments [75, 76].

3.8.4 INTERMEDIATE PROTEASOMES

The discovery that few tissues possess some but not all proteasome immunosubunits suggested the creation of proteasomes (intermediate) made up of a fusion of conventional and immunosubunits [77]. By recognizing proteasomes in their native structure and utilizing an exceptional set of Abs targeted against proteasome catalytic subunits, it was shown that intermediate proteasomes with 1 (β5i) or 2 (β5i β1i) immuno-subunits exist. According to the principles of assembly of the proteasome, which depend on the character of the propeptide subclass, there are no other forms of intermediate proteasomes [78].

3.8.5 ADDITIONAL PROTEASES THAT GENERATE ANTIGENIC PEPTIDES

Although the majority of antigen peptides are believed to be produced by the proteasome, some are also degraded by cytosolic proteases such as insulin-degrading enzymes or TPPII [79]. Recent studies have shown that the proteasome is required for cleaving the N-terminus of a PRAME peptide, while thimet oligopeptidase and nardilysin are needed for cleaving the C-terminus in sequence. It has been suggested that the proteasome-assisted modification of the carboxyl-terminus of MHC class I peptides is facilitated by angiotensin-converting enzyme (ACE). In the context of immunotherapy, antigenic peptides generated through proteasome-independent mechanisms may present attractive targets, as some of these antigens could be continuously expressed by tumors despite their environment [80, 81].

3.8.6 HUMAN TUMOR ANTIGEN CENTRAL TOLERANCE

Cancer patients' blood displays CTL targeted against differentiation antigens or antigens produced by cancer-germline genes, indicating that innate immune tolerance to these antigens is either nonexistent or inadequate. It's interesting to note that the immuno- or intermediate proteasomes typically produce antigenic peptides coded by MAGE genes better than the regular proteasome [82]. In contrast, immuno- or intermediate proteasomes do not create variation antigens, which are typically produced by the regular proteasome. Although the precise proteasome composition of thymic epithelial cells is unknown, it was hypothesized to primarily comprise immunoproteasome

[83, 84]. The paucity of T lymphocytes that recognize immunoproteasome-dependent peptides in the tumor or blood of cancer patients likely reflects their effective central resistance to differentiation antigens: Thymic selection does not spare T lymphocytes that are directed against immunoproteasome-dependent differentiation peptides. For antigens that are coded by cancer germline genes, the situation is more complicated. As previously mentioned, the immunoproteasome or intermediate proteasomes are typically used to process these antigens, which suggests an absence of central persistence to these antigens. The lack of central persistence was formerly attributed to the lack of these genes' expression in the thymus. However, research revealed that medullary thymic epithelial cells (mTEC), cells that offer antigens for negative choice, expressed a variety of cancer-germline genes at low concentrations [85, 86]. Reduced expression of these genes in mTEC was therefore predicted to cause partial tolerance and aid in the negative choice of only high avidity T-cells. The low-avidity T-cells that have escaped negative selection in the thymus would then be the source of the antitumor T lymphocytes found in the blood of cancer patients. This idea was supported by a study using knockout mice for the murine cancer-germline gene P1A [87]. These mice were supposed to have a non-selective anti-P1A T-cell repertoire that included T-cells. Unexpectedly, despite the observable expression of P1A in mTEC, the P1A-KO mice's anti-P1A T-cell count was only marginally higher and their repertoire little distinguished from that of wild mice [88]. This suggests that WT mice only have minimal tolerance to this antigen. The apparent absence of central tolerability to the antigen P1A may be due to mTEC's inability to degrade the P1A peptide. The particular proteasome type expressed in epithelial cells of the thymic should therefore be determined to better recognize the absence of central tolerability to cancer germline genes. Additionally, it should be able to pinpoint the antigens for which peripheral T-cells should have a relatively strong affinity and for which central tolerance is limited [87].

3.9 TELOMERASE AS A UNIVERSAL TUMOR ANTIGEN

Human telomerase reverse transcriptase (hTERT), the catalytic subunit of telomerase, has been proposed as a prototype for a common tumor antigen [89, 90]. MDM2, CYP 1B1, and survivin are additional classic tumor antigens that have been previously identified. Human telomeres consist of repetitive DNA sequences, non-coding regions located at the ends of chromosomes with the nucleotide sequence TTAGGG [91, 92]. The telomerase holoenzyme plays a role in maintaining chromosomal stability by enzymatically preserving telomeric DNA, which would otherwise be progressively lost through multiple cell divisions, particularly in cancer cells. The biology of telomerase and its associated proteins, such as hTERT, has been extensively studied, as previously described [93, 94]. Most human cells do not possess telomerase activity and undergo a limited number of divisions, leading to telomere shortening. In contrast, the majority of human cancers exhibit hTERT expression, have significant telomerase activity, and maintain telomere length [95, 96]. The rate-limiting component of the telomerase complex, hTERT, shows a strong correlation between telomerase activity and its expression levels [97, 98].

By enabling unlimited potential for replication, hTERT expression in tumor cells has been associated with tumor growth and makes a significant contribution to oncogenic transformation. For illustration, it was shown that one of the genetic elements in humans called hTERT allows healthy people's fibroblast and epithelial cells to develop into tumor cells [99]. In addition, suppressing telomerase action in hTERT-positive tumor cells causes gradual telomere length reduction and ultimately culminates in apoptosis [100, 101].

3.9.1 IMMUNOTHERAPY USING HTERT IN CLINICAL TRIALS

The clinical assessment of hTERT vaccines for cancer patients results from translational and basic immunological research of hTERT as a TAA. There have been several reported trials with close to 200 patients (Table 3.2) and the outcomes have been thoroughly examined elsewhere [89].

In a clinical study, 19 individuals with metastatic, HLA-A2-positive breast malignancy were examined. They received up to eight subcutaneous (SC) injections of the hTERT I540 peptide along with GM-CSF [110, 111]. Peripheral blood samples taken before and after therapy were analyzed *in vitro*, revealing that 68% of patients showed an immunological response. More than 4% of $CD8^+$ tumor-infiltrating lymphocytes (TILs) were specific for I540 hTERT after immunization but not before. Clinical signs of TILs induction included pain and pruritus at the tumor site, while pathological signs included significant changes in the tumor microenvironment (TME) like necrosis, edema, and histiocytic infiltration. Furthermore, it was observed that peripheral blood hTERT-specific $CD8^+$ T-cells could be activated, proliferate, secrete IFN-ɤ, and eliminate tumors. An exploratory landmark analysis

TABLE 3.2 Clinical Trials of hTERT-based Vaccine in Cancer Patients

Vaccines	Type of Cancer	Specific Immune Response Rate (%)	Clinical Results	References
I540 peptide in dendritic cells.	Prostate and breast (n = 7)	57	No results	[102]
I540 peptide in montanide.	Colon and kidney (n = 14)	50	No results	[103]
Dendritic cells transduced with hTERT (full-length) mRNA.	Prostate (n = 20)	95	Molecular clearance of circulating micro-metastases and a decrease in the velocity of prostate specific antigen.	[104]
GM-CSF and I540 and E611 peptides.	NSCLC (n = 26)	46	One result	[105]
GM-CSF and E611 peptide.	Pancreas (n = 48)	63	Relationship between overall survival and immunological reaction.	[106]
572 peptide and 572Y peptide in montanide.	Multiple (n = 19)	95	No results	[107]
572 peptide and 572Y peptide in montanide.	NSCLC (n = 22)	76	No measurable results, relationship between overall survival and immunological reaction.	[108]
B cells loaded with 572Y peptide and I540 peptide.	Prostate (n = 15)	Varying with respect to dose	No results	[109]
I540 peptide in montanide with GM-CSF.	Breast (n = 19)	68	No measurable results; relationship between overall survival and immunological reaction.	[110]

Abbreviations: hTERT: Human telomerase reverse transcriptase; GM-CSF: granulocyte-macrophage colony-stimulating factor; and NSCLC: non-small cell lung cancer.

revealed a correlation between the overall survival (OS) of these patients and a specific $CD8^+$ T-cell immune response to hTERT [110].

Only a few of the various vaccine compositions have been thoroughly researched, such as B cells or dendritic cells (DCs) loaded with the hTERT antigen, additional adjuvants, other peptides, peptides with cytokines, and others. A wide range of tumor types, including breast, pancreatic, lung, and prostate malignancies, have been studied due to the nearly universal expression of hTERT. Despite low response rates by objective criteria, individuals in these studies often had large, advanced tumors and had previously refused treatment. Despite changes in preparation and trial design, as well as the majority of patients being in advanced stages, several consistent results have been observed in these initial trials. HTERT immunotherapies have shown that immune responses can be elicited without harm, can impact clinical outcomes, and can modulate changes in the TME [89].

3.9.2 POTENTIAL SIDE EFFECTS FROM USING VACCINATIONS TO BLOCK HTERT

Even though hTERT vaccinations have not been associated with any major adverse events (AEs) in patients, there is currently no accessible long-term safety information. It remains unknown whether stronger hTERT immune responses could have a negative impact on normal cells expressing hTERT. While telomerase activity is absent in most vital organs, HTERT mRNA expression has been detected in hematopoietic stem cells (HSCs), epithelial cells, stimulated gonadal cells, lymphocytes, and basal keratinocytes [112–114]. Preclinical studies indicate that CD34+ stimulated T lymphocytes are not lysed *in vitro* by hTERT-specific CTLs. Similarly, immunodeficient animals restored with patient hematopoietic progenitors did not show CTL-mediated toxicity towards healthy bone marrow cells after exposure to hTERT-specific CTLs [114]. However, individuals with hereditary mutations in hTERT or related genes may experience significant clinical effects due to long-term telomerase insufficiency. These observations are crucial when planning and conducting clinical trials targeting hTERT [114, 115].

3.9.3 SECOND-GENERATION HTERT VACCINATION

The immunological response elicited *in vivo* by hTERT vaccination is being enhanced using several strategies. One approach utilizes full-length hTERT recombinant constructs instead of individual peptides to simultaneously target multiple CD4 and CD8 epitopes. In one therapeutic trial, patients with prostate malignancy received autologous DCs transfected with mRNA expressing HTERT. The treatment was well tolerated in this study. Following immunization, both $CD8^+$ and $CD4^+$ T-cells with hTERT specificity were detected. These cells were associated with the molecular clearance of circulating micrometastases and an increase in the doubling time of antigen-specific to the prostate. Moreover, peripheral blood from 95% of patients showed an increase in hTERT-specific $CD8^+$ T-cells [104].

A viral vector with full-length hTERT and direct injection of plasmid DNA encoding full-length HTERT are two other alternatives. Both of these options would avoid the complex manufacturing and medical issues linked to producing antigen-loaded DCs [104, 116].

The ultimate goal is to develop preparations capable of eliciting specific immune responses that are as potent and robust as those triggered by viruses. The frequency of T-cell responses achieved through immunization against various tumor antigens remains significantly lower than T-cell responses known to confer immunity against viral infections. This observation may be elucidated by several well-established immune evasion mechanisms activated during tumor progression, involving both host-derived factors and factors originating from the tumor (such as GF, IL-10, and PD-L1) that dampen cellular immunity (e.g., myeloid-derived suppressor cells (MDSCs), PD-1, regulatory T-cells (TReg), and CTLA-4). With a better understanding of the molecular and cellular underpinnings of these processes, targeting these mechanisms clinically has become a key focus in advancing state-of-the-art cancer immunotherapy [117, 118].

3.9.4 FUTURE DIRECTIONS AND EFFECTS OF THE HTERT VACCINE

Overall, it is evident that efficacious hTERT-specific immunotherapy strategies must maximize immune reaction induction and avoid immune resistance posed by both the host and the tumor; however, the potential use of such a common antigen of tumor for prevention of melanoma, rather than the vaccine, may eventually be an approach with the greatest scientific potential. Since more than 85% of patient malignancies overexpress hTERT, it is reasonable to see patients receiving vaccinations against hTERT and other common tumor antigens as a way to stop cancer from growing or returning. This strategy would ultimately evaluate the hTERT vaccine in healthy persons without a cancer diagnosis and in the presence of minor remaining disease (e.g., following adjuvant treatment). This may include people who have genetic conditions like BRCA1 or BRCA2 mutations that put them at high genetic risk for developing cancer [119].

It is significant to emphasize that, despite the confirmed production of particular immune reactions, therapeutic cancer immunotherapy has been an experimental failure for the majority of patients. Therapeutic cancer immunotherapy has not outperformed controls in numerous Phase III studies of clinical end goals such as OS. However, tumor burden and preceding immunotoxic therapy may severely impact patients' immunological competence that attempts to use therapeutic vaccinations may be ineffective in all but the most isolated cases. Therefore, once safety is demonstrated, researchers are aggressively transferring innovative vaccine techniques from metastatic cancer to individuals with minimal residual illness following adjuvant therapy. There is also a case to be made that the initial in-human trials of cancer immunotherapies ought to be conducted in adjuvant or other environments with low residual disease, both ethically and scientifically. The time may be right to test this idea with hTERT vaccinations; however, it has not yet been done. The development of long-term immunological retention and the creation of a specific toxicity profile are two obstacles that must be overcome for hTERT vaccinations to succeed. However, there are currently resources and opportunities to assess telomerase vaccine outside of the context of advanced cancer patients with bulky illness [120, 121].

3.10 IMPLICATIONS FOR THE FUTURE OF CANCER VACCINES

3.10.1 PROSPECTS FOR CANCER VACCINES

Recent research on the adoptive transmission of antitumor T-cells or the usage of T-cell activating Abs has shown the effectiveness of cancer vaccination techniques [122, 123]. Many patients who receive these medicines have remarkable clinical outcomes, and few of these individuals even seem to be disease-free years after therapy began. However, as was already mentioned, both treatments frequently result in the emergence of potent autoimmune adverse effects that must be managed. Many patients continue to be resistant to adoptive transfer and stimulating antibody therapy, even though the response to these treatments shows considerable potential. Therapeutic CVs are made to stimulate antitumor CTL already found in the tumors or blood of cancer individuals. Additionally, the vaccine's target T-cells have experienced thymic collection, which should reduce the risk of unfavorable immune reactions against self-antigens released by healthy organs. These two factors together make vaccination particularly effective at activating anti-vaccine T-cells. Together, these factors make vaccination appear safer and better controlled. To be more precise, antigenic peptides found in mixed-tumor lymphocyte cultures or when T-cells from cancer individuals are induced with autologous cells of a tumor *in vitro* should be harmless because they frequently aim antigens against which a prior impulsive reaction has been displayed, generally without any clear adverse effects. However, since it is not necessarily clear if the immune system can successfully attack these antigens, it is important to properly assess the safety of the antigens discovered using the reverse vaccination method, which entails *in vitro* priming of T-cells from cancer individuals with pulsed DCs [25, 124].

Even in the few individuals who responded to the MAGE-1 or MAGE-3-containing peptide, full-length protein, or viral vaccination, no adverse reactions were observed in the cohorts of cancer patients. In contrast to anti-vaccine T-cells, individuals who responded to the immunotherapy exhibited a significantly higher presence of antitumor T-cells in their metastases post-vaccination. This discovery suggests the potential for a small number of anti-vaccine CTL to prime or reactivate T-cells that target tumor antigens unrelated to the immunotherapy [125, 126].

3.10.2 CHALLENGES OF VACCINE: THE IMMUNOSUPPRESSION AND TUMOR ESCAPE

Most individuals with melanoma seem to develop a spontaneous T-cell response to their tumor, which is now well understood. While the presence of TIL at the tumor site is usually associated with a favorable prognosis, in many cases, these TIL are overwhelmed by tumor growth and remain inactive and energized at the tumor site, likely explaining why, so far, cancer patient vaccination has only resulted in 5–10% clinical responses [127, 128]. Several factors may contribute to the inability of antitumor lymphocytes to eliminate the tumor. One of these reasons is the loss of antigens that T-cells previously recognized on tumors. The consequences of this antigen loss are due to genetic defects affecting the antigenic peptide, any of the proteins involved in the HLA molecule, and the machinery for processing MHC class I. An extreme example is the reduction of the TAP transporter, also known as tapasin, found in various cancer types, which affects the processing and presentation of antigenic peptides [129, 130]. Cytokines like IFN, produced by lymphocytes or NK cells that have infiltrated the tumor, can also alter the repertoire of tumor peptides. IFN decreases the presentation of peptides that are poorly synthesized or degraded by the normal proteasome while enhancing the presentation of certain antigenic peptides by upregulating the expression of heavy chains of HLA class I, TAP2, Tapasin, TAP1, or 2m [131]. It is true, for instance, that the immunoproteasome does not handle peptides produced from melanoma differentiation antigens; instead, they are digested by the regular proteasome. Consequently, the following series of events may occur anticancer lymphocytes assault the cells of the tumor at the initial stage of T-cell tumor infiltration. These lymphocytes' ability to detect antigens is most likely due to the steady-state expression of standard or intermediate proteasomes in malignancies. These T-cells secrete INF, which changes the proteasome composition of the cells of tumor nearby. Consequently, tumor cells start expressing immunoproteasome-dependent antigens that may be identified by additional anticancer T-cells instead of the antigens dependent on the standard proteasome that the immune reaction was first established against. By reducing the expression of genes essential for the appearance of peptides to antitumor T-cells, the tumor may periodically escape the immune reaction during this entire pathway under the selective pressure of antitumor CTL. Additionally, the production of immunomodulatory substances that tamp down the immune response and hinder the activity of antitumor CTL is induced by the IFN developed by stimulated cells at the tumor site. One of these immune-modulating substances is the indoleamine 2,3-dioxygenase enzyme, which deteriorates tryptophan and produces metabolites that restrict the proliferation of T-cell and activity while enhancing tumor survival and movement [132, 133]. Another instance is the programmed death ligand-1 (PD-L1), which acts with its T-cell receptor PD-1 to reduce T-cell adhesion, cytokine, and proliferation, released in tumor cells, which is also triggered by interferon [134]. In that situation, reversing local immunosuppression by using CVs and neutralizing anti-PD-L1 or anti-PD-1 Abs simultaneously may help sustain the efficacy of anticancer T-cell responses. By extending the lifetime of animals that bear tumors and boosting antigen-specific immune reactions, PD-1 inhibition improved the efficiency of GM-CSF-releasing tumor cell vaccination in preclinical colon carcinoma and melanoma models. When PD-1 inhibition and 4-1BB stimulation were used in a different mouse melanoma model, the immunization with a recombinant human adenovirus that expressed the dopachrome tautomerase antigen was significantly improved and the tumor completely disappeared. Last but not least, a different study found that combining GM-CSF expressing tumor cell immunotherapy with both PD-1 and CTLA-4 inhibiting Abs effectively enhanced tumor refusal in animal ovarian and colon cancer (CC) models. The proliferation, functionality, and tumor infiltration of tumor-specific CTL increased as a result. Conversely, suppressive Tregs were reduced in quantity and blocked at the tumor site. These findings demonstrate how combination therapy may improve the efficiency of cancer vaccinations [135].

3.10.3 TUMOR VACCINE PERSPECTIVES

The current obstacle to the development of more effective immunotherapeutic vaccines lies in the challenge of tumor evasion and immunosuppression. Cancer immunotherapy should aim to activate T-cells against a broad spectrum of antigens to minimize the risk of antigen variant escape. Moreover, vaccines targeting antigens with high tumor specificity and universal presentation by tumors should be prioritized. Considering that intermediate proteasomes are prevalent in most cancers in a steady state and that immunoproteasomes play a crucial role in tumors exposed to an inflammatory environment, tumor antigens processed by both immunoproteasomes and intermediate proteasomes seem to offer the most promising approach in this context. It is also essential to consider relevant targets as antigens that can still be expressed even when components of the antigen-processing machinery (such as ERAP, tapasin, and TAP) are compromised. The signal sequences of ER-targeted proteins serve as the primary source of peptides for TAP-independent antigenic peptides, which are subsequently released in the ER after being cleaved by signal peptide peptidase and peptidase [136, 137]. Antigens generated specifically in cells lacking MHC class I processing machinery fall into a category of antigens referred to as T-cell epitopes associated with impaired peptide processing. T-cell epitopes linked with impaired peptide processing antigens are absent from the surface of TAP-positive cells because these peptides are excluded from the ER lumen compared to peptides transported by TAP. Though the parent protein was effectively expressed in thymic cells, thymic epithelial cells from wild mice were unable to digest the T-cell epitopes linked with impaired peptide processing peptides, in contrast to what was seen in TAP knockout animals. Negative selection of T-cells that recognize such TEIPP antigens should thus not occur in wild animals, and T-cell epitopes linked with impaired peptide processing T-cells are anticipated to exhibit greater avidity than other antitumor T-cells that underwent negative selection. To circumvent tumor escape mechanisms, it is now possible to identify novel tumor antigens that are delivered in the absence of MHC class I machinery components or TAP [138, 139].

The distribution method of CVs will also significantly impact their effectiveness. Ideally, vaccinations should elicit potent CTL reactions. Peptide- or protein-based vaccinations are not optimal in this regard. Conversely, vaccinations based on viral vectors that produce desired antigens are highly immunogenic as they stimulate both the adaptive and innate immune systems.

Specifically, viral-based vaccinations effectively infect DC, which is essential for priming antitumor CTL. Recombinant viral vectors have been shown to induce very potent CTL responses linked to anticancer effects. However, to mitigate the adverse effects of vector-specific neutralizing Abs caused by some vectors after a single shot, heterologous prime-boost methods are typically required, where a different vector is used for the boost and priming [140]. Lastly, reversing local immunosuppression will be crucial for the development of cancer immunotherapy. Anticancer vaccines may be most effective in individuals whose tumors exhibit minimal immunosuppression. Research has focused extensively on developing effective inhibitors of immunomodulatory enzymes such as arginase, tryptophan 2,3-dioxygenase (TDO), and IDO. Various other immunosuppressive factors could also be targeted, including soluble substances like interleukin-10 (IL-10), galectins, and transforming growth factor-β (TGF-β), as well as suppressive cells such as MDSCs, TRegs, and tolerogenic DCs present at the tumor site. Future treatments hold significant promise when adoptive T-cell transfer cancer immunotherapies and checkpoint blockade inhibitors are combined with novel medications that target these immunosuppressive elements [141, 142].

3.11 CONCLUSION

The effectiveness of vaccines in preventing viral and bacterial diseases has been demonstrated. Since the first vaccination was developed more than 200 years ago, it has helped save the lives of hundreds of millions of people by averting some of the worst diseases of the 20th century. Even though preclinical data often shows promising results, clinical trials have not yet yielded sufficient data. The main cause of these setbacks is the trouble in identifying tumor-specific antigens, which are supposed to be overexpressed by tumor cells relative to healthy cells. Most tumor antigens used in cancer immunotherapy are non-mutated self-antigens and overexpressed self-antigens that primarily stimulate T-cells with insufficient T-cell binding sites, preventing them from effectively triggering an anti-tumor response. The objective tumor regression after adoptively transferring IL and lymphocytes that infiltrated tumors into people with autologous cancer demonstrates that T-cells are essential for tumor regression. So far, the three types of tumor antigens that have been identified are tissue-specific antigens, tumor-specific shared antigens, and tumor-specific differentiation antigens. $CD4^+$ T-cells are essential for planning the host reaction response to cancer, infectious infections, and autoimmune diseases. Since it is expressed in the vast majority of cancers and plays an essential practical function in malignant transformation, telomerase is a template for an all-encompassing cancer antigen. The possibility of utilizing the telomerase catalytic subtype, hTERT, as a widely applicable immunological target has been explored in view of these favorable characteristics. Initial clinical trials of different immunotherapy agents have indicated that it is feasible to induce immune responses specific to hTERT in individuals, which could potentially influence clinical results. hTERT has been shown to be immunogenic through basic immunological tests.

KEYWORDS

- **antigens**
- **cancer antigens**
- **CD4+ T-cells**
- **human telomerase reverse transcriptase**
- **passenger mutations**
- **reverse immunology**
- **T-cells**
- **telomerase**
- **tumor-associated antigens**
- **vaccine**

REFERENCES

1. Rahib, L., Smith, B. D., Aizenberg, R., Rosenzweig, A. B., Fleshman, J. M., & Matrisian, L. M., (2014). Projecting cancer incidence and deaths to 2030: The unexpected burden of thyroid, liver, and pancreas cancers in the United States. *Cancer Res., 74*(11), 2913–2921.
2. Schlom, J., Hodge, J. W., Palena, C., Tsang, K. Y., Jochems, C., Greiner, J. W., et al., (2014). Therapeutic cancer vaccines. *Advances in Cancer Research, 121*, 67–124.
3. *Cancer Fact Sheet: WHO,* (2015). [Available from: www.who.int/mediacentre/factsheets/fs297/en/ (accessed on 11 March 2024).
4. Chiang, C. L., Coukos, G., & Kandalaft, L. E., (2015). Whole tumor antigen vaccines: Where are we? *Vaccines (Basel), 3*(2), 344–372.

5. Liu, S., Wu, X., Chandra, S., Lyon, C., Ning, B., Jiang, L., et al., (2022). Extracellular vesicles: Emerging tools as therapeutic agent carriers. *Acta Pharmaceutica Sinica B., 12*(10), 3822–3842.
6. Chiang, C. L. L., Benencia, F., & Coukos, G., (2010). Whole tumor antigen vaccines. *Seminars in Immunology, 22*(3), 132–143.
7. Alharbi, N., Skwarczynski, M., & Toth, I., (2022). The influence of component structural arrangement on peptide vaccine immunogenicity. *Biotechnol. Adv., 60*, 108029.
8. Hollingsworth, R. E., & Jansen, K., (2019). Turning the corner on therapeutic cancer vaccines. *NPJ Vaccines, 4*, 7.
9. Buonaguro, L., & Tagliamonte, M., (2020). Selecting target antigens for cancer vaccine development. *Vaccines (Basel), 8*(4).
10. Orange, M., Reuter, U., & Hobohm, U., (2016). Coley's lessons remembered: Augmenting mistletoe therapy. *Integr. Cancer Ther., 15*(4), 502–511.
11. Kakimi, K., Karasaki, T., Matsushita, H., & Sugie, T., (2017). Advances in personalized cancer immunotherapy. *Breast Cancer (Tokyo, Japan), 24*(1), 16–24.
12. Vigneron, N., Stroobant, V., Van, D. E. B. J., & Van, D. B. P., (2013). Database of T cell-defined human tumor antigens: The 2013 update. *Cancer Immunity, 13*, 15.
13. Parkhurst, M. R., Yang, J. C., Langan, R. C., Dudley, M. E., Nathan, D. A., Feldman, S. A., et al., (2011). T cells targeting carcinoembryonic antigen can mediate regression of metastatic colorectal cancer but induce severe transient colitis. *Molecular Therapy: The Journal of the American Society of Gene Therapy, 19*(3), 620–626.
14. Gettinger, S., Choi, J., Hastings, K., Truini, A., Datar, I., Sowell, R., et al., (2017). Impaired HLA class I Antigen processing and presentation as a mechanism of acquired resistance to immune checkpoint inhibitors in lung cancer. *Cancer Discovery, 7*(12), 1420–1435.
15. Gandhi, L., Rodríguez-Abreu, D., Gadgeel, S., Esteban, E., Felip, E., De Angelis, F., et al., (2018). Pembrolizumab plus chemotherapy in metastatic non–small-cell lung cancer. *New England Journal of Medicine, 378*(22), 2078–2092.
16. Germeau, C., Ma, W., Schiavetti, F., Lurquin, C., Henry, E., Vigneron, N., et al., (2005). High frequency of antitumor T cells in the blood of melanoma patients before and after vaccination with tumor antigens. *J. Exp. Med., 201*(2), 241–248.
17. Vigneron, N., (2015). Human tumor antigens and cancer immunotherapy. *Biomed Res. Int., 2015*, 948501.
18. Job, S., Rapoud, D., Dos, S. A., Gonzalez, P., Desterke, C., Pascal, G., et al., (2020). Identification of four immune subtypes characterized by distinct composition and functions of tumor microenvironment in intrahepatic cholangiocarcinoma. *Hepatology, 72*(3), 965–981.
19. Rizvi, S., Khan, S. A., Hallemeier, C. L., Kelley, R. K., & Gores, G. J., (2018). Cholangiocarcinoma – evolving concepts and therapeutic strategies. *Nat. Rev. Clin. Oncol., 15*(2), 95–111.
20. Vigneron, N., Ooms, A., Morel, S., Ma, W., Degiovanni, G., & Van, D. E. B. J., (2005). A peptide derived from melanocytic protein gp100 and presented by HLA-B35 is recognized by autologous cytolytic T lymphocytes on melanoma cells. *Tissue Antigens, 65*(2), 156–162.
21. Engelhard, V. H., Obeng, R. C., Cummings, K. L., Petroni, G. R., Ambakhutwala, A. L., Chianese-Bullock, K. A., et al., (2020). MHC-restricted phosphopeptide antigens:

Preclinical validation and first-in-humans clinical trial in participants with high-risk melanoma. *Journal for Immunotherapy of Cancer, 8*(1).

22. Dalet, A., Robbins, P. F., Stroobant, V., Vigneron, N., Li, Y. F., El-Gamil, M., et al., (2011). An antigenic peptide produced by reverse splicing and double asparagine deamidation. *Proceedings of the National Academy of Sciences of the United States of America, 108*(29), E323–331.
23. Vigneron, N., & Van, D. E. B. J., (2011). Insights into the processing of MHC class I ligands gained from the study of human tumor epitopes. *Cell Mol. Life Sci., 68*(9), 1503–1520.
24. Heather, J. M., Myers, P. T., Shi, F., Aziz-Zanjani, M. O., Mahoney, K. E., Perez, M., et al., (2019). Murine xenograft bioreactors for human immunopeptidome discovery. *Sci. Rep., 9*(1), 18558.
25. Coulie, P. G., Van, D. E. B. J., Van, D. B. P., & Boon, T., (2014). Tumor antigens recognized by T lymphocytes: At the core of cancer immunotherapy. *Nature Reviews Cancer, 14*(2), 135–146.
26. Melssen, M. M., Petroni, G. R., Chianese-Bullock, K. A., Wages, N. A., Grosh, W. W., Varhegyi, N., et al., (2019). A multipeptide vaccine plus toll-like receptor agonists LPS or polyICLC in combination with incomplete Freund's adjuvant in melanoma patients. *Journal for Immunotherapy of Cancer, 7*(1), 163.
27. Murin, C. D., Wilson, I. A., & Ward, A. B., (2019). Antibody responses to viral infections: A structural perspective across three different enveloped viruses. *Nature Microbiology, 4*(5), 734–747.
28. Kenter, G. G., Welters, M. J., Valentijn, A. R., Lowik, M. J., Berends-Van, D. M. D. M., Vloon, A. P., et al., (2009). Vaccination against HPV-16 oncoproteins for vulvar intraepithelial neoplasia. *The New England Journal of Medicine, 361*(19), 1838–1847.
29. Zhao, Y., Baldin, A. V., Isayev, O., Werner, J., Zamyatnin, A. A. Jr., & Bazhin, A. V., (2021). Cancer vaccines: Antigen selection strategy. *Vaccines (Basel), 9*(2).
30. Lennerz, V., Fatho, M., Gentilini, C., Frye, R. A., Lifke, A., Ferel, D., et al., (2005). The response of autologous T cells to a human melanoma is dominated by mutated neoantigens. *Proceedings of the National Academy of Sciences of the United States of America, 102*(44), 16013–16018.
31. Khan, S. T., Montroy, J., Forbes, N., Bastin, D., Kennedy, M. A., Diallo, J. S., et al., (2020). Safety and efficacy of autologous tumor cell vaccines as a cancer therapeutic to treat solid tumors and haematological malignancies: A meta-analysis protocol for two systematic reviews. *BMJ Open, 10*(6), e034714.
32. Jiang, T., Shi, T., Zhang, H., Hu, J., Song, Y., Wei, J., et al., (2019). Tumor neoantigens: From basic research to clinical applications. *J. Hematol. Oncol., 12*(1), 93.
33. Chomez, P., De Backer, O., Bertrand, M., De Plaen, E., Boon, T., & Lucas, S., (2001). An overview of the MAGE gene family with the identification of all human members of the family. *Cancer Research., 61*(14), 5544–5551.
34. Florke, G. R. R., Chen, H., Lee, A. K., Daly, C. A., Wilander, B. A., Fon, T. K., et al., (2020). Emerging roles of the MAGE protein family in stress response pathways. *The Journal of Biological Chemistry, 295*(47), 16121–16155.
35. Tischler, J., Crew, K. D., & Chung, W. K., (2019). Cases in precision medicine: The role of tumor and germline genetic testing in breast cancer management. *Annals of Internal Medicine, 171*(12), 925–930.

36. Zappasodi, R., Merghoub, T., & Wolchok, J. D., (2018). Emerging concepts for immune checkpoint blockade-based combination therapies. *Cancer Cell, 33*(4), 581–598.
37. Mandal, R., Samstein, R. M., Lee, K. W., Havel, J. J., Wang, H., Krishna, C., et al., (2019). Genetic diversity of tumors with mismatch repair deficiency influences anti-PD-1 immunotherapy response. *Science (New York, NY), 364*(6439), 485–491.
38. Grosser, R., Cherkassky, L., Chintala, N., & Adusumilli, P. S., (2019). Combination immunotherapy with CAR T cells and checkpoint blockade for the treatment of solid tumors. *Cancer Cell, 36*(5), 471–482.
39. Almeida, L. G., Sakabe, N. J., deoliveira, A. R., Silva, M. C., Mundstein, A. S., Cohen, T., et al., (2009). CTdatabase: A knowledge-base of high-throughput and curated data on cancer-testis antigens. *Nucleic Acids Research, 37*(Database issue), D816–819.
40. Kenter, G. G., Welters, M. J., Valentijn, A. R., Lowik, M. J., Berends-Van, D. M. D. M., Vloon, A. P., et al., (2008). Phase I immunotherapeutic trial with long peptides spanning the E6 and E7 sequences of high-risk human papillomavirus 16 in end-stage cervical cancer patients shows low toxicity and robust immunogenicity. *Clinical Cancer Research: An Official Journal of the American Association for Cancer Research, 14*(1), 169–177.
41. Vonderheide, R. H., & Nathanson, K. L., (2013). Immunotherapy at large: The road to personalized cancer vaccines. *Nature Medicine, 19*(9), 1098–1100.
42. Cheever, M. A., Allison, J. P., Ferris, A. S., Finn, O. J., Hastings, B. M., Hecht, T. T., et al., (2009). The prioritization of cancer antigens: A national cancer institute pilot project for the acceleration of translational research. *Clinical Cancer Research: An Official Journal of the American Association for Cancer Research, 15*(17), 5323–5337.
43. Butterfield, L. H., Ribas, A., Dissette, V. B., Amarnani, S. N., Vu, H. T., Oseguera, D., et al., (2003). Determinant spreading associated with clinical response in dendritic cell-based immunotherapy for malignant melanoma. *Clinical Cancer Research: An Official Journal of the American Association for Cancer Research, 9*(3), 998–1008.
44. DeLucia, D. C., & Lee, J. K., (2022). Development of cancer immunotherapies. *Cancer Treatment and Research, 183*, 1–48.
45. Disis, M. L., (2011). Immunologic biomarkers as correlates of clinical response to cancer immunotherapy. *Cancer Immunology, Immunotherapy: CII, 60*(3), 433–442.
46. Zhang, H., Ye, L., Yu, X., Jin, K., & Wu, W., (2022). Neoadjuvant therapy alters the immune microenvironment in pancreatic cancer. *Front. Immunol., 13*, 956984.
47. Gogas, H., Ioannovich, J., Dafni, U., Stavropoulou-Giokas, C., Frangia, K., Tsoutsos, D., et al., (2006). Prognostic significance of autoimmunity during treatment of melanoma with interferon. *The New England Journal of Medicine, 354*(7), 709–718.
48. Darbeheshti, F., (2022). The immunogenetics of melanoma. *Advances in Experimental Medicine and Biology, 1367*, 383–396.
49. Monzavi-Karbassi, B., Pashov, A., & Kieber-Emmons, T., (2013). Tumor-associated glycans and immune surveillance. *Vaccines (Basel), 1*(2), 174–203.
50. Zhao, Y., Li, S., Lv, J., Liu, Y., Chen, Y., Liu, Y., et al., (2021). Generation of triacyl lipopeptide-modified glycoproteins by metabolic glycoengineering as the neoantigen to boost anti-tumor immune response. *Theranostics, 11*(15), 7425–7438.
51. Hurley, L. C., Levin, N. K., Chatterjee, M., Coles, J., Muszkat, S., Howarth, Z., et al., (2020). Evaluation of paraneoplastic antigens reveals TRIM21 autoantibodies as biomarker for early detection of ovarian cancer in combination with autoantibodies to NY-ESO-1 and TP53. *Cancer Biomark., 27*(3), 407–421.

52. Goodenough, E., Robinson, T. M., Zook, M. B., Flanigan, K. M., Atkins, J. F., Howard, M. T., et al., (2014). Cryptic MHC class I-binding peptides are revealed by aminoglycoside-induced stop codon read-through into the 3' UTR. *Proceedings of the National Academy of Sciences of the United States of America, 111*(15), 5670–5675.
53. Wagner, E. K., Qerqez, A. N., Stevens, C. A., Nguyen, A. W., Delidakis, G., & Maynard, J. A., (2019). Human cytomegalovirus-specific T-cell receptor engineered for high affinity and soluble expression using mammalian cell display. *The Journal of Biological Chemistry, 294*(15), 5790–5804.
54. Vigneron, N., & Van, D. E. B. J., (2014). Proteasome subtypes and regulators in the processing of antigenic peptides presented by class I molecules of the major histocompatibility complex. *Biomolecules, 4*(4), 994–1025.
55. Abi Habib, J., De Plaen, E., Stroobant, V., Zivkovic, D., Bousquet, M. P., Guillaume, B., et al., (2020). Efficiency of the four proteasome subtypes to degrade ubiquitinated or oxidized proteins. *Sci. Rep., 10*(1), 15765.
56. Gonzalez, C. C. M., Merino, B., Casanueva, A. E., Postigo, C. T., Camara, T. P., Fernandez, D. C. M., et al., (2021). Modulation of insulin sensitivity by insulin-degrading enzyme. *Biomedicines, 9*(1).
57. Morozov, A. V., & Karpov, V. L., (2018). Biological consequences of structural and functional proteasome diversity. *Heliyon, 4*(10), e00894.
58. Thibaudeau, T. A., & Smith, D. M., (2019). A practical review of proteasome pharmacology. *Pharmacological Reviews, 71*(2), 170–197.
59. Harris, J. L., Alper, P. B., Li, J., Rechsteiner, M., & Backes, B. J., (2001). Substrate specificity of the human proteasome. *Chemistry & Biology, 8*(12), 1131–1141.
60. Serrano-Aparicio, N., Ferrer, S., & Świderek, K., (2022). Covalent inhibition of the human 20S proteasome with Homobelactosin C inquired by QM/MM studies. *Pharmaceuticals (Basel, Switzerland), 15*(5).
61. Toes, R. E., Nussbaum, A. K., Degermann, S., Schirle, M., Emmerich, N. P., Kraft, M., et al., (2001). Discrete cleavage motifs of constitutive and immunoproteasomes revealed by quantitative analysis of cleavage products. *J. Exp. Med., 194*(1), 1–12.
62. Prinz, J. C., (2022). Antigen processing, presentation, and tolerance: Role in autoimmune skin diseases. *J. Invest. Dermatol., 142*(3 Pt B), 750–759.
63. Michaux, A., Larrieu, P., Stroobant, V., Fonteneau, J. F., Jotereau, F., Van, D. E. B. J., et al., (2014). A spliced antigenic peptide comprising a single spliced amino acid is produced in the proteasome by reverse splicing of a longer peptide fragment followed by trimming. *Journal of Immunology (Baltimore, Md: 1950), 192*(4), 1962–1971.
64. Mansurkhodzhaev, A., Barbosa, C. R. R., Mishto, M., & Liepe, J., (2021). Proteasome-Generated cis-spliced peptides and their potential role in CD8(+) T cell tolerance. *Front. Immunol., 12*, 614276.
65. Admon, A., (2021). Are there indeed spliced peptides in the immunopeptidome? *Molecular & Cellular Proteomics: MCP., 20*, 100099.
66. Oliveri, F., Basler, M., Rao, T. N., Fehling, H. J., & Groettrup, M., (2022). Immunoproteasome inhibition reduces the T helper 2 response in mouse models of allergic airway inflammation. *Front. Immunol., 13*, 870720.
67. Guillaume, B., Chapiro, J., Stroobant, V., Colau, D., Van, H. B., Parvizi, G., et al., (2010). Two abundant proteasome subtypes that uniquely process some antigens presented by HLA class I molecules. *Proceedings of the National Academy of Sciences of the United States of America, 107*(43), 18599–18604.

68. Erokhov, P. A., Kulikov, A. M., Karpova, Y. D., Rodoman, G. V., Sumedi, I. R., Goncharov, A. L., et al., (2021). Proteasomes in patient rectal cancer and different intestine locations: Where does proteasome pool change? *Cancers (Basel), 13*(5).
69. Guimaraes, G., Gomes, M. T. R., Campos, P. C., Marinho, F. V., De Assis, N. R. G., Silveira, T. N., et al., (2018). Immunoproteasome subunits are required for CD8(+) T cell function and host resistance to brucella abortus infection in mice. *Infection and Immunity, 86*(3).
70. de Verteuil, D., Muratore-Schroeder, T. L., Granados, D. P., Fortier, M. H., Hardy, M. P., Bramoulle, A., et al., (2010). Deletion of immunoproteasome subunits imprints on the transcriptome and has a broad impact on peptides presented by major histocompatibility complex I molecules. *Molecular & Cellular Proteomics: MCP, 9*(9), 2034–2047.
71. Tripathi, S. C., Vedpathak, D., & Ostrin, E. J., (2021). The functional and mechanistic roles of immunoproteasome subunits in cancer. *Cells, 10*(12).
72. Morel, S., Levy, F., Burlet, S. O., Brasseur, F., Probst-Kepper, M., Peitrequin, A. L., et al., (2000). Processing of some antigens by the standard proteasome but not by the immunoproteasome results in poor presentation by dendritic cells. *Immunity, 12*(1), 107–117.
73. Chapiro, J., Claverol, S., Piette, F., Ma, W., Stroobant, V., Guillaume, B., et al., (2006). Destructive cleavage of antigenic peptides either by the immunoproteasome or by the standard proteasome results in differential antigen presentation. *Journal of Immunology (Baltimore, Md: 1950), 176*(2), 1053–1061.
74. Guillaume, B., Stroobant, V., Bousquet, D. M. P., Colau, D., Chapiro, J., Parmentier, N., et al., (2012). Analysis of the processing of seven human tumor antigens by intermediate proteasomes. *Journal of Immunology (Baltimore, Md: 1950), 189*(7), 3538–3547.
75. Dalet, A., Stroobant, V., Vigneron, N., & Van, D. E. B. J., (2011). Differences in the production of spliced antigenic peptides by the standard proteasome and the immunoproteasome. *Eur. J. Immunol., 41*(1), 39–46.
76. Specht, G., Roetschke, H. P., Mansurkhodzhaev, A., Henklein, P., Textoris-Taube, K., Urlaub, H., et al., (2020). Large database for the analysis and prediction of spliced and non-spliced peptide generation by proteasomes. *Sci. Data, 7*(1), 146.
77. Rackova, L., & Csekes, E., (2020). Proteasome biology: Chemistry and bioengineering insights. *Polymers, 12*(12).
78. De, M., Jayarapu, K., Elenich, L., Monaco, J. J., Colbert, R. A., & Griffin, T. A., (2003). Beta 2 subunit propeptides influence cooperative proteasome assembly. *The Journal of Biological Chemistry, 278*(8), 6153–6159.
79. Guil, S., Rodriguez-Castro, M., Aguilar, F., Villasevil, E. M., Anton, L. C., & Del, V. M., (2006). Need for tripeptidyl-peptidase II in major histocompatibility complex class I viral antigen processing when proteasomes are detrimental. *The Journal of Biological Chemistry, 281*(52), 39925–39934.
80. Kessler, J. H., Khan, S., Seifert, U., Le Gall, S., Chow, K. M., Paschen, A., et al., (2011). Antigen processing by nardilysin and thimet oligopeptidase generates cytotoxic T cell epitopes. *Nat. Immunol., 12*(1), 45–53.
81. Santos, N. B. D., Franco, R. D., camarini, R., Munhoz, C. D., Eichler, R. A. S., Gewehr, M. C. F., et al., (2019). Thimet oligopeptidase (EC 3.4.24.15) Key functions suggested by knockout mice phenotype characterization. *Biomolecules, 9*(8).

82. Vigneron, N., & Van, D. E. B. J., (2012). Proteasome subtypes and the processing of tumor antigens: Increasing antigenic diversity. *Current Opinion in Immunology, 24*(1), 84–91.
83. Nil, A., Firat, E., Sobek, V., Eichmann, K., & Niedermann, G., (2004). Expression of housekeeping and immunoproteasome subunit genes is differentially regulated in positively and negatively selecting thymic stroma subsets. *Eur. J. Immunol., 34*(10), 2681–2689.
84. Limanaqi, F., Biagioni, F., Gaglione, A., Busceti, C. L., & Fornai, F., (2019). A sentinel in the crosstalk between the nervous and immune system: The (immuno)-proteasome. *Front. Immunol., 10*, 628.
85. Gotter, J., Brors, B., Hergenhahn, M., & Kyewski, B., (2004). Medullary epithelial cells of the human thymus express a highly diverse selection of tissue-specific genes colocalized in chromosomal clusters. *J. Exp. Med., 199*(2), 155–166.
86. Benlaribi, R., Gou, Q., & Takaba, H., (2022). Thymic self-antigen expression for immune tolerance and surveillance. *Inflamm. Regen., 42*(1), 28.
87. Huijbers, I. J., Soudja, S. M., Uyttenhove, C., Buferne, M., Inderberg-Suso, E. M., Colau, D., et al., (2012). Minimal tolerance to a tumor antigen encoded by a cancer-germline gene. *Journal of Immunology (Baltimore, Md: 1950), 188*(1), 111–121.
88. Cane, S., Van, S. J., Uyttenhove, C., Pilotte, L., & Van, D. E. B. J., (2021). TGFβ1 neutralization displays therapeutic efficacy through both an immunomodulatory and a non-immune tumor-intrinsic mechanism. *Journal for Immunotherapy of Cancer, 9*(2).
89. Vonderheide, R. H., (2008). Prospects and challenges of building a cancer vaccine targeting telomerase. *Biochimie, 90*(1), 173–180.
90. Mizukoshi, E., & Kaneko, S., (2019). Telomerase-targeted cancer immunotherapy. *Int. J. Mol. Sci., 20*(8).
91. Gordan, J. D., & Vonderheide, R. H., (2002). Universal tumor antigens as targets for immunotherapy. *Cytotherapy, 4*(4), 317–327.
92. De Jonge, H., Iamele, L., Maggi, M., Pessino, G., & Scotti, C., (2021). Anti-cancer auto-antibodies: Roles, applications, and open issues. *Cancers (Basel), 13*(4).
93. Hahn, W. C., & Meyerson, M., (2001). Telomerase activation, cellular immortalization, and cancer. *Annals of Medicine, 33*(2), 123–129.
94. Keshavarz, M., Xie, K., Schaaf, K., Bano, D., & Ehninger, D., (2022). Targeting the "hallmarks of aging" to slow aging and treat age-related disease: Fact or fiction? *Molecular Psychiatry, 2022.*
95. Harley, C. B., Futcher, A. B., & Greider, C. W., (1990). Telomeres shorten during ageing of human fibroblasts. *Nature, 345*(6274), 458–460.
96. Mishra, R., Haldar, S., Biondi, S., Bhari, V. K., Singh, G., & Bhowmick, N. A., (2022). TGF-β controls stromal telomere length through epigenetic modifications. *3 Biotech, 12*(11), 290.
97. Vaziri, H., & Benchimol, S., (1998). Reconstitution of telomerase activity in normal human cells leads to elongation of telomeres and extended replicative life span. *Current Biology: CB., 8*(5), 279–282.
98. Guo, D., Zhang, L., Wang, X., Zheng, J., & Lin, S., (2022). Establishment methods and research progress of livestock and poultry immortalized cell lines: A review. *Front Vet Sci., 9*, 956357.
99. Sato, M., Shay, J. W., & Minna, J. D., (2020). Immortalized normal human lung epithelial cell models for studying lung cancer biology. *Respiratory Investigation, 58*(5), 344–354.

100. Herbert, B., Pitts, A. E., Baker, S. I., Hamilton, S. E., Wright, W. E., Shay, J. W., et al., (1999). Inhibition of human telomerase in immortal human cells leads to progressive telomere shortening and cell death. *Proceedings of the National Academy of Sciences of the United States of America, 96*(25), 14276–14281.
101. Borges, G., Criqui, M., & Harrington, L., (2022). Tieing together loose ends: Telomere instability in cancer and aging. *Molecular Oncology, 16*(18), 3380–3396.
102. Vonderheide, R. H., Domchek, S. M., Schultze, J. L., George, D. J., Hoar, K. M., Chen, D. Y., et al., (2004). Vaccination of cancer patients against telomerase induces functional antitumor CD8+ T lymphocytes. *Clinical Cancer Research: An Official Journal of the American Association for Cancer Research, 10*(3), 828–839.
103. Parkhurst, M. R., Riley, J. P., Igarashi, T., Li, Y., Robbins, P. F., & Rosenberg, S. A., (2004). Immunization of patients with the hTERT:540-548 peptide induces peptide-reactive T lymphocytes that do not recognize tumors endogenously expressing telomerase. *Clinical Cancer Research: An Official Journal of the American Association for Cancer Research, 10*(14), 4688–4698.
104. Su, Z., Dannull, J., Yang, B. K., Dahm, P., Coleman, D., Yancey, D., et al., (2005). Telomerase mRNA-transfected dendritic cells stimulate antigen-specific CD8+ and CD4+ T cell responses in patients with metastatic prostate cancer. *Journal of Immunology (Baltimore, Md: 1950), 174*(6), 3798–3807.
105. Brunsvig, P. F., Aamdal, S., Gjertsen, M. K., Kvalheim, G., Markowski, G. C. J., Sve, I., et al., (2006). Telomerase peptide vaccination: A phase I/II study in patients with non-small cell lung cancer. *Cancer Immunology, Immunotherapy: CII, 55*(12), 1553–1564.
106. Bernhardt, S. L., Gjertsen, M. K., Trachsel, S., Moller, M., Eriksen, J. A., Meo, M., et al., (2006). Telomerase peptide vaccination of patients with non-resectable pancreatic cancer: A dose escalating phase I/II study. *British Journal of Cancer, 95*(11), 1474–1482.
107. Mavroudis, D., Bolonakis, I., Cornet, S., Myllaki, G., Kanellou, P., Kotsakis, A., et al., (2006). A phase I study of the optimized cryptic peptide TERT(572y) in patients with advanced malignancies. *Oncology, 70*(4), 306–314.
108. Bolonaki, I., Kotsakis, A., Papadimitraki, E., Aggouraki, D., Konsolakis, G., Vagia, A., et al., (2007). Vaccination of patients with advanced non-small-cell lung cancer with an optimized cryptic human telomerase reverse transcriptase peptide. *Journal of Clinical Oncology: Official Journal of the American Society of Clinical Oncology, 25*(19), 2727–2734.
109. Cortez-Gonzalez, X., & Zanetti, M., (2007). Telomerase immunity from bench to bedside: Round one. *Journal of Translational Medicine, 5*, 12.
110. Domchek, S. M., Recio, A., Mick, R., Clark, C. E., Carpenter, E. L., Fox, K. R., et al., (2007). Telomerase-specific T-cell immunity in breast cancer: Effect of vaccination on tumor immunosurveillance. *Cancer Res., 67*(21), 10546–10555.
111. Jiang, C., Li, J., Zhang, W., Zhuang, Z., Liu, G., Hong, W., et al., (2022). Potential association factors for developing effective peptide-based cancer vaccines. *Front Immunol., 13*, 931612.
112. Shay, J. W., & Wright, W. E., (2002). Telomerase: A target for cancer therapeutics. Cancer *Cell, 2*(4), 257–265.
113. Diaz, A. R., Lopez, N. M., Colon, B. E. S., Cabrera, C. R., Cunci, L., Gonzalez, C. I., et al., (2022). Test strip platform spin-off for telomerase activity detection: Development of an electrochemical biosensor. *ACS Omega, 7*(11), 9964–9972.

114. Danet, D. G. A., Luongo, J. L., Bonnet, D. A., Domchek, S. M., & Vonderheide, R. H., (2005). Telomerase vaccination has no detectable effect on SCID-repopulating and colony-forming activities in the bone marrow of cancer patients. *Exp. Hematol., 33*(11), 1275–1280.
115. Singh, T., Fatehi, H. M., & Fatehi, H. A., (2021). Non-small cell lung cancer: Emerging molecular targeted and immunotherapeutic agents. *Biochim. Biophys. Acta Rev. Cancer, 1876*(2), 188636.
116. Yang, L., Tang, L., Zhang, M., & Liu, C., (2022). Recent advances in the molecular design and delivery technology of mRNA for vaccination against infectious diseases. *Front. Immunol., 13*, 896958.
117. Drake, C. G., Jaffee, E., & Pardoll, D. M., (2006). Mechanisms of immune evasion by tumors. *Advances in Immunology, 90*, 51–81.
118. Castillo, S. R., Churruca, S. A., Martinez, I. M., & Salazar, E. P., (2022). Cancer associated fibroblasts communicate with breast tumor cells through extracellular vesicles in tumor development. *Technology in Cancer Research & Treatment, 21*, 15330338221131647.
119. Kerr, J. R., Freeman, A. L. J., Marteau, T. M., & Van, D. L. S., (2021). Effect of information about COVID-19 Vaccine effectiveness and side effects on behavioral intentions: Two online experiments. *Vaccines (Basel), 9*(4).
120. Vonderheide, R. H., (2007). Universal tumor antigens for cancer vaccination: Targeting telomerase for immunoprevention. *Discovery Medicine, 7*(39), 103–108.
121. Dosset, M., Castro, A., Carter, H., & Zanetti, M., (2020). Telomerase and CD4 T cell immunity in cancer. *Cancers (Basel), 12*(6).
122. Topalian, S. L., Hodi, F. S., Brahmer, J. R., Gettinger, S. N., Smith, D. C., McDermott, D. F., et al., (2012). Safety, activity, and immune correlates of anti-PD-1 antibody in cancer. *The New England Journal of Medicine, 366*(26), 2443–2454.
123. Wu, X., Deng, Z., & Zhao, Q., (2022). Immunotherapy improves disease prognosis by affecting the tumor microenvironment: A bibliometric study. *Front. Immunol., 13*, 967076.
124. Liu, Y., Tian, S., Ning, B., Huang, T., Li, Y., & Wei, Y., (2022). Stress and cancer: The mechanisms of immune dysregulation and management. *Front. Immunol., 13*, 1032294.
125. Carrasco, J., Van, P. A., Neyns, B., Lethe, B., Brasseur, F., Renkvist, N., et al., (2008). Vaccination of a melanoma patient with mature dendritic cells pulsed with MAGE-3 peptides triggers the activity of nonvaccine anti-tumor cells. *Journal of Immunology (Baltimore, Md: 1950), 180*(5), 3585–3593.
126. Lu, L. C., Hsu, C., Shao, Y. Y., Chao, Y., Yen, C. J., Shih, I. L., et al., (2019). Differential organ-specific tumor response to immune checkpoint inhibitors in hepatocellular carcinoma. *Liver Cancer, 8*(6), 480–490.
127. Van, H. I. S., Sluijter, B. J., Moesbergen, L. M., Vos, W. M., De Gruijl, T. D., Molenkamp, B. G., et al., (2008). Favorable outcome in clinically stage II melanoma patients is associated with the presence of activated tumor infiltrating T-lymphocytes and preserved MHC class I antigen expression. *International Journal of Cancer, 123*(3), 609–615.
128. Rossi, F., Fredericks, N., Snowden, A., Allegrezza, M. J., & Moreno-Nieves, U. Y., (2022). Next generation natural killer cells for cancer immunotherapy. *Front Immunol., 13*, 886429.

129. Vigneron, N., Peaper, D. R., Leonhardt, R. M., & Cresswell, P., (2009). Functional significance of tapasin membrane association and disulfide linkage to ERp57 in MHC class I presentation. *Eur. J. Immunol., 39*(9), 2371–2376.
130. Poorebrahim, M., Abazari, M. F., Sadeghi, S., Mahmoudi, R., Kheirollahi, A., Askari, H., et al., (2020). Genetically modified immune cells targeting tumor antigens. *Pharmacology & Therapeutics, 214*, 107603.
131. Willimsky, G., Beier, C., Immisch, L., Papafotiou, G., Scheuplein, V., Goede, A., et al., (2021). *In vitro* proteasome processing of neo-splicetopes does not predict their presentation *in vivo. eLife*, 10.
132. Platten, M., Wick, W., & Van, D. E. B. J., (2012). Tryptophan catabolism in cancer: Beyond IDO and tryptophan depletion. *Cancer Res., 72*(21), 5435–5440.
133. Hubkova, B., Valko, R. M., Cizmarova, B., Zabavnikova, M., Marekova, M., & Birkova, A., (2022). Tryptophan: Its metabolism along the kynurenine, serotonin, and indole pathway in malignant melanoma. *Int. J. Mol. Sci., 23*(16).
134. Moore, E. K., Strazza, M., & Mor, A., (2022). Combination approaches to target PD-1 signaling in cancer. *Front. Immunol., 13*, 927265.
135. Jia, L., Zhang, Q., & Zhang, R., (2015). PD-1/PD-L1 pathway blockade works as an effective and practical therapy for cancer immunotherapy. *Cancer Biology & Medicine, 15*(2), 116–123.
136. El Hage, F., Stroobant, V., Vergnon, I., Baurain, J. F., Echchakir, H., Lazar, V., et al., (2008). Preprocalcitonin signal peptide generates a cytotoxic T lymphocyte-defined tumor epitope processed by a proteasome-independent pathway. *Proceedings of the National Academy of Sciences of the United States of America, 105*(29), 10119–10124.
137. Durgeau, A., Virk, Y., Gros, G., Voilin, E., Corgnac, S., Djenidi, F., et al., (2018). Human preprocalcitonin self-antigen generates TAP-dependent and -independent epitopes triggering optimised T-cell responses toward immune-escaped tumors. *Nat. Commun., 9*(1), 5097.
138. Oliveira, C. C., Querido, B., Sluijter, M., Derbinski, J., Van, D. Burg, S. H., & Van, H. T., (2011). Peptide transporter TAP mediates between competing antigen sources generating distinct surface MHC class I peptide repertoires. *Eur. J. Immunol., 41*(11), 3114–3124.
139. Sun, T., Li, Y., Yang, W., Wu, H., Li, X., Huang, Y., et al., (2019). Histone deacetylase inhibition up-regulates MHC class I to facilitate cytotoxic T lymphocyte-mediated tumor cell killing in glioma cells. *Journal of Cancer, 10*(23), 5638–5645.
140. Bonam, S. R., Renia, L., Tadepalli, G., Bayry, J., & Kumar, H. M. S., (2021). *Plasmodium falciparum* malaria vaccines and vaccine adjuvants. *Vaccines (Basel), 9*(10).
141. Larocca, C., & Schlom, J., (2011). Viral vector-based therapeutic cancer vaccines. *Cancer Journal (Sudbury, Mass), 17*(5), 359–371.
142. Finn, O. J., (2018). The dawn of vaccines for cancer prevention. *Nat. Rev. Immunol., 18*(3), 183–194.

CHAPTER 4

DNA-Based Cancer Vaccine: Advancement and Challenges

SAGAR PANDAV,[1] VIPUL P. PATEL,[1] ARDE GAYATRI,[1] and BHUPENDRA PRAJAPATI[2]

[1]*Sanjivani College of Pharmaceutical Education and Research, Kopargaon, Maharashtra, India*

[2]*S.K. Patel College of Pharmaceutical Education and Research, Ganpat University, Kherva, Gujarat, India*

ABSTRACT

A DNA vaccine is a novel immunotherapeutic strategy utilized to combat cancer and diseases caused by various antigens by inducing immunity against the specific antigen. DNA vaccines are considered safe, easy to produce, cost-effective, and exhibit *in-vivo* stability. The activation of the cytotoxic T-cell immunological pathway can be enhanced by incorporating a suitable sequence from an antigen gene into the carrier or vector of the vaccine. Through the use of cytokines and a xenogenic p53 version present in the antigen, T-cell stimulation by the vaccine reacts against viral or mutated proteins. Despite their numerous advantages, vaccines containing genes like DNA and RNA still face challenges in eliciting robust immune responses against their antigens, although solutions are being developed. This chapter provides a brief overview of DNA-based vaccines for cancer, including progress in DNA-based vaccines, challenges, and insights into the various immunological mechanisms contributing to their ability to trigger immune responses.

Cancer Vaccination and Challenges, Volume 1: Strategies for Therapeutic Cancer Vaccine Development. Rishabha Malviya, Bhupendra Prajapati, and Sonali Sundram (Eds.)

4.1 INTRODUCTION

A variety of diseases are characterized by the abnormal proliferation of cells, which exceeds the normal rate. This condition, known as cancer, adversely affects and invades surrounding cells and tissues, leading to an increased risk of mortality for the affected individual. This, in turn, impacts the annual mortality rate [1]. The treatment of such diseases poses a significant and formidable challenge. Innovative treatments and therapies, such as chemo-therapy, hormone therapy, hyperthermia, immunotherapy, photodynamic therapy, tumor removal surgery, and site-specific treatments, play a crucial role in combating cancer [2]. Immunotherapy stands out as one of the most effective and commonly employed approaches in cancer treatment. Vaccines are pivotal in immunotherapies as they enhance immunity against cancer antigens. Vaccines contain antigens that stimulate the immune system to develop immunity against specific target antigens [3]. The immune system serves as the body's intrinsic defense mechanism against illnesses [4]. Since Edward Jenner's discovery of vaccinations in 1796, this preventive immu-notherapy method has demonstrated success. While vaccines have tradition-ally been used to prevent infectious diseases, recent years have seen their application in stimulating cytotoxic T-cells against infected or cancerous cells. For example, intramuscular injection of an antigen encoded in plasmid DNA has effectively boosted the antibody-mediated and adaptive immune response to the antigen [5]. The identification of these antigens is crucial in vaccine development. In contrast to most tumors, where the antigen is endogenous and improperly recognized, DNA vaccines are designed to cure cancer, which is most efficient when such cancer-causing agents or genes are of foreign viral origin [3]. The ability of DNA vaccines to boost immunity can be enhanced by genetic modification of the DNA in such a way that they have the ability to encode secreted homodimeric fusion proteins focused on antigen-contaminated cells (APCs). DNA vaccines have the advantage of activating all pathways of immunity. To achieve this, immune-activating segments must be encoded into the vaccine's DNA. As a result, two approaches are used when formulating a DNA vaccine. Selecting the suit-able antigen to install within the plasmid DNA in such a way as to enhance immunogenicity is the first step in the strategy. By using complementary therapy, the second strategy aims to increase immune cell activity at the site of action and/or immunosuppress immune cells there [3]. The development of genetic information-carrying material (DNA/RNA) based vaccines, also known as "third generation vaccines," was observed in the final years of the 19th century. Genetic information-carrying materials such as DNA and RNA have the ability to provide instructions for producing specific proteins that the body's immune system can recognize, thereby triggering an immune response against self or non-self-antigens. This valuable genetic information is translated into a protein antigen once it is introduced into host cells, subsequently eliciting an immune response. Research in gene therapy explored the possibility of replacing viral vectors with direct injections of DNA or RNA for transcription and translation. It was later found that introducing non-replicative DNA or RNA expression vectors into muscle cells through lipoplexes influenced the behavior of genetic material in the cells. Surprisingly, it was subsequently observed that this effect also occurred with plasmid DNA constructs, even in the absence of a lipid-based delivery method [6].

In-depth research has been conducted on cancer immunotherapy due to its emergence as a significant anticancer strategy that can halt tumor growth by eliciting an immune response against the tumor. Cancer vaccines (CV), as an additional immunotherapy approach, hold the potential to provoke immune reactions specific to tumors and establish immunological memory, presenting a promising therapeutic avenue. However, certain drawbacks, such as the toxicity affecting multiple organs or body parts and the inadequate immune response to tumor antigens, limit the efficacy of vaccines in treating such conditions. Ideally, CVs should be able to trigger tumor-specific immune responses without causing severe side effects. To induce robust immune responses, antigen-presenting cells (APCs) need to be provided with an adequate supply of tumor-associated antigens (TAAs). Additionally, an immune booster is administered alongside TAAs to activate dendritic cells (DCs). Subsequently, cytotoxic T cells and T helper lymphocytes are activated after DCs present antigens to T cells, which have been previously activated, directly stimulating B cells to produce numerous antibodies (Abs) and induce humoral immunity. Natural killer (NK) cells, a type of innate immunity, can be activated by DCs through the secretion of interferon-alpha. Finally, cytotoxic T lymphocytes (CTLs) and NK cells infiltrate the tumor, overcoming the immunosuppressive tumor microenvironment (TME) to eliminate tumor cells. Managing systemic toxicity is crucial throughout this process [8]. DNA-based vaccines are considered a promising solution for enhancing immunity against cancer, offering a straightforward, safe, and effective approach to leveraging the immune system against both infectious diseases and various types of cancer. However, there are numerous drawbacks to producing vaccines that contain plasmid DNA, such as long lead times and the requirement to clean bacterial cultures of impurities [7]. Internal pathogenic infections have been found to trigger spontaneous tumor remission. This conclusion demonstrated the immune system's potential anti-tumor abilities and served as the impetus for the creation of strong immunity, as well as therapeutic vaccines to cure cancer. The ease of use, stability, and safety make DNA vaccination a popular immunotherapeutic approach to treat cancer. Findings from numerous clinical trials show that the majority of patients accept the DNA vaccine well, with no notable adverse reactions. The DNA-based vaccine is affordable and can be administered repeatedly for lifetime immunity. DNA vaccines face challenges like strong antigen-targeting cell-mediated immunity due to the immune system's ability to identify endogenic self-antigens in tumors, despite all the practical benefits. Strategies such as encoding heterogenic antigens, fusing antigens to molecules that stimulate T cells and promote collaborative identification, utilizing viral vectors to enhance the effects of the DNA vectors, and using molecules or substances that support the immune system are all methods used to enhance the ability of DNA vaccines to boost immunity against self-antigens [4].

This chapter will focus on discussing cancer antigens, DNA vaccine mechanism, advantages and drawbacks, approaches for the development of DNA-based vaccines, optimal priming approaches, DNA vaccine and immune-tolerance, nanocarriers, and future prospects of vaccines that include DNA. One of the most significant achievements in medicine is considered to be the DNA vaccine. The use of vaccines has led to the eradication of diseases that once claimed millions of lives worldwide. Vaccines provide immunity by eliciting an appropriate immune response in the recipient. When a person is re-exposed to the same or a related disease, their immune system will respond rapidly due to the memory established by B and T cells. By utilizing the intrinsic machinery of host cells, DNA vaccines lower the costs associated with developing, purifying, and delivering a pathogen or antigenic protein. The DNA vaccination strategy emerged concurrently with advancements in molecular biology, providing researchers with new tools to investigate and manipulate the fundamental components of an organism's genome [9].

4.2 NEED OF DNA VACCINES

Numerous studies have examined and described how DNA can protect chickens or murine models against a broad range of viruses, bacteria, and protozoa. Cancer and autoimmune diseases have both been studied as potential targets for DNA immunization therapy. In addition to demonstrating efficacy as a vaccine against a wide range of veterinary and human diseases, such as malaria, tuberculosis, SARS, and HIV-1, the DNA vaccine has been approved for use by veterinarians on horses infected with the West Nile virus. Clinical studies have shown the effectiveness of plasmid DNA for immunization against various diseases like influenza, malaria, hepatitis B, human papillomavirus (HPV), and infectious diseases caused by HIV, as well as neoplastic diseases such as melanoma. These research projects and clinical studies have highlighted the effectiveness of the DNA vaccine against infectious illnesses [9].

4.3 CANCER ANTIGEN

Vaccination against cancer should target tumors specifically by expressing only malignant cells and being highly immunogenic to elicit significant immune responses. Due to negative selection, these cells present self-antigens and lack phenotypes that can activate immune cells [4]. The breakdown of proteins produced inside the cells occurs in the cytoplasm by proteasomes, initiating antigen processing. The transporter associated with antigen processing delivers the digested polypeptides to the endoplasmic reticulum (ER) (TAP). ER-resident amino peptidase associated with antigen processing (ERAP1 or ERAAP), located in the ER, trims the pro-peptides to optimize HLA class I binding. The peptide loading complex (PLC), composed of tapasin, ERp57, calreticulin, and TAP, generates stable peptide loading. Peptide exchange is likely facilitated by the PLC [10]. Cancer Testis Antigens are typically restricted to the testes and placenta in normal human tissues, making their presence in other contexts mostly impermissible to malignant tumors. CTAs are excellent candidates for immunotherapy due to their expression, which is only found in tumors, and their immunogenic properties. Previous research has shown that 14.8% of early primary triple-negative breast cancer tumors and 43% of locally progressed primary TNBC4 express the helicase antigen HAGE (DEADbox protein DDX43). HAGE has also been shown to play a role in predicting the outcome of adjuvant treatment for patients with breast cancer and as a potential prognostic marker [11].

DNA vaccines have poor immunogenicity. It has been particularly challenging to apply positive results in rodents to larger animals and humans. The poor immunogenicity of naked DNA is primarily mediated by administration using electropermeabilization (EP) or non-invasive needle-free jet injector methods [5]. In our nanodevices, peptides (OVA257-264, gp10025-33, or Adpgk) were utilized as antigens. Cancer/testis (CT) antigens, due to their limited expression profile, are promising immunotherapy targets. Testicular germ cells are the only healthy adult tissues that express CT antigens, making the testis a uniquely protected organ by the immune system. Tumors with various histological origins exhibit abnormal expression of CT antigens. CT antigens released in tumors, referred to as neoantigens, can trigger an immune reaction that is highly cancer-targeting. Trials of adoptive T-cell treatments and vaccinations targeting CT antigens have shown excellent results, indicating efficient action, but are limited by the diverse behavior strategy associated with these antigens. Only a small proportion of patients with a specific tumor type and often only a small percentage of cancer cells in tumors express Cancer/testis antigens (CTA). As a result, methods that target several CT antigens may be necessary to produce positive results in a variety of patients. Characterizing the expression profiles of additional CT antigens can lead to the identification of highly cancer-specific targets for cancer immunotherapy treatments [12].

Previous research has shown that immunodominant antigen epitopes can be more effective than the entire protein vaccine. Immunoinformatic techniques were used in our earlier study to identify the BORIS antigen regions that were most immunodominant. To create the finalized CVs containing DNA and peptide builds, the chosen areas were joined with the proper linkers and underwent numerous modifications. Bioinformatic analyzes revealed that the finalized vaccine design has been an immunogenic, an allergic, steady protein that contained a large number of cytotoxic T cells (CTL) epitopes, IFN-inducing epitopes, and straight and conformational B cell epitopes [13, 18]. Due to their strengths *in vivo* immunogenicity and relatively limited distribution in normal tissues, CT antigens may be satisfactory fragments to develop a vaccine for tumors against hematological malignancies [14]. The CT antigens, which were discovered to be expressed in several cancers including MM, seem to be of particular interest. In humans, CT genes exhibit a pattern of normal expression that may only be found in the testis' germ line cells. Some CTAs are tumor-specific due to the testis' immune-privileged location. CTAs are converted into intracellular antigens and presented as epitopes by HLA class I molecules [15].

In contrast to the testis, normal tissues do not express the carcinoma testis antigen (CTA) genes, making them appealing targets for vaccination treatment. CTLs obtained from the same person or animal, which are constrained by Human Leucocyte antigen (HLA) belonging to class I, can identify these antigens. For the treatment of various cancers, numerous vaccination treatments depend on Cancer/Testis antigen-derived peptides that have been created. CTA-negative neoplastic cells inside the tumor mass can also avoid immune detection by CTA-specific molecules, according to reports. However, CTAs are said to express themselves in tumor tissues in a localized or heterogeneous manner [16]. Numerous tumor types exhibit a class of tumor antigens commonly called CTAs, which are not found in usual tissues [17]. CT antigens are excellent choices for targeted cancer immunity because of their immunogenicity and limited availability. However, because the identified CT antigens are only seen in a small percentage of instances of a particular kind of tumor and because immune selection during vaccination causes the tumor antigens to gradually disappear, it is urgently necessary to find new CT antigens to create polyvalent tumor vaccines that will increase the effectiveness of immunotherapy [18].

A highly immunogenic antigen with targeted tumor-killing properties is appropriate to be included in CVs. However, negative selection has resulted in the self-antigens lacking tumor antigen characteristics because cancer cells cannot control their proliferation. To avoid autoimmunity, lymphocytes which possess a strong affection for the self-antigens of the host are removed during lymphocyte maturation. One of the first cancer antigens that were discovered to be identified by T cells belonged to a class of antigens termed MAGE-A1. They have been viewed as the best candidates for the creation of a cancer DNA vaccine since CT antigens are only expressed on particular tumor cells [19].

4.3.1 IDENTIFICATION OF CT ANTIGENS

The genetic copy of MAGE-1 by van der Bruggen et al. in 1991 represented a significant advance in the determination of tumor antigens recognized by T cells (CTL) of a host. MAGE-1, later called MAGEA1, melanoma antigen A1, was discovered as the target antigen for particular CTL copy utilizing the melanoma cell line MZ2-MEL and autologous CTL osmotic lytic to the line. Moreover, this was the first immunogenic tumor antigen that has been demonstrated to have caused the same cytotoxic T-lymphocyte behavior in a patient who suffers from cancer [20].

4.4 DNA CANCER VACCINE MECHANISM

Since mammalian cells can easily express genes encoded on plasmid DNA after transfection, plasmid DNA is used in DNA CVs. Plasmid DNA allows for longer gene expression and adaptive and cell-mediated immune responses to the encoded antigen. When introduced into the host's body and transferred within cells, a plasmid DNA containing an antigen and additional suitable genes under the control of a mammalian promoter in the appropriate conditions enables *in vivo* generation and expression with the assistance of the host's protein expression machinery. DNA vaccines have been shown to elicit a fundamental immune response. Depending on their composition and the site of vaccine administration, DNA vaccines can induce a unique adaptive and cell-mediated immune response specific to the antigen [4]. The appropriate gene can encode a therapeutic protein to directly target cancer cells. Examples include a cytotoxic peptide, a prodrug-activating enzyme, a proapoptotic protein, or bacterial toxins (B.T.). Plasmids encoding specific small segments of RNA (siRNA) molecules can be used for cancer gene therapy. In this scenario, transgenic expression in cancer cells can be controlled by a tumor-specific promoter. Another approach involves the target gene producing cytokines or antigens that activate immune cells, particularly lymphocytes or antigen-presenting cells (APCs), to combat cancer growth. The therapeutic protein can also be expressed by any cell that acquires the plasmid [21]. The fundamental goal of cancer immunotherapy is to introduce multiple tumor antigens into the host to stimulate an immune response for eliminating tumor cells [4]. Tumor antigens, primarily proteins highly expressed in tumor tissues, play a critical role in tumor development and progression. Tumor-associated antigens (TAAs) and mutational antigens (TSAs) are the two main categories of TAs [58]. One of the delivery methods involves transferring the desired genetically engineered gene sequence to the intradermal (ID) layer of the skin, subcutaneously, and intramuscularly. The plasmids enter the nuclei of nearby cells such as muscle cells or basal cells, as well as antigen-presenting cells (APCs), through the host cellular machinery. This process leads to the production of foreign antigens through plasmid-mediated gene expression. Upon interaction with a cluster of differentiation (CD289) or Toll-like receptor 9 (TLR9) present on APCs, DNA plasmids derived from bacteria trigger an immune response. With the help of identifying pathogen-associated chemical patterns, TLR9, which belongs to the category of transmembrane proteins on DC, plays a significant role in the body's defense mechanism. TLR9 interacts with a motif of hypomethylated CpG dinucleotides, which is abundant in the DNA of bacteria but uncommon in DNA found in mammals. In other words, unmethylated CG islands, which are often concentrated in DNA found in bacteria, including numerous plasmids, are recognized by TLR9. Type I IFNs are released when unmethylated DNA binds to TLR9, and in rare circumstances, proinflammatory cytokines are also released. Studies investigating TLR9 signaling contribution to DNA vaccine immunogenicity revealed that while it is not required for the development of robust immune reactions, it may contribute to their augmentation. Zahm et al. [22] used DNA vaccines encoding tumor antigens in preclinical and clinical studies, and they improved TLR9 signaling by utilizing a plasmid vector (pTVG4) with a modified CpG-rich sequence backbone. Nevertheless, TLR9 is one of several innate DNA sensors.

Many cells related to immunity, such as DCs, B cells, and NK cells, express the TLR9 receptor, which is activated when the inserted DNA is taken up through transfection or phagocytosis. TLR9 stimulation initiates a cascade of pro-inflammatory reactions, leading to the release of numerous cytokines. Localized inflammation and increased cytokine production resulting from the innate immune response can attract and activate additional immune cells, such as T cells, subsequently enhancing targeted immune responses. TLR9 activation specifically triggers Interferon Regulatory Factor (IRF) 7 *via* MyD88 signaling, promoting the production of Type I interferons (IFNs). Additionally, Tank-binding kinase 1 (TBK1) was identified as mediating the adjuvant effect of plasmid DNA. DNA sensors like DLM-1, Histone H2B, IFI16, Bead Box Helicase 41, LRRFIP1, and cGAS can detect intracellular DNA plasmids within the cytoplasmic matrix. These sensors then activate TBK1 and Stimulator of Interferon Genes (STING), which in turn activates IRF3, facilitating the production of Type I IFNs. These signaling pathways are crucial for the activation of T and B cells responsive to specific antigens [4]. Direct and indirect exposure of antigens to APCs enhances the antigen-specific immune response. The plasmid's antigen is produced in host cells and specialized APCs, leading to immediate immune response priming or transfer to APCs, resulting in cross-priming in non-professional cells. Both of these main models have been proposed. The DNA delivery mechanism has an impact on the types of transfected cells, according to early research. Epidermal keratinocytes and Langerhans cells were frequently directly transfected using the Gene Gun, which bombarded the epidermis with plasmid. These cells were then demonstrated to move quickly to nearby lymph nodes. Immature DCs called Langerhans cells, which are detected within the skin, play a role in the collection and processing of antigens. Antigens are expressed and processed when DNA is transfected into Langerhans cells, and they then enter the presentation route. After migrating to the lymph nodes to serve as specialized antigen-presenting cells (APCs), Langerhans cells expose naive T cells to antigens. Major histocompatibility complex (MHC) class I molecules on Langerhans cells deliver these endogenously generated antigens to activate CD8C cytotoxic lymphocytes (CTLs) [4]. Professional APCs were directly transfected in this instance and act as an origin of Ag presentation.

Plasmid introduction into the muscle mostly causes myocytes to become transfected. The direct priming of T lymphocytes by myocytes is not anticipated because they do not express MHC class II or costimulatory molecules. Since they are not professional APCs, they are unable to trigger robust immune responses with a specific focus. Therefore, after vaccination, dendritic cells migrate to the site of DNA infusion in response to inflammatory or chemotactic signals. Here, DCs are likely responsible for immunological priming. At this stage, APCs utilize processes like phagocytosis to capture antigens produced by transfected cells as illustrated in (Figure 4.1). When these foreign antigens interact with $CD4^+$ helper T cells through DCs' presentation via MHC class II molecules, a humoral response is initiated. Moreover, it is believed that these DCs present antigens through direct transfection of plasmid DNA or by cross-presenting extracellular antigens. Through cross-presentation to $CD8^+$ cytotoxic T cells, the acquired foreign antigens can be displayed in the context of MHC class I molecules, leading to the initiation of a cell-mediated immune response. Type I IFNs facilitate this process. Direct DNA transfection may also occur in APCs, which can prime $CD8^+$ T lymphocytes through direct antigen presentation via MHC class I. Antitumor immunity is induced through the stimulation of cellular immune responses. Notably, a strong correlation

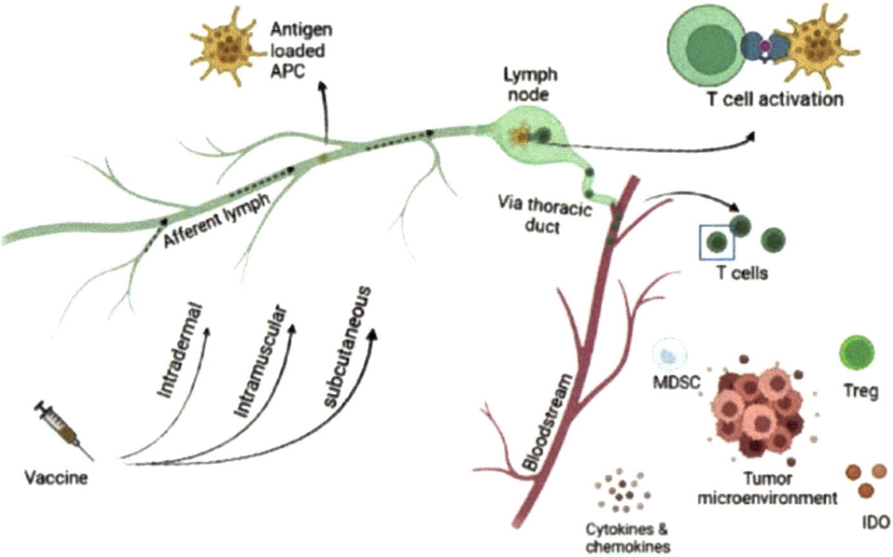

FIGURE 4.1 Route of administration and migration of immune cells.

has been observed between a robust $CD8^+$ cytotoxic T-cell response and a favorable prognosis for tumor control and clearance. Since DNA vaccines can generate antigens intracellularly, activating the MHC class I antigen presentation pathway, they are particularly effective in eliciting a $CD8^+$ T cell response [4, 23].

4.5 ADVANTAGES AND DRAWBACKS OF DNA VACCINES

4.5.1 ADVANTAGES

Long-term antibody resistance – a broad-spectrum vaccination could be created using a combination of plasmids [24]. Due to their simplicity, ease of manufacture, and safety, DNA vaccines offer several advantages over existing anti-tumor vaccine techniques [22]. The low toxicity of DNA vaccines encourages their ongoing development; however, for successful integration into clinical settings, additional methods to enhance their effectiveness are necessary [25]. Plasmid DNA particles encoding an important antigen introduce cell-mediated as well as antibody immune reactions, which is likely the most significant aspect. The *in vivo* synthesis of the encoded antigen DNA Vaccines 1190 allows the protein to be processed for presentation on the MHC class I complex, promoting the development of class I-limited cytolytic T cells (CTL). CV strategies based on the activation of CTL against tumor antigens are effective because CTLs are recognized as significant mediators of the anticancer immune response [25]. DNA vaccines are significantly more cost-effective and easily accessible. CVs work by triggering an immune response targeted towards tumor antigens, ideally long-lasting. They are easily modifiable in design, scalable for broad-scale manufacture, non-infectious, non-specific to individuals with a particular MHC type, and thermally stable for simple distribution and transport [26]. Plasmid DNA vaccines are advantageous for their simplicity in production and development. Additionally, a substantial body of evidence supports the safety of plasmid DNA-based products, with several products currently in late-stage clinical trials [27]. Genetically engineered lentiviruses, adenoviruses, adeno-associated viruses, and retroviruses are valuable due to their high transfection ability and stability [28]. The significant transfection effectiveness and stability of genetically engineered lentiviruses, adenoviruses, adeno-associated viruses, and retroviruses make them beneficial [29]. Furthermore, additional advantages are summarized in (Table 4.1).

4.5.2 *DRAWBACKS*

Only protein immunogens are not utilizable for non-protein-based antigens such as bacterial glycan. For instance, pneumococcal and meningococcal vaccines utilize protective polysaccharide antigens [24]. When administered as an unformulated intramuscular injection, plasmid DNA vaccines exhibit weak immunogenicity, which is their primary drawback. To achieve only mild immunogenicity, large doses (5–10 mg) are required [30]. This is undesirable due to their high cost, adverse side effects, limitations on gene transfer amounts, and potential for insertional mutagenesis [31]. The main limitation of DNA vaccines has been the inefficiency of directly injecting plasmid DNA into host cells, especially in primates compared to rodents. Consequently, DNA vaccinations result in weaker immunogenic effects than viral vaccines [32]. Additionally, several other disadvantages are outlined in Table 4.1.

TABLE 4.1 DNA Vaccine Advantages and Disadvantages [4, 57, 23]

Advantages	Disadvantages
Flexible DNA vector design is made possible by molecular recombinant DNA technology, which can encode a variety of antigens and immunomodulatory compounds.	Low immunogenicity.
The DNA plasmid has inherent capacities to activate innate immune responses.	*In vivo*, naked DNA is difficult to transmit from one cell to another.
DNA vaccines have the ability to polarize the resulting immune response, leading to either humoral or cellular immune behavior. This can be achieved by utilizing different administration routes and modifying the antigen processing pathways to selectively activate Th1 or Th2 Helper T cells.	APCs are not readily able to take up expressed antigens and trigger adequate immune responses.
DNA vaccines induce immunological responses that are mediated by $CD8^+$ cytotoxic T lymphocytes (CTLs), which play a crucial role in tumor destruction.	May require a special administration device.
They are simple to engineer. DNA vectors are extremely stable and can be effortlessly stored, making it possible to transfer them anywhere.	Potential integration into the human genome.
Risk-free from pathogenic infection.	Low transfection efficiency.
DNA vaccination is a feasible alternative due to its safety, simplicity of synthesis, and low cost.	–

4.6 APPROACHES FOR DEVELOPMENT OF DNA VACCINE

DNA vaccines are a promising strategy with significant potential for clinical use. They are more cost-effective compared to alternative vaccinations like recombinant proteins, tumor cells, or viral vectors. Recent advancements in molecular biology and recombinant techniques have enabled plasmid gene manipulation, along with an increased understanding of tumor antigens. To regulate immunological responses, DNA vaccines can incorporate various antigens and immunomodulatory substances as sequences. Furthermore, DNA vaccines offer the advantage of multiple doses, and numerous studies have demonstrated their safety profile [4].

4.6.1 VARIOUS APPROACHES PHYSICAL DNA DELIVERY

A unique approach to gene transfer called magnetofection combines biochemical and physical transfection elements. Due to the use of cationic magnetic nanoparticles (NPs), it provides high transfection efficiency. Nucleic acids are transported into cells by an external magnetic field while they are attached to magnetic NPs. Magnetofection has been successfully employed for transferring into both cell lines and primary cells. Sonoporation, also known as cellular sonication, involves the temporary modification of the cell membrane's permeability by ultrasound to facilitate the transfer of mostly nucleic acids. Sonoporation occurs when nearby cells undergo sonic cavitation. The force generated by ultrasound and gas cavities within microbubbles can induce the formation of tiny pores in the cell membrane [33].

4.6.2 APPROACHES FOR DNA DELIVERY THROUGH DEVICES

For the delivery of DNA-coated metal particles to create genetically modified sets, a gene gun was used to administer DNA as well as conventional immunizations. DNA delivery techniques mediated by devices have been investigated in DNA vaccination studies. A biolistic system known as a "gene gun" was initially used to deliver metal particles coated with DNA to create genetically modified sets. Both DNA and conventional immunizations were administered via the gene gun [33].

4.6.3 APPROACHES FOR DNA DELIVERY THROUGH 'DANGER-SIGNAL'

Numerous mini-chaperones have recently been used in the creation of innovative vaccine methods. By binding antigenic peptides through CD91 and other receptors, heat-shock proteins (HSPs) function as molecular chaperones and transfer them to antigen-presenting cells. HSPs transport peptides made inside of cells and mediate their transfer to the MHC class I molecule of that cell; HSPs released as a result of cell lysis transport chaperoned peptides to the MHC molecules of nearby APCs. HSPs can be exploited as potent adjuvants in innovative ways in cancer immunotherapy, according to several studies that have found that they have immune-stimulating function. Numerous HSPs have been shown to activate NK cells as well as toll-like receptors (TLR) on DCs in part [33].

4.7 OPTIMAL PRIMING APPROACHES

The first process in the immunity-stimulating cascade of DNA vaccines is the delivery of the encrypted DNA vector. Because DNA vaccines are generally less immunogenic, it is crucial to select the administration route that yields the highest immunity. The majority of vectors used to deliver DNA plasmids to cells end up in the extracellular space and are typically injected intramuscularly or intradermally. The minimal transfection efficiency is due to the fact that many cells at the injection site are ineffective at absorbing the injected DNA vectors. To achieve adequate DNA uptake, large amounts of DNA would need to be administered, which would reduce the cost-effectiveness of DNA vaccines [4]. Priming is the initial and most essential step in the development of immunity and involves direct collaboration between antigens and helper T-cells, showing optimal DNA vaccine activity. However, cells near the injection site are generally incapable of effectively absorbing DNA vectors, resulting in poor transfection. In such cases, a large amount of DNA must be administered to ensure effective cellular uptake, making the medication less cost-effective. To deliver DNA directly to APCs, a delivery system must be in place. One such strategy is utilizing the Gene Gun, a delivery system where heavy metal nanomaterials (typically gold) coated with a plasmid DNA vector are used to overload the keratinocytes with the device. Pressurized helium is often used to aid the process. Consequently, the antigen can be taken up by the DC and transported to the lymph nodes [19].

One of the most challenging tasks in developing a CV is determining the appropriate antigens to elicit a specific antitumor immune response. Tumor antigens that are ideal for vaccines are those that can distinguish between normal and cancerous tissues. These antigens can be self or non-self (neo-antigens) based on their functionality. Self-tumor antigens are generated by unaltered genes that meet certain criteria, such as limited expression to fetal tissues, heightened expression in cancer, or cancer-specific post-translational modifications. Non-self-antigens can originate from exogenous or endogenous sources. Exogenous non-self-antigens may include peptides from cancer-causing viruses, while endogenous non-self-antigens can arise from mutated proteins due to genetic alterations in malignancies. Cheever et al. identified 75 types of cancer antigens under investigation, along with nine criteria for assessing their suitability. These criteria encompass factors like therapeutic efficacy, immunogenicity, oncogenic potential, specificity, expression levels and patterns, stem cell expression, patient groups with antigen-positive cancers, epitope quantity, and cellular localization of expression [34].

There are some additional components that may be required to optimize immune cell vaccines, especially DC vaccines. The first of these components is of the DC type. The most commonly used DCs, MoDCs, have some flaws due to their slow generation time and limited function. However, naturally occurring DCs have been found to be better at delivering antigens since they express more MHC molecules. The efficacy of the DC vaccine has been shown to be influenced by the type of DC used. Myeloid/plasmacytoid DCs (mDCs/pDCs) and conventional DCs (cDCs) are often used as substitutes for MoDCs. Patients with stage III melanoma and metastatic castration-resistant prostate cancer (PC) were treated with natural DC, which includes mDCs, cDCs, or a combination of the two (NCT02692976, NCT02574377). Since mDCs and cDCs secrete different cytokines and have different TLRs, they have different actions. Therefore, their combined application may be beneficial in treating murine lung cancer (LC) [35].

4.8 CANCER VACCINES AND IMMUNOTOLERANCE

Immunotherapy for cancer is a challenging process because the immune system can become tolerant to its own antigens. Developing a tumor vaccine using plasmid DNA presents challenges, particularly overcoming immunologic tolerance. The immune system can tell the difference between self- and non-self-antigens, which is essential to its functioning. It must be able to mount an effective immune response against infections and cancer cells while also maintaining tolerance to autoantigens. However, if the balance between tolerance and immunity tips too far in either direction, it can lead to undesirable health outcomes such as infections, cancer, or autoimmune diseases [36]. Immunological tolerance is a significant barrier to DNA vaccination because many tumor antigens are self-antigens and unable to stimulate a robust immune response. DNA vaccines also face challenges in activating tumor-specific immune function since most TAAs are transfected self-antigens. Central and peripheral tolerance mechanisms and tumor-induced immunosuppression regulate the activation of individuals to make T-cell responses against these low-immunogenic antigens [37]. The immune system is incredibly adaptable, but sometimes it stops responding to specific antigens. T-cell tolerance results from two distinct processes: the reduction of self-reactive T cells during thymic T cell maturation (central tolerance) and the inhibition or inactivation of peripheral self-reactive mature T cells [38].

The tumor environment is a complex interaction of various cells such as tumor-bearing host cells, cancer cells, TILs, fibroblasts, MDSCs, iDCs, and stromal cells. This interaction produces proinflammatory and tolerogenic cytokines that enhance the tumor's ability to promote immunological tolerance [39].

Although early stages of carcinoma may see some cancer cells cleared by innate and adaptive immune responses, melanoma cells (CiC) quickly give invading immune cells instructions to modify regulatory phenotypes. This results in an intense immune-suppressive response and a build-up of Tregs, B cells, DCs, and M2 macrophages within the TME. These cells may provide immuno-regulatory mediators like IL-10 or VEGF, and B cells could produce IgG4 and IgA with a different Th2-biased molecule [40]. Professional antigen-presenting cells, known as DCs, have the rare capacity to trigger the distinctions of naive T cells into CTLs that are specifically directed against tumor antigens. After DC-mediated antigen presentation, pro-inflammatory cytokines (IL-1, IL-12, IL-6, and TNF) must be secreted in order for proliferative T cell responses to be effective (please refer to Figure 4.2 for a detailed description) [41, 47].

DCs not only act as immune system sentinels, but they also perform legislative functions. Myeloid DCs undergo changes within the TME, transforming into tolerogenic phenotypes that support regulatory T cells (TRegs) instead of T-effector cells as a means of evading immune clearance. This

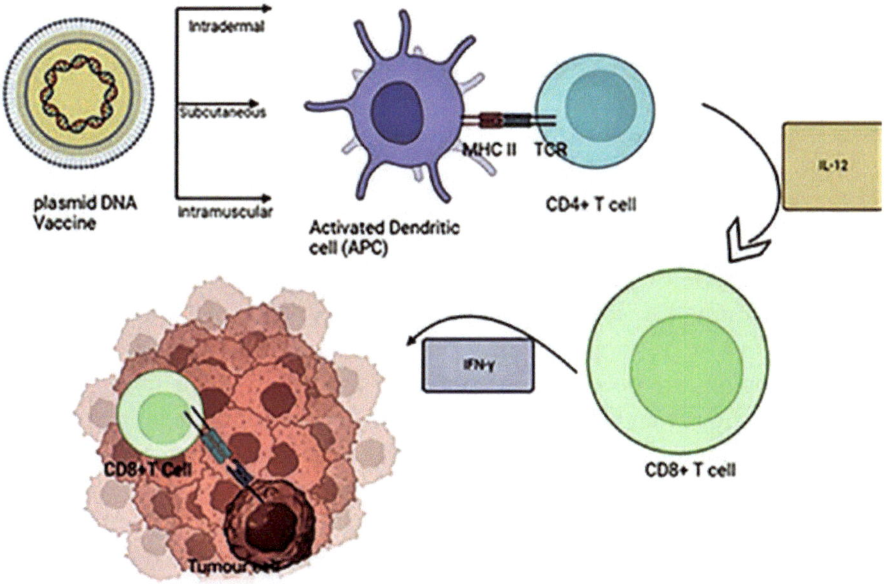

FIGURE 4.2 Tumor antigen-specific immune response.

escape strategy is mediated by transforming growth factor (TGF) and interleukin-10 (IL-10). Tolerogenic DCs produce low levels of IL-12 and IL-10, express low levels of co-stimulatory molecules, and generate IL-10. Cancer patients should avoid DCs that promote antigen presentation as it may inhibit the anti-tumor T cell response [40]. Tregs, a subset of $CD4^+$ T cells, are an immune-suppressive cell type that is present in the TME. Tregs can inhibit the immune system both in tumor-specific and non-tumor-specific ways, and their frequency has been associated with poor tumor prognosis. Pre-clinical and clinical studies have shown that anti-cancer vaccination is less effective in the presence of higher Treg frequency [14]. Solid human tumors, such as non-small cell lung cancer (NSCLC), melanoma, breast, ovarian, cervical, and colorectal cancer (CRC), contain malignant B cells (TiBCs) that can influence immune reactions against cancer cells through various mechanisms. TiBCs can promote TLSs or antigen presentation, which can modulate the immune systems of other TILs. TiBCs may also directly exert cytotoxicity through B cell-secreted mediators; however, Bregs may function similarly to promote immunological tolerance or amplify Treg responses, thereby accelerating tumor growth [40].

4.8.1 BREAKING IMMUNOTOLERANCE

DCs must process TAAs before presenting them to antigen-specific T cells for stimulation and growth. This process must continue for as long as it takes the T cells to completely destroy the malignancy. Suppression can occur at any point during this process and result in immune tolerance. Immune alteration can lead to a significant decrease in T cells capable of responding to specific tumor antigens. Insufficient co-stimulatory signals can impede the full activation and development of malignant lymphocytes. While a robust immune response is initiated, it may not persist long enough to induce tumor regression. Effective trafficking and accumulation of activated T lymphocytes are necessary at the tumor site. Once in the TME, they must overcome fatigue and immunosuppression. Tumor cells can induce tolerance and reduce immune efficacy through various mechanisms, such as altering antigen presentation machinery or secreting immunosuppressive substances that cause lymphocyte cell death or activate adverse regulatory pathways. The growth of tumor cells is facilitated by a selective survival advantage enjoyed by those cells that directly or indirectly enhance immunogenicity. Depending on the stage of the procedure at which the therapy is applied, cancer immunotherapy techniques aimed at breaking tolerance can be broadly categorized. These techniques are discussed in subsections.

4.8.1.1 IMMUNE EFFECTOR ADOPTION

4.8.1.1.1 Monoclonal Antibodies (mAb) Those Are Tumor-Specific

MAbs can induce anticancer effects through specific mechanisms such as cell-mediated immunity, cytotoxicity, ingestion, and complement-dependent neurotoxicity. They can also modify tumor cell signaling pathways and disrupt melanoma connections. When tumor antigens are cross-presented by mAb-induced tumor cell lysis, they can be more efficiently absorbed and presented by DCs, leading to the activation of adaptive immune system responses. This example illustrates how altering the immune system in one way can impact the immunological response in multiple ways.

Current research is focused on modifying the structure of mAbs to enhance the interaction between effector cells like DCs and NK-cells. Immunoconjugates, which combine the effectiveness of anticancer agents with the selectivity of mAbs, are being explored as a valuable tool to overcome immunological tolerance. Immunoconjugates do not rely on immune identification of the target antigen. For instance, Trastuzumab Emtansine is used in breast cancer, LC, and CRC, while Brentuximab Vedotin is used for relapsed or refractory systemic anaplastic large-cell lymphoma (ALCL), CD30+ peripheral T cell lymphoma (PTCL), and Hodgkin's lymphoma. For example, in the case of Trastuzumab Emtansine, it is used for breast cancer, LC, and CRC. Brentuximab Vedotin is used for relapsed or refractory systemic ALCL, $CD30^+$ PTCL, and Hodgkin's lymphoma. Trastuzumab Deruxtecan is used for $ErbB2^+$ advanced breast cancer, ErbB2-expressing NSCLC, and ErbB2-positive gastric cancer. Nivolumab immunotherapy combo is used for advanced breast and urothelial malignancies, among others [42].

Bispecific mAbs, by attaching to tumor antigens with one arm and an antigen on immune effector cells with the other, exploit the cytolytic ability of T cells. This redirects the T cell, regardless of its native specificity, towards cancer cells that present the targeted antigen. Bispecific mAbs have shown clinical effectiveness in certain cancers and growth. Some cancers and developmental processes have demonstrated therapeutic efficacy with bispecific mAbs [43].

4.8.1.1.2 Transfer of Adoptive T-Cells (ACT)

By extracting lymphocytes from patients with tumors, cultivating them *ex vivo* in the presence of specific growth factors, and subsequently reintroducing them into the patient, this procedure is referred to as adaptive T cell transfer. The addition of interleukin 2 (IL-2) co-infusion was implemented to enhance T cell activation. Carcinoma cells lymphocytes were utilized with the expectation that their elevated anticancer T cell counts would enhance the selectivity and effectiveness of the transplanted cells. Chimeric antigen receptor expressing T cells that have undergone adaptive transfer (CARs) are employed. The complementarity-determining region (CDR) of Abs is fused with the signaling domain of a T cell receptor (TCR) to create chimeric single-chain constructs known as CARs. When CAR genes are transferred to homologous T cells, the result is activated T cells that proliferate *in vivo* upon interaction with their target antigen. Clinically, this can lead to the eradication of a significant tumor burden and the development of acquired immunity against specific targeted antigens. For instance, CD19-based CARs show great promise in the treatment of B cell malignancies [44].

4.8.1.2 VACCINATION

Therapeutic CVs provide tumor antigens, peptides, or entire tumor cells to elicit an immune response in individuals. Immune tolerance is commonly observed in patients with tumors expressing the antigen. Various techniques are employed to overcome this tolerance and enhance the efficacy of vaccinations, including antigen selection and modifications, the use of antigen presentation cells such as DCs, and the application of immunological adjuvants. However, factors such as immune modification, reduced antigen processing and presentation, and the challenge of maintaining a robust immune system can all diminish the effectiveness of CVs. Cell-based vaccinations allow the host to generate immune responses against a wide range of malignant cell antigens, resulting in a more diverse immune response. This can be advantageous if the immunological response is stronger for one antigen compared to another. To enhance the ability to break tolerance, strategies such as altering autologous cells, genetically modifying tumor cells, fusing malignant cells with APCs, and utilizing multiple immune adjuvants can be employed [45].

Immunization with undamaged peptides derived from malignant cell antigens has the potential to overcome immune tolerance by bypassing antigen processing. These peptides can bind to MHC molecules, typically class I, on DCs, thereby delivering a strong activating signal to T lymphocytes. The use of immunological enhancers may further enhance the activation of peptide-specific T lymphocytes. However, there are several challenges that must be addressed for peptide-based vaccinations to be considered successful. These include selecting the appropriate peptide, dealing with the heterogeneous expression of the peptide by tumor cells, enhancing DC stimulation at the immunization site to optimize T cell activation, attempting to prevent peripheral tolerance even after an immune response has been generated, and blocking tumor immunomodulatory processes. If peptide vaccines are to be clinically successful, they will likely need to be combined with alternative techniques to overcome immunogenicity that go beyond the use of typical immunological additives. DC-based vaccinations were developed in response to the discovery that while DCs are crucial for processing and presenting tumor antigens, they may be functionally deficient or insufficiently present in cancer patients [46, 47].

4.8.1.3 IMMUNOMODULATORY THERAPY

Myeloid-derived suppressor cells (MDSCs) are widely present within the TME and exert potent inhibitory effects on T lymphocytes. MDSCs also stimulate T regulatory cells (Tregs), which suppress effector T cell function through various mechanisms. In various types of cancer, the presence of MDSCs and Tregs at the tumor site or in peripheral blood has been associated with a poor prognosis. Other tumor-produced chemicals, such as IL-10 and VEGF, can hinder the development of myeloid DCs and result in the accumulation of immature DCs with reduced activation of costimulatory molecules (CD80/CD86), leading to T cell anergy. To shift the balance from immune tolerance to an immunological response, a wide range of drugs that can modulate immune regulation within the TME have been employed. IL, which has been studied both as a standalone agent and in combination with other vaccine components, is a factor that influences T cell activation. Another requirement for immunogenic T cell activation is receiving co-activatory signals from DCs, which would trigger T cell proliferation and activation. Clinical trials are underway for mAbs that can provide costimulatory signals. GITR, CD40, OX40 (CD134), and 4-1BB (CD137) are other examples. Immune checkpoints tightly regulate the strength of T cell responses and are crucial for preventing autoimmunity. Checkpoint blockade is a promising approach to unleash the strength and durability of beneficial anticancer immune responses. Cytotoxic T-lymphocyte antigen 4 (CTLA-4), programmed death 1 (PD-1), T-cell immunoglobulin and mucin domain-containing protein 3 (Tim-3), and lymphocyte activation gene 3 (LAG-3) are essential T-cell membrane components involved in regulatory checkpoint control. These molecules function to inhibit effector T cells and are highly expressed on Tregs. Therefore, they are highly desirable as strategies for reversing immunological tolerance.

In situ immunization, specifically 3C *in situ* immunization, modifies the TME to promote a systemic anticancer response and overcome immune resistance. By altering the TME, the anticancer response can be enhanced. There are several TLR agonists in malignancies that induce systemic anticancer responses. However, to successfully achieve *in situ* immunization and modify multiple aspects of the TME, therapies are likely required. An ideal strategy would involve treatments that promote antigen uptake and presentation, induce immunomodulatory cell apoptosis, stimulate T-cell activation, and sustain T-cell responses. Although complex, early research suggests that this approach may be beneficial [48–51].

4.9 FORCES IMPACTING ON DNA CANCER VACCINE

The forces impacting on DNA cancer vaccine are listed in Table 4.2.

TABLE 4.2 Forces Impacting on DNA Cancer Vaccine

Forces Impacting on DNA Cancer Vaccine			References
Force I: Entry Barriers (Unquestionably, the first step in the development and administration of a successful DNA vaccine is the wise selection of the antigen and the route of entry within the host body. DNA-based vaccines are an appealing option because antigens expressed in cancers with non-viral etiologies can elicit an immune response directly targeted against cancer cells.)	Tumor-associated antigens.	These antigens can be either expressed during tumor development, highly expressed by oncogenes, or chosen as distinguishing antigens.	[4, 58]
	Neoantigens	The much more powerful antigen to be used in the creation of a DNA vaccine will be completely foreign to the patient because there is no central tolerance already present. Neo-antigens, also referred to as self-antigens, mutate to produce fresh non-self-antigens.	[21, 52, 58]
	Plasmid backbone	A DNA vaccine plasmid backbone must contain a strong promoter sequence, an antibiotic selection marker, and a poly-A sequence to stabilize mRNA transcription.	[21, 23]
	Administration site	The TAA gene must be delivered to the DCs using DNA vaccine in order to ensure efficient antigen presentation.	[23, 59]
Force II: Direct Stimulator of T-Cells Response	Duties of APCs	DCs receive concentrated antigen delivery as a result of the DNA vaccine design. Together, $CD4^+$ and $CD8^+$ T-cell responses are upregulated by these activated DCs.	[23, 66]
	Role of $CD8^+$ T-cells.	After the administration of a DNA vaccination, $CD8^+$ T cells are activated whenever the expressed TAAs are displayed alongside MHC class I proteins. These newly stimulated CD8 T-cells carry out their effector function by lysing the tumor cells when they come into contact with them, through the secretion of the pore-forming protein perforin in a calcium-dependent manner into the target cell membrane. The stimulated $CD8^+$ T-cells also release large quantities of inflammatory cytokines such as interferon (IFN) and tumor necrosis factor (TNF) to promote an anti-tumor response.	[66]
	Role of $CD4^+$ T-cells.	$CD4^+$ T-cells are essential for effective and durable effector and cellular memory responses. B cells generate antibodies, and $CD4^+$ T-helper cells, which are primarily activated by epitopes in the context of MHC class II molecules, promote memory $CD8^+$ T-cell responses.	[66]
	Duties of fusion genes on T-cell activation.	Many research teams have combined the TAA gene with the single-chain Fv (scFv) sequence of the immunoglobulin differential region genes in recombinant plasmids to cause the activation of $CD4^+$ T cells from the non-tolerized $CD4^+$ T cells nave repertoire.	[63]

TABLE 4.2 *(Continued)*

Forces Impacting on DNA Cancer Vaccine			References
Force III: The Immune Avoidance Threat	Downregulation of MHC Class I on tumor cell.	MHC class I molecules, which are essential for the innate immunity and stimulation of cytotoxic $CD8^+$ T cells, are expressed by all nucleated cells. Numerous investigations have shown that tumors specifically exhibit decreased cell surface expression of MHC class I proteins. The transcript stages of several other proteins necessary for the class I presentation route have also been decreased in tumor cells, in addition to the particular decrease in MHC class I protein expression.	[64]
	Tumor associated macrophages	Immune-suppressive, pro-tumor, alternatively activated M2 macrophages are more prevalent in the tumor microenvironment. These M2-type macrophages contribute to TH2 cell responses, reduce inflammation, inhibit immunity, and accelerate tumor growth. In many tumors, a significant infiltration of M2-type macrophages is related to a poor prognosis.	[64]
	Myeloid-derived suppressor cells (MDSCs)	In large quantities, the blood and tumors of cancer patients contain immature myeloid cells known as MDSCs. The production of the cytokine GM-CSF by tumor cells promotes the accumulation of MDSCs within tumors. These MDSCs associated with tumors hinder the maturation of naive T-cells into cancer-fighting cells and their activation in the tumor microenvironment.	[64]
	Regulatory T-cell (Tregs)	In the tumor microenvironment, the immune-suppressive Treg subset of $CD4^+$ T cells is present. Furthermore, it has been demonstrated that Tregs can suppress the immune system in both tumor-specific and tumor-nonspecific ways. Numerous studies have shown a correlation between Treg frequency and tumor prognosis.	[64, 65]
Force IV: Secondary Stimulator	Inflammatory cytokines	For optimal stimulation of $CD8^+$ T-cell and effector cytotoxic responses, the cytokines type I interferons (IFNs) and interleukin-12 (IL-12) are required. Both the innate and adaptive immune systems produce these inflammatory cytokines. T-cell activation by IL-12 has been demonstrated, while type I IFNs are known to directly stimulate $CD8^+$ T-cells and other cell types, including APCs. It has been shown that the pluripotent cytokine IL-2 promotes T-cell maturation and proliferation. Future cytokine-DNA vaccination strategies may have greater theoretical achievements.	[61]

TABLE 4.2 *(Continued)*

Forces Impacting on DNA Cancer Vaccine			References
	Adjuvants	The adjuvants function by attaching to their ligands, such as poly I:C and CpG islands on cellular DNA, in order to induce costimulatory molecules like toll-like receptor factors (TLR 3 and 9). Gene fusion with TLR ligand pattern recognition peptides can further enhance DNA vaccines by increasing APC activation.	[23, 61]
Force V: Supplements	With chemotherapy and radiotherapy	Chemotherapy targets tumors depending on the distinct genetic makeup of cancer cells, which favors rapid cell growth compared to typical terminally differentiated cells. Additionally, numerous studies have demonstrated that these anti-cancer medications can stimulate the reactivation of immune responses specific to tumors by enhancing T-cell infiltration into the tumor microenvironment and increasing the susceptibility of tumor cells to cytotoxic $CD8^+$ T-cell effector responses. Combining chemotherapy and DNA vaccination can not only effectively eliminate cancer cells but also induce effector immune responses in the tumor microenvironment.	[62]
	With immune-check point inhibitors	Co-stimulatory molecules such as CD28 and OX-40 mediate signaling pathways that are crucial for the merger of T-cells with APCs and subsequent stimulation and maturation. T-cell anergy is caused by a significantly low density of ligands for co-activatory molecules in the tumor microenvironment. However, stimulation with agonistic mAbs that are specifically designed to bind to the co-activatory molecules and act as surrogate ligands can overcome this T-cell insensitivity.	[60, 61]

4.10 NANOPARTICLES FOR ENHANCING DNA VACCINE EFFICACY

The major barrier to the development of DNA vaccines in humans is their weak immunogenicity. In this context, the majority of DNA vaccine-related activities at the moment are concentrated on the use of molecular adjuvants and delivery methods that can greatly enhance the immunogenicity and effectiveness of DNA vaccines. To achieve optimal results from the DNA vaccination method, this complementary modality could be crucial. The introduction of foreign DNA into cells using NP-mediated plasmid delivery is particularly helpful for easy and affordable immunization. In terms of stability, solubility, and pharmacology, this can be a good approach to overcome the drawbacks of conventional administration methods [9]. There is no one-size-fits-all method for delivering genes, thus novel approaches, chemicals, and physical or biological vectors (such as viruses and bacteria) are regularly proposed. The long-term objective of gene therapy is the creation of effective, non-toxic gene delivery systems and methods that can transport and distribute foreign DNA to specific cell types. Recombinant viruses and bacteria have shown promise and are currently the subject of extensive research. The most explored approach in viral vectors that activates host tumor immune reaction via T-cell-specific immune responses is vaccination with tumor-specific antigens [54].

Oncolytic viruses (OVs) are used in cancer treatment to directly destroy cancer cells while also inducing an immune reaction by releasing antigens unique to the tumor, causing inflammation. It has been demonstrated that oncolytic therapy has a synergistic impact in the treatment of cancer when combined with other treatments such as radiation therapy, immunotherapy, chemotherapy, and thermal radiofrequency. The effectiveness of cancer therapy can be enhanced by genetically modifying the bacterial and viral vectors. The clinical safety and efficacy of recombinant bacterial and viral vectors could potentially be improved, making it easier to utilize these vaccines therapeutically [1].

4.10.1 NANOCARRIER FOR DNA VACCINE

Nanocarriers are particles with a size ranging from 1 to 1,000. They are typically multilayered and made of a variety of materials. The potential of NPs to enhance the development of nanovaccines is under investigation. Nanotherapeutics with specific targeting capabilities, improved systemic availability, and controlled drug release patterns are particularly useful for cancer therapy. The shape, charge, size, and lipid solubility of the nanocarrier (NC) significantly influence its interaction with the cell membrane. The NC can be internalized by the cell through receptor-mediated endocytosis or membrane permeation, facilitated by transient pore formation. To enhance cellular uptake and endosomal release of NC-complexed DNA, cell-penetrating peptides (CPP) are often included in the formulation. The NC surface has been modified to minimize serum protein adsorption, such as by densely decorating it with polyethylene glycol (PEG), which helps maintain the behavior of CPP [3]. The creation of engineered objects or substances at the nanometer scale, at the molecular level, is known as nanotechnology. They are involved in the development of analytical tools, physical therapy equipment, contrast agents, diagnostic tools, drug delivery systems, and drug discovery. DNA vaccines, which can induce adaptive and cell-mediated immune responses and are currently considered the safest vaccines, have emerged as the most effective treatment for challenging diseases. Conventional vaccines have not been effective against a significant number of infectious diseases. Particularly for cancer and viral illnesses, DNA vaccines appear to be the preferred choice. However, the level of immunogenicity they generate in primates is insufficient for use in humans. There is evidence to suggest that these immune responses can be enhanced through the use of appropriately designed NPs, which can prevent DNA degradation and enable targeted and controlled delivery to antigen-presenting cells. DNA formulations can be stabilized through adsorption, formulation, or particle entrapment. NP delivery for vaccines is a versatile technology that has been successfully used for various vaccines. Lipid-based dendrimer-calcium-phosphate (TT-LDCP) NPs, in combination with thymine-functionalized dendrimers, demonstrate enhanced gene delivery capability and immune combination therapy potential by activating the STING-cGAS route. Hepatocellular carcinoma (HCC) was treated with TT-LDCP NPs, which delivered siRNA targeting the immune checkpoint ligand PD-L1 and the immune activator IL-2-encoding plasmid DNA. This approach promoted tumor infiltration, activated $CD8^+$ T cells, improved the efficacy of CVs in boosting immunity, and inhibited HCC progression. Exosomes, microparticles (or microvesicles), and apoptotic bodies are examples of extracellular vesicles (EVs) that cells can produce. These EVs are classified based on their size and mechanism. Exosomes, small multivesicular bodies (30–100 nm) released from endosomes into the extracellular environment after fusion with the plasma membrane, play a crucial role. In response to various stimuli, cells modulate their cytoskeletal structure by releasing Ca^{2+} from the endoplasmic reticulum (ER), leading to the encapsulation of cytosolic components within the plasma membrane and the release of vesicles into the extracellular space [55].

By focusing on the C-type lectin receptors (CLRs) found on antigen-presenting cells, mannose- and bisphosphonate-modified calcium phosphate NPs act as efficient carriers for vaccine delivery to antigen-presenting cells (APCs) [56].

4.10.1.1 LIPID-BASED SYSTEMS FOR DNA ADMINISTRATION

Among the most innovative delivery strategies are lipid-based systems for DNA administration. These systems increase DNA intracellular uptake, improve delivery to target APCs, and protect the loaded DNA. Cationic solid-lipid nanoparticles, known as SLN, are one of the NPs approved for pharmaceutical use in humans. SLN are lipid-based colloidal delivery systems that are physiologically acceptable. The protective efficacy of cationic SLN in treating NSCLC cells has been investigated. It has been discovered that cationic SLN-mediated p53 gene delivery has clinical capacity and is effective for inducing apoptosis and inhibiting tumor growth [9]. Lipids have been studied for gene introduction to the body for several decades. Felgner and colleagues were the first to show that N-[1-(2,3-dioleoyl oxy) propyl]-N,N,N-trimethylammonium chloride (DOTAP) could transfer DNA within cells. *In vivo* transfection utilization was high when DOTAP was complexed with protamine-condensed DNA. Cationic lipid complexes serve as an ideal platform to deliver lipids with high improved efficacy due to the anionic nature of cellular membranes. Cationic lipids are attached to the cargo, which is DNA, to prevent its degeneration. A neutral supporter lipid stabilizes the lipid bilayer and may even facilitate the delivery of the DNA by buffering the pH of protonatable groups or interacting with the endosomal membrane. The main mechanism of uptake for DOTAP-based NPs, which include cholesterol as a supporter lipid, was found to be micropinocytosis. The liposomal coat of these NPs has been enhanced with coiled lipopeptides to enable efficient delivery of cargo into the cytosol. Liposomes' ability to reorganize into nanostructures and effectively encapsulate protein antigens has made them valuable in adjuvant delivery systems [3]. The activation of interferon genes (STING)-cGAS route, along with TT-LDCP NPs containing thymine-functionalized dendrimers, not only demonstrates improved gene delivery capability but also immune adjuvant properties. In HCC, si-RNA targeting the immune checkpoint ligand PD-L1 and immunostimulatory IL-2-encoding plasmid DNA were delivered using TT-LDCP NPs. This approach enhanced tumor infiltration, stimulated $CD8^+$ T cells, improved the effectiveness of CV immunity enhancement therapy, and inhibited HCC progression [54].

4.10.1.2 INORGANIC NANOPARTICLES FOR DNA VACCINE DELIVERY

The emergence of a wide range of inorganic NPs and their potential applications in material sciences has been observed. Inorganic particles offer numerous advantages over organic particles. For example, they possess good storage properties and are less susceptible to microbial infections. Additionally, they are relatively inexpensive and easy to prepare. These inorganic substances have been extensively studied for their potential use in various biomedical applications, including drug delivery, gene therapy, vaccine adjuvants, carriers for different proteins, and dental hygiene agents. Among these inorganic substances, commonly used salts such as calcium phosphate and metals like gold, silver, and iron are notable [9].

4.11 CHALLENGES

CVs have been researched for years with sporadic achievements, but they have not yet reached the mainstream of oncology. In comparison to infectious diseases, where prophylactic vaccines are more common, CVs are often used as a therapeutic approach. Tumor-derived antigens (TAs) are the target of CVs, aiming to stimulate an immune response [67].

- TAs may play a crucial role in the development, growth, and metastasis of tumors. We outline the challenges in creating a DNA vaccine [53, 67, 68].
- Mutated self-antigens, known as tumor-specific antigens (TSAs) or neoantigens, can be costly and time-consuming to identify in each patient's tumor. However, they have proven to be more effective in clinical trials compared to TAAs [37, 69].
- The aforementioned delivery techniques and virus-like particles (VLPs) have shown promising results, but each has its drawbacks and risks. These include the potential for an unfavorable immune reaction, especially with the constituents of the vectors (particularly

viruses), and the challenges in manufacturing the carriers, requiring extra caution.

- When DNA vaccines are injected, they primarily remain extracellular, where nucleases quickly degrade them. This necessitates repeated injections over time.
- The DNA CV only needs to function for a specific duration within the body. They are unable to generate a sustained immune response due to their rapid breakdown and excretion. While DNA vaccines are more stable, there is still room for improvement in terms of formulation and carrier loading [70].
- The objective of the cancer DNA Vaccine is to elicit an immune response (both adaptive and innate) specifically against the encoded tumor antigens, rather than against the delivery vehicle. However, it may also lead to undesired immune reactions targeting the vaccine itself, which could diminish their efficacy and hinder the production of the target protein [71].
- Some studies have linked excessive innate immune stimulation to inhibition of cellular translation and antigen expression. For instance, excessive stimulation of type I interferon (IFN) receptors on T cells by Type I IFNs can impede T cell function, despite their potential benefits.
- Additionally, type I IFNs are produced as a result of the innate immune response triggered by the double-stranded structure of DNA through non-TLR signaling pathways. The right-handed helical B form of DNA can also activate TLR-9, which in turn stimulates B and T cells, as well as DC maturation and activation. Depending on the type of vaccine, the targeted host cells, the level of type I IFN production, and other unidentified factors, TLR-9 stimulation may have either positive or negative effects.
- DNA vaccines with fewer prokaryotic sequences have shown greater effectiveness, highlighting the need for improvement in the hybrid structure of DNA vaccines. It has been hypothesized that reducing the prokaryotic nature of DNA vaccines could result in smaller vectors and simpler transfection into host cells. Moreover, DNA vaccines with shorter bacterial sequences, such as minicircle vectors instead of plasmid vectors, may have fewer inhibitory effects on TA expression in eukaryotic host cells.
- Some toxicities have been associated with DNA vaccines, including insertional mutagenesis and DNA plasmid integration issues. The

utilization of DNA vaccines may be affected by the presence of contaminated microbes during their preparation.

Clinical studies on cancer DNA vaccines have primarily focused on cervical, prostate, and breast cancer. Despite ongoing initiatives to enhance cancer DNA vaccines, there are still several challenges that must be overcome before fully functional DNA vaccines for cancer immunotherapy can be developed. However, given sufficient time, these issues may be resolved.

The following are some hypothesized explanations for why there isn't strong proof that using current NAVs is beneficial:

1. One reason is the biology of cancer cells, which makes it challenging to find TAs that can trigger a strong immune response. Further research is still needed in this area. It appears that NAVs created to treat infectious diseases have outperformed NAVs created to treat cancer due to more precise and well-defined TAs. Additionally, as cancer progresses, tumor cells may stop expressing a specific neoantigen for which a tailored vaccine had been developed. Therefore, it is often assumed that using multiple TAs at once will be advantageous when designing NAVs.

   Since we still don't fully understand the factors that influence immunogenicity, such as peptide processing by the proteasome complex and stability of epitope binding to MHC I, there are no fully valid criteria for the selection and identification of appropriate immunogenic TAs. Furthermore, tumor heterogeneity makes this issue even more challenging.
2. The immunosuppressive nature of tumors is considered a significant obstacle to the success of NAVs, especially in patients with advanced stages of cancer. The TME contains various immune suppressor cells, such as immature myeloid cells or TRegs, as well as inhibitory cell surface molecules expressed on cancer cells, such as programmed death-ligand 1 (PD-L1), and immunosuppressive cytokines (TME).
3. Previous research has demonstrated that anticancer drugs that promote immunogenic cell death (as opposed to non-immunogenic cell death) are more effective in inducing antitumor immunity. Treatments that promote immunogenic cell death may help reduce immunosuppressive cells within the TME and enhance antigen

presentation to T cells [44]. Therefore, combining immunological modification of the TME with TA-targeting NAVs could potentially lead to a powerful cancer treatment. NAVs designed to combat infectious diseases do not have to overcome a specific immunosuppressive environment.

4. Systemic reactions, such as fever and cytokine release syndrome, can be induced by human responses to NAVs.
5. One potential safety concern is that certain individuals may be prone to autoimmune reactions triggered by the type I INF response caused by NAVs.
6. There is uncertainty surrounding the immune signaling mechanisms in response to NAVs because, in some cases, these signaling mechanisms are believed to enhance adjuvant effects, but they may actually be unnecessary inflammatory signaling.
7. Lastly, the existence of differences between humans and animal models may have contributed to some exaggerated expectations. If humans were identical to mice, cancer would have already been cured, according to some commentators [72].

4.12 CONCLUSION

Immune tolerance to self-antigens is mediated by several mechanisms, which poses challenges in achieving cancer immunotherapy. The design of cancer immunotherapy is based on a deeper understanding of immunological tolerance. DNA vaccination has shown promising results as an immunotherapeutic approach for cancer treatment. The goal of DNA vaccines is to develop effective vaccination techniques against existing cancers. Ongoing efforts are focused on improving the DNA vaccine technology platform to overcome tolerance to tumor antigens. Extensive research is being conducted to enhance antigen expression, immunogenicity, and the utilization of next-generation delivery systems. Future DNA vaccines should aim to significantly enhance antitumor immunity by avoiding immune tolerance, disrupting immunosuppressive networks in the TME, and establishing long-lasting memory. DNA plasmids should be constructed to maximize immune response priming. It is crucial to control the activities of immunosuppressive elements such as TRegs and MDSCs in the TME. Combining immunotherapy with conventional chemotherapy and other therapeutic strategies may yield additive or synergistic therapeutic benefits.

4.13 FUTURE OUTLOOK

Future clinical trials for CVs should fully consider the activity of the patient's defense mechanism and tumor burden. CVs are compelling immune therapeutics that can be utilized to establish immune surveillance and stimulate the immune system to eliminate tumors. However, there is still much work to be done in identifying neoantigens, developing combination therapies, and improving vaccine platforms in order for CVs to become a potent immunotherapy approach [50].

Recent advancements in genetic analysis have made it possible to easily assess the immunogenicity of tumor cells and the immune status of the TME in specific cancer patients. This knowledge is expected to contribute to the identification of predictive biomarkers for selecting patients for cancer immunotherapy treatment, as well as the development of personalized peptide-based CVs, which may enhance the effectiveness of this immunotherapy [51].

Clinical studies have been conducted and are still ongoing for DNA vaccines. The use of a codon-optimized, polyepitopic DNA vaccine encoding TAAs or neoantigens in combination with other therapies appears to be the most relevant option for modulating the immunosuppressive TME. However, there are still numerous issues that need to be resolved [52].

Significant advancements are expected in high-output sequencing, antigen identification algorithms, modeling and activation efforts, production, development, and regulatory guidelines in the coming decade. Conventional quantitative clinical pharmacology techniques can contribute to a more rational and personalized selection of dosage and time frames for TCV studies as part of these efforts. Quantifiable modeling and systematic platform testing are innovative strategies emerging to support clinical development. Overcoming significant challenges that distinguish vaccine-based therapy from other immunotherapies remains a remarkable task. In addition to the difficulties described in this chapter, issues related to patient access and affordability due to production costs also need to be addressed [53].

Recent advancements in cancer therapy research have been impressive, leading to less invasive, more precise, and potent cancer treatments. Targeted therapy and nanomedicine have improved the absorption and bioavailability of experimental or new chemotherapy agents around the targeted region. Additionally, gene therapy, siRNA delivery, and immunotherapy have also been successful, providing cancer patients with new options. Antioxidant substances offer further promise. Magnetic hyperthermia and thermal ablation show potential as substitutes for tumor removal. Furthermore, methods utilizing radiomics and pathomics aid in the management of large cancer datasets, improving patients' prognosis and results [2]. Combining immunotherapy with conventional chemotherapy and other therapy strategies may produce additive or synergistic therapeutic benefits.

KEYWORDS

- **anticancer**
- **cancer antigen**
- **cancer immunotherapy**
- **DNA vaccine**
- **immunotolerance**
- **nanotechnology**
- **nucleic acid-based vaccines**
- **plasmid DNA**

REFERENCES

1. Shanmugaraj, B., Priya, L. B., Mahalakshmi, B., Subbiah, S., Hu, R. M., Velmurugan, B. K., & Baskaran, R., (2020). Bacterial and viral vectors as vaccine delivery vehicles for breast cancer therapy. In: *Life Sciences* (Vol. 250). Elsevier Inc. https://doi.org/10.1016/j.lfs.2020.117550.
2. Pucci, C., Martinelli, C., & Ciofani, G., (2019). Innovative approaches for cancer treatment: Current perspectives and new challenges. In: *Ecancermedicalscience* (Vol. 13). E-cancer Global Foundation. https://doi.org/10.3332/ecancer.2019.961.
3. Sabya Sachi Das, Ishan Moitra, Sandeep Kumar Singh, P. R. P. Verma, & Suryakanta Swain, (2022). Chapter 13 – DNA vaccines for cancer treatment: challenges and promises. In: Mahfoozur Rahman, Sarwar Beg, Waleed H. Almalki, Nabil A. Alhakamy, Hani Choudhry, (eds.). *Nanotherapeutics in Cancer Vaccination and Challenges*, Academic Press, pp. 259–278, https://doi.org/10.1016/B978-0-12-823686-4.00017-3.
4. Yang, B., Jeang, J., Yang, A., Wu, T. C., & Hung, C. F., (2014). DNA vaccine for cancer immunotherapy. *Human Vaccines and Immunotherapeutics*, *10*(11), 3153–3164. https://doi.org/10.4161/21645515.2014.980686.
5. Braathen, R., Spång, H. C. L., Hinke, D. M., Blazevski, J., Bobic, S., Fossum, E., & Bogen, B., (2020). A DNA Vaccine that encodes an antigen-presenting cell-specific heterodimeric protein protects against cancer and influenza. *Molecular Therapy*

– *Methods and Clinical Development*, *17*, 378–392. https://doi.org/10.1016/j.omtm.2020.01.007.
6. Conforti, A., Aurisicchio, L., & Palombo, F., (2022). Nucleic acid-based vaccines. In: *Cancer Vaccines as Immunotherapy of Cancer* (pp. 227–245). Elsevier. https://doi.org/10.1016/B978-0-12-823901-8.00002-9.
7. Conforti, A., Salvatori, E., Lione, L., Compagnone, M., Pinto, E., Shorrock, C., Hayward, J. A., et al., (2022). Linear DNA amplicons as a novel cancer vaccine strategy. *Journal of Experimental and Clinical Cancer Research*, *41*(1). https://doi.org/10.1186/s13046-022-02402-5.
8. Li, S., Wu, J., Li, X., Chen, J., & Wang, C., (2022). Biomaterial-enhanced cancer vaccines. *Materials and Design*, *218*. https://doi.org/10.1016/j.matdes.2022.110720.
9. Shah, M. A. A., He, N., Li, Z., Ali, Z., & Zhang, L., (2014). Nanoparticles for DNA vaccine delivery. In: *Journal of Biomedical Nanotechnology* (Vol. 10, No. 9, pp. 2332–2349). American Scientific Publishers. https://doi.org/10.1166/jbn.2014.1981.
10. Kanaseki, T., & Torigoe, T., (2019). Proteogenomics: Advances in cancer antigen research. In: *Immunological Medicine* (Vol. 42, No. 2, pp. 65–70). Taylor and Francis Ltd. https://doi.org/10.1080/25785826.2019.1640500.
11. Nagarajan, D., Pearson, J., Brentville, V., Metheringham, R., Pockley, A. G., Durrant, L., & McArdle, S. E., (2021). Helicase antigen (HAGE)-derived vaccines induce immunity to HAGE and immunobody®-HAGE DNA vaccine delays the growth and metastasis of HAGE-expressing tumors *in vivo*. *Immunology and Cell Biology*, *99*(9), 972–989. https://doi.org/10.1111/imcb.12485.
12. Jakobsen, M. K., Traynor, S., Stæhr, M., Duijf, P. G., Nielsen, A. Y., Terp, M. G., Pedersen, C. B., et al., (2021). The cancer/testis antigen gene VCX2 Is rarely expressed in malignancies but can be epigenetically activated using DNA methyltransferase and histone deacetylase inhibitors. *Frontiers in Oncology*, *10*. https://doi.org/10.3389/fonc.2020.584024.
13. Mahdevar, E., Kefayat, A., Safavi, A., Behnia, A., Hejazi, S. H., Javid, A., & Ghahremani, F., (2021). Immunoprotective effect of an in silico designed multiepitope cancer vaccine with BORIS cancer-testis antigen target in a murine mammary carcinoma model. *Scientific Reports*, *11*(1). https://doi.org/10.1038/s41598-021-01770-w.
14. Meklat, F., Li, Z., Wang, Z., Zhang, Y., Zhang, J., Jewell, A., & Lim, S. H., (2007). Cancer-testis antigens in haematological malignancies. In: *British Journal of Haematology* (Vol. 136, No. 6, pp. 769–776). https://doi.org/10.1111/j.1365-2141.2006.06484.x.
15. Joseph-Pietras, D., Gao, Y., Zojer, N., Ait-Tahar, K., Banham, A. H., Pulford, K., Rice, J., et al., (2010). DNA vaccines to target the cancer testis antigen PASD1 in human multiple myeloma. *Leukemia*, *24*(11), 1951–1959. https://doi.org/10.1038/leu.2010.196.
16. Yawata, T., Nakai, E., Park, K. C., Chihara, T., Kumazawa, A., Toyonaga, S., Masahira, T., et al., (2010). Enhanced expression of cancer testis antigen genes in glioma stem cells. *Molecular Carcinogenesis*, *49*(6), 532–544. https://doi.org/10.1002/mc.20614.
17. Mahdevar, E., Safavi, A., Abiri, A., Kefayat, A., Hejazi, S. H., Miresmaeili, S. M., & IranpurMobarakeh, V., (2022). Exploring the cancer-testis antigen BORIS to design a novel multi-epitope vaccine against breast cancer based on immunoinformatics approaches. *Journal of Biomolecular Structure and Dynamics*, *40*(14), 6363–6380. https://doi.org/10.1080/07391102.2021.1883111.

18. Yin, Y. H., Li, Y. Y., Qiao, H., Wang, H. C., Yang, X. A., Zhang, H. G., Pang, X. W., et al., (2005). TSPY is a cancer testis antigen expressed in human hepatocellular carcinoma. *British Journal of Cancer*, *93*(4), 458–463. https://doi.org/10.1038/sj.bjc.6602716.
19. Kaneda, Y., & Hoon, D. S. B., (2001). DNA vaccines for cancer treatment. *IDrugs*, *4*(3), 301–311. https://doi.org/10.1016/b978-0-12-823686-4.00017-3.
20. Caballero, O. L., & Chen, Y. T., (2009). Cancer/testis (CT) antigens: Potential targets for immunotherapy. In: *Cancer Science* (Vol. 100, No. 11, pp. 2014–2021). https://doi.org/10.1111/j.1349-7006.2009.01303.x.
21. Martínez-Puente, D. H., Pérez-Trujillo, J. J., Zavala-Flores, L. M., García-García, A., Villanueva-Olivo, A., Rodríguez-Rocha, H., Valdés, J., et al., (2022). Plasmid DNA for therapeutic applications in cancer. In: *Pharmaceutics* (Vol. 14, No. 9). MDPI. https://doi.org/10.3390/pharmaceutics14091861.
22. Zahm, C. D., Colluru, V. T., & McNeel, D. G., (2017). DNA vaccines for prostate cancer. In: *Pharmacology and Therapeutics* (Vol. 174, pp. 27–42). Elsevier Inc. https://doi.org/10.1016/j.pharmthera.2017.02.016.
23. Fioretti, D., Iurescia, S., Fazio, V. M., & Rinaldi, M., (2010). DNA vaccines: Developing new strategies against cancer. In: *Journal of Biomedicine and Biotechnology* (Vol. 2010). https://doi.org/10.1155/2010/174378.
24. Hayat, K. K., (2013). *DNA Vaccines: Roles Against Diseases* (Vol. 3, No. 1). In www.germs.ro • GERMS. www.germs.ro.
25. Shaw, D. R., & Strong, T. V., (2006). DNA vaccines for cancer. In: *Frontiers in Bioscience* (Vol. 11).
26. Gamat-Huber, M., Jeon, D., Johnson, L. E., Moseman, J. E., Muralidhar, A., Potluri, H. K., et al., (2020). Treatment combinations with DNA vaccines for the treatment of metastatic castration-resistant prostate cancer (mCRPC). In: *Cancers* (Vol. 12, No. 10, pp. 1–23). MDPI AG. https://doi.org/10.3390/cancers12102831.
27. Anderson, R. J., & Schneider, J., (2007a). Plasmid DNA and viral vector-based vaccines for the treatment of cancer. In: *Vaccine* (Vol. 25, No. SUPPL. 2). https://doi.org/10.1016/j.vaccine.2007.05.030.
28. Fioretti, D., Iurescia, S., Fazio, V. M., & Rinaldi, M., (2010a). DNA vaccines: Developing new strategies against cancer. In: *Journal of Biomedicine and Biotechnology* (Vol. 2010). https://doi.org/10.1155/2010/174378.
29. Bei, R., & Scardino, A., (2010). TAA polyepitope DNA-based vaccines: A potential tool for cancer therapy. In: *Journal of Biomedicine and Biotechnology* (Vol. 2010). https://doi.org/10.1155/2010/102758.
30. Anderson, R. J., & Schneider, J., (2007b). Plasmid DNA and viral vector-based vaccines for the treatment of cancer. In: *Vaccine* (Vol. 25, No. SUPPL. 2). https://doi.org/10.1016/j.vaccine.2007.05.030.
31. Fioretti, D., Iurescia, S., Fazio, V. M., & Rinaldi, M., (2010b). DNA vaccines: Developing new strategies against cancer. In: *Journal of Biomedicine and Biotechnology* (Vol. 2010). https://doi.org/10.1155/2010/174378.
32. McNeel, D. G., Becker, J. T., Johnson, L. E., & Olson, B. M., (2012). DNA vaccines for prostate cancer. *Current Cancer Therapy Reviews*, *8*(4), 245–253. https://doi.org/10.2174/157339412804143113.
33. Tiptiri-Kourpeti, A., Spyridopoulou, K., Pappa, A., & Chlichlia, K., (2016). DNA vaccines to attack cancer: Strategies for improving immunogenicity and efficacy. *Pharmacology and Therapeutics*, *165*, 32–49. https://doi.org/10.1016/j.pharmthera.2016.05.004.

34. Johnston, S. A., Kumar, S., Miller, L., Stafford, P., & Sykes, K., (2012). *Identification of Neo-antigens for a Cancer Vaccine*. https://core.ac.uk/download/pdf/79563686.pdf.
35. Liu, J., Fu, M., Wang, M., Wan, D., Wei, Y., & Wei, X., (2022). Cancer vaccines as promising immuno-therapeutics: Platforms and current progress. In: *Journal of Hematology and Oncology* (Vol. 15, No. 1). BioMed Central Ltd. https://doi.org/10.1186/s13045-022-01247-x.
36. Pan, P. Y., Ozao, J., Zhou, Z., & Chen, S. H., (2008a). Advancements in immune tolerance. In: *Advanced Drug Delivery Reviews* (Vol. 60, No. 2, pp. 91–105). https://doi.org/10.1016/j.addr.2007.08.025.
37. Herrada, A. A., Rojas-Colonelli, N., González-Figueroa, P., Roco, J., Oyarce, C., Ligtenberg, M. A., & Lladser, A., (2012). Harnessing DNA-induced immune responses for improving cancer vaccines. In: *Human Vaccines and Immunotherapeutics* (Vol. 8, No. 11, pp. 1682–1693). https://doi.org/10.4161/hv.22345.
38. Alpdogan, O., & Van, D. B. M. R. M., (2012). Immune tolerance and transplantation. *Seminars in Oncology, 39*(6), 629–642. https://doi.org/10.1053/j.seminoncol.2012.10.001.
39. Pan, P. Y., Ozao, J., Zhou, Z., & Chen, S. H., (2008b). Advancements in immune tolerance. In: *Advanced Drug Delivery Reviews* (Vol. 60, No. 2, pp. 91–105). https://doi.org/10.1016/j.addr.2007.08.025.
40. Jensen-Jarolim, E., Bax, H. J., Bianchini, R., Crescioli, S., Daniels-Wells, T. R., Dombrowicz, D., Fiebiger, E., et al., (2018). Allergo oncology: Opposite outcomes of immune tolerance in allergy and cancer. *Allergy: European Journal of Allergy and Clinical Immunology, 73*(2), 328–340. https://doi.org/10.1111/all.13311.
41. Amara, S., & Tiriveedhi, V., (2017). The five immune forces impacting DNA-based cancer immunotherapeutic strategy. In: *International Journal of Molecular Sciences* (Vol. 18, No. 3). MDPI AG. https://doi.org/10.3390/ijms18030650.
42. Hafeez, U., Parakh, S., Gan, H. K., & Scott, A. M., (2020). Antibody – drug conjugates for cancer therapy. In: *Molecules* (Vol. 25, No. 20). MDPI AG. https://doi.org/10.3390/molecules25204764.
43. Antonarelli, G., Giugliano, F., Corti, C., Repetto, M., Tarantino, P., & Curigliano, G., (2021). Research and clinical landscape of bispecific antibodies for the treatment of solid malignancies. In: *Pharmaceuticals* (Vol. 14, No. 9). MDPI. https://doi.org/10.3390/ph14090884.
44. Choi, B. D., Maus, M. V., June, C. H., & Sampson, J. H., (2019). Immunotherapy for glioblastoma: Adoptive T-cell strategies. *Clinical Cancer Research, 25*(7), 2042–2048. https://doi.org/10.1158/1078-0432.CCR-18-1625.
45. Morse, M. A., Gwin, W. R., & Mitchell, D. A., (2021). Vaccine therapies for cancer: Then and now. In: *Targeted Oncology* (Vol. 16, No. 2, pp. 121–152). Adis. https://doi.org/10.1007/s11523–020–00788-w.
46. Liu, W., Tang, H., Li, L., Wang, X., Yu, Z., & Li, J., (2021). Peptide-based therapeutic cancer vaccine: Current trends in clinical application. In: *Cell Proliferation* (Vol. 54, No. 5). Blackwell Publishing Ltd. https://doi.org/10.1111/cpr.13025.
47. Palucka, K., & Banchereau, J., (2013). Dendritic-cell-based therapeutic cancer vaccines. In: *Immunity* (Vol. 39, No. 1, pp. 38–48). https://doi.org/10.1016/j.immuni.2013.07.004.
48. Lemke-Miltner, C. D., Blackwell, S. E., Yin, C., Krug, A. E., Morris, A. J., Krieg, A. M., & Weiner, G. J., (2020). Antibody opsonization of a TLR9 agonist–containing

virus-like particle enhances *in-situ* immunization. *The Journal of Immunology, 204*(5), 1386–1394. https://doi.org/10.4049/jimmunol.1900742.
49. Locy, H., De Mey, S., De Mey, W., De Ridder, M., Thielemans, K., & Maenhout, S. K., (2018). Immunomodulation of the tumor microenvironment: Turn foe into friend. In: *Frontiers in Immunology* (Vol. 9). Frontiers Media S.A. https://doi.org/10.3389/fimmu.2018.02909.
50. Syed, S., Hasson, A. A., Khalifa, J., Al-Busaidi, Z., Sallam, T. A., & Hasson, S., (2015). The past, current and future trends in DNA vaccine immunizations. In: *Arabia Asian Pac. J. Trop. Biomed.* (Vol. 5, No. 5). www.elsevier.com/locate/apjtb (accessed on 11 March 2024).
51. Hirayama, M., Nishimura, Y., & Nishimura, Y., (n.d.). *The Present Status and Future Prospects of Peptide-Based Cancer Vaccines.* http://intimm.oxfordjournals.org/ (accessed on 11 March 2024).
52. Lopes, A., Vandermeulen, G., & Préat, V., (2019). Cancer DNA vaccines: Current preclinical and clinical developments and future perspectives. In: *Journal of Experimental and Clinical Cancer Research* (Vol. 38, No. 1). BioMed Central Ltd. https://doi.org/10.1186/s13046-019-1154-7.
53. Shemesh, C. S., Hsu, J. C., Hosseini, I., Shen, B. Q., Rotte, A., Twomey, P., Girish, S., & Wu, B., (2021). Personalized cancer vaccines: Clinical landscape, challenges, and opportunities. In: *Molecular Therapy* (Vol. 29, No. 2, pp. 555–570). Cell Press. https://doi.org/10.1016/j.ymthe.2020.09.038.
54. Huang, K. W., Hsu, F. F., Qiu, J. T., Chern, G. J., Lee, Y. A., Chang, C. C., Huang, Y. T., et al., (2020). *Highly Efficient and Tumor-Selective Nanoparticles for Dual-Targeted Immunogene Therapy Against Cancer.* http://advances.sciencemag.org/ (accessed on 11 March 2024).
55. Ma, J., Zhang, H., Tang, K., & Huang, B., (2020). Tumor-derived microparticles in tumor immunology and immunotherapy. In: *European Journal of Immunology* (Vol. 50, No. 11, pp. 1653–1662). Wiley-VCH Verlag. https://doi.org/10.1002/eji.202048548.
56. Sun, B., Zhao, X., Wu, Y., Cao, P., Movahedi, F., Liu, J., Wang, J., et al., (2021). Mannose-functionalized biodegradable nanoparticles efficiently deliver DNA vaccine and promote anti-tumor immunity. *ACS Applied Materials and Interfaces, 13*(12), 14015–14027. https://doi.org/10.1021/acsami.1c01401.
57. Colluru, V. T., Johnson, L. E., Olson, B. M., & McNeel, D. G., (2016). Preclinical and clinical development of DNA vaccines for prostate cancer. In: *Urologic Oncology: Seminars and Original Investigations* (Vol. 34, No. 4, pp. 193–204). Elsevier Inc. https://doi.org/10.1016/j.urolonc.2013.09.014.
58. Valilou, S. F., & Rezaei, N., (2018). Tumor antigens. In: *Vaccines for Cancer Immunotherapy: An Evidence-Based Review on Current Status and Future Perspectives* (pp. 61–74). Elsevier. https://doi.org/10.1016/B978-0-12-814039-0.00004-7.
59. Melief, C. J. M., Van, H. T., Arens, R., Ossendorp, F., & Van, D. B. S. H., (2015). Therapeutic cancer vaccines. *Journal of Clinical Investigation* (Vol. 125, pp. 3401–3412). American Society for Clinical Investigation.
60. Makkouk, A., & Weiner, G. J., (2015). *Cancer Immunotherapy and Breaking Immune Tolerance: New Approaches to an Old Challenge* (Vol. 75, pp. 5–10). Cancer Research. American Association for Cancer Research Inc.
61. Miri, S. M., Pourhossein, B., Hosseini, S. Y., Keshavarz, M., Shahmahmoodi, S., Zolfaghari, M. R., et al., (2022). Enhanced synergistic antitumor effect of a DNA

vaccine with anticancer cytokine, MDA-7/IL-24, and immune checkpoint blockade. *Virol. J., 19*(1).
62. Chopra, A., Kim, T. S., O-Sullivan, I. S., Martinez, D., & Cohen, E. P., (2006). Combined therapy of an established, highly aggressive breast cancer in mice with paclitaxel and a unique DNA-based cell vaccine. *Int. J. Cancer, 118*(11), 2888–2898.
63. Stevenson, F. K., Ottensmeier, C. H., Johnson, P., Zhu, D., Buchan, S. L., Mccann, K. J., et al., (2004). *DNA Vaccines to Attack Cancer [Internet].* Available from: doi:10.1073/pnas.0404896101 (accessed on 11 March 2024).
64. Pan, P. Y., Ozao, J., Zhou, Z., & Chen, S. H., (2008). Advancements in immune tolerance. *Advanced Drug Delivery Reviews, 60*, 91–105.
65. Alpdogan, O., & Van, D. B. M. R. M., (2012). Immune tolerance and transplantation. *Semin. Oncol., 39*(6), 629–642.
66. Shen, Y., & Qiu, L., (2019). Effective oral delivery of gp100 plasmid vaccine against metastatic melanoma through multi-faceted blending-by-blending nanogels. *Nanomedicine, 22.*
67. Thomas, S., & Prendergast, G. C., (2016). Cancer vaccines: A brief overview. In: *Methods in Molecular Biology* (pp. 755–761). Humana Press Inc.
68. Mastelic-Gavillet, B., Balint, K., Boudousquie, C., Gannon, P. O., & Kandalaft, L. E., (2019). Personalized dendritic cell vaccines-recent breakthroughs and encouraging clinical results. *Frontiers in Immunology* (Vol. 10). Frontiers Media S.A.
69. Peng, M., Mo, Y., Wang, Y., Wu, P., Zhang, Y., Xiong, F., et al., (2019). Neoantigen vaccine: An emerging tumor immunotherapy. *Molecular Cancer* (Vol. 18). BioMed Central Ltd.
70. Dhar, R., Seethy, A., Singh, S., Pethusamy, K., Srivastava, T., Talukdar, J., et al., (2021). Cancer immunotherapy: Recent advances and challenges. *Journal of Cancer Research and Therapeutics* (Vol. 17, pp. 834–844). Wolters Kluwer Medknow Publications.
71. Pejawar-Gaddy, S., & Finn, O. J., (2008). Cancer vaccines: Accomplishments and challenges. *Critical Reviews in Oncology/Hematology* (Vol. 67, pp. 93–102).
72. Jahanafrooz, Z., Baradaran, B., Mosafer, J., Hashemzaei, M., Rezaei, T., Mokhtarzadeh, A., et al., (2020). Comparison of DNA and mRNA vaccines against cancer. *Drug Discovery Today* (Vol. 25, pp. 552–560). Elsevier Ltd.

CHAPTER 5

RNA-Based Cancer Vaccines: Advancement and Challenges

PALAK A. SHAH,[1] RASHMI A. SHUKLA,[2] MEHUL R. CHORAWALA,[2] NIMISHA P. KAKADIA,[1] BHUPENDRA PRAJAPATI,[3] and SHRIKALP S. DESHPANDE[1]

[1]*Department of Pharmacology and Pharmacy Practice, K. B. Institute of Pharmaceutical Education and Research, Gandhinagar, Gujarat, India*

[2]*Department of Pharmacology, L. M. College of Pharmacy, Ahmedabad, Gujarat, India*

[3]*Shree S.K. Patel College of Pharmaceutical Education and Research, Ganpat University, Mehsana, Gujarat, India*

ABSTRACT

Biological response modifier therapy has proven to be a significant advancement in the management of cancer in the current era. The use of messenger RNA (mRNA) in cancer immunotherapy is gaining widespread acceptance due to its ability to stimulate the production of target-specific antibodies (Abs). mRNA-based vaccines aim to generate proteins that combat cancer and related conditions. These vaccines offer notable advantages such as high efficacy, targeted action, reduced toxicity, ease of manufacturing, and convenient administration, making them superior to established and conventional vaccines. mRNA-based vaccines have demonstrated exceptional success in preventing COVID-19 infection and other similar conditions. Numerous clinical trials have been conducted in recent years to evaluate the efficacy of mRNA vaccines in patients with various types of cancer. However, to date, no mRNA vaccines have been approved by regulatory bodies worldwide, either as standalone treatments or as adjuvant therapies for cancer management.

Cancer Vaccination and Challenges, Volume 1: Strategies for Therapeutic Cancer Vaccine Development.
Rishabha Malviya, Bhupendra Prajapati, and Sonali Sundram (Eds.)

This is primarily due to the challenges associated with the variability and ineffective delivery of mRNA, which hinders the widespread application of this technology. In order to address these issues, significant modifications have been made to the structure and administration methods of mRNA over the years. This review will provide an overview of the current developments and improvements in mRNA vaccines and their delivery in cancer immunotherapy, as well as the challenges and regulatory hurdles that need to be addressed for clinical use. It will also highlight the approaches towards optimizing mRNA-based vaccines for the effective and successful treatment of cancer.

5.1 INTRODUCTION

Immunization is one of the most significant advancements in global health, contributing to the extension of human life. The eradication of smallpox, as well as the significant reduction in cases of polio and measles, are major milestones in the field of immunization. While live-attenuated or inactivated conventional vaccines have proven effective against various infections, they are not suitable for non-infectious diseases such as cancer [1]. An alternative approach to traditional vaccines is nucleic acid therapeutics, specifically DNA or RNA-based vaccines. These vaccines stimulate the immune response by promoting protein translation within cells, making them a form of immunotherapy. Nucleic acid-derived vaccines are preferred due to their non-infectious nature, lack of contamination, and good tolerance for both prophylaxis and treatment [2]. Although both DNA and RNA vaccines share similar characteristics, RNA vaccines offer additional benefits. Firstly, they do not integrate into the genome, and secondly, they only need to cross the plasma membrane, unlike DNA vaccines that must cross the nuclear membrane [3]. Recently, mRNA vaccines have gained significant attention during the severe acute respiratory syndrome coronavirus 2 (SARS-CoV-2) pandemic, showcasing their therapeutic potential and boosting mRNA-based therapeutics [4]. However, the history of mRNA vaccines dates back to 1961 when they were first discovered by Brenner et al. [5]. Subsequent studies in mice demonstrated that intramuscular administration of unformulated mRNA led to the expression of the encoded protein within a few days [6]. The studies provided evidence that genetic information could be transferred through *in-vitro* transcribed mRNA, without the need for any vector, eliminating the speculation of mRNA instability *in-vivo* [7]. In addition to replacing missing proteins, it has also been proven that mRNA can conveniently provide desired antigenic information to antigen-presenting cells (APCs) [8]. In 1995, the first preclinical study revealed the anticancer effect of the mRNA vaccine

[9]. To date, research and development have generated expectations for a new candidate in mRNA therapeutics. A brief overview of the evolution of the mRNA vaccine is presented in Table 5.1. In this chapter, we will discuss the mechanism followed by mRNA-based vaccines and the opportunities and challenges they offer as advantageous cancer immunotherapeutics.

TABLE 5.1 Timeline for mRNA-based Vaccine Discoveries

Year of Research	Findings	References
1961	Discovery of mRNA.	[5]
1975	Discovery of 5' cap modification of mRNA.	[10]
1984	*In-vitro* transcription of mRNA.	[11]
1989	*In-vitro* transfection of cationic liposome-mediated mRNA (concept proposal of mRNA-based drug).	[12]
1990	*In-vivo* transfection of pure mRNA (concept proposal of mRNA vaccine).	[6]
1993	mRNA-induced cellular immunity.	[8]
1995	mRNA-induced humoral immunity (First preclinical study: Proof of concept of mRNA vaccine).	[9]
1996	Preclinical study: Tumor-derived mRNA pulsed dendritic cell (DC) cancer vaccine (CV).	[13]
2002	A first clinical trial of PSA mRNA-transfected DC cancer vaccine.	[14]
2005	Nucleoside-modified mRNA has limited immunogenicity.	[15]
2009	Utilization of direct intranodal injection of protamine-protected mRNA in the clinic.	[16]
2010	Preclinical study: Naked mRNA cancer vaccine.	[17]
2015	Preclinical study: Neoantigen mRNA cancer vaccine.	[18]
2017	A clinical trial neoantigen mRNA.	[19]
2021	FDA approval of the first mRNA vaccine, approval of a fast-track designation for the BNT111 cancer vaccine.	[20]

5.2 HOW MRNA-BASED VACCINE WORKS?

The process of developing an mRNA vaccine involves sequencing the genes and replicating the gene into the template strand from the pathogenic antigen of interest. This is followed by the transcription of mRNA *in vitro* and its delivery into humans.

The host cell machinery then translates the mRNA from the vaccine into the corresponding antigen. This process naturally occurs when an actual viral infection elicits an immune response in an infected person [21]. From virtual sequence designing *in silico*, synthesizing, and delivering the mRNA vaccine to testing it *in vivo*, it becomes an easy and effective therapeutic option. The Zika mRNA vaccine developed by Richner et al. in 2017 [22] has also shown enhanced neutralization capacity, optimal codon usage, and reduced undesired cross-reactivity, which is an additional advantage [22]. The detailed mechanism of the mRNA vaccine is depicted in Figure 5.1.

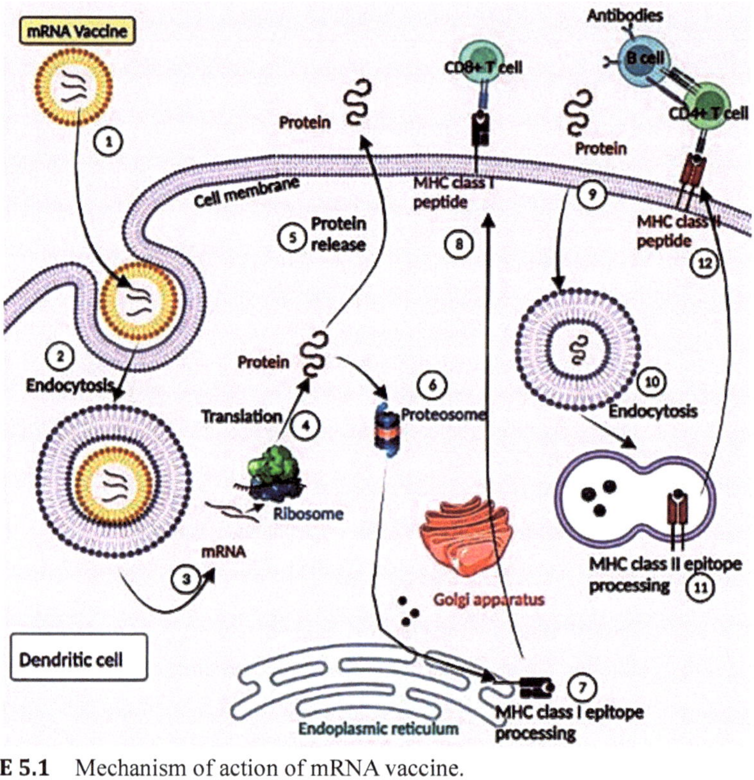

FIGURE 5.1 Mechanism of action of mRNA vaccine.

Note: The multistep process of mRNA vaccine mechanism in DCs is as follows: (1) Antigen-presenting cells, such as DCs, take up the mRNA vaccine. (2) The uptake of the mRNA vaccine is facilitated through endocytosis. (3) The mRNA is released into the cytoplasm from the delivery system. (4) The mRNA is then translated into protein. (5) This protein is released outside the cell and acts as an exogenous antigen for neighboring cells. (6) The translated protein inside the cytosol is processed by the proteosomes to release peptides. (7) MHC class I epitope processing occurs in the endoplasmic reticulum. (8) Antigen presentation on the MHC class I at the cell surface activates CD8+ T-cells. (9) Antigen-presenting cells take up exogenous proteins. (10) This uptake occurs through the process of endocytosis. (11) Extracellular proteins are taken up for MHC Class II epitope processing. And (12) MHC Class II presentation on the cell surface activates CD4+ T-cells and co-activates B-cells to produce specific antibodies. [MHC: major histocompatibility complex; CD4: cluster of differentiation 4; and CD8: cluster of differentiation 8].

5.2.1 INDUCTION OF INNATE IMMUNE RESPONSE BY MRNA VACCINE

Pathogen-associated molecular patterns (PAMPs) are detected by pattern recognition receptors (PRRs), which ultimately lead to the activation of the host immune machinery [23]. After the administration of mRNA in humans, the host body identifies the vaccine as an exogenous substance or PAMPs through Toll-like receptors (TLRs) located on antigen-presenting cells (APCs). Specifically, TLR3, TLR7, and TLR8 are responsible for recognizing PAMPs from various sites such as the cytoplasm, endosomes, and the cell membrane, ultimately eliciting an immune response [24]. TLR3 receptors can recognize double-stranded RNA (dsRNA) formed by single-stranded RNA (ssRNA), thereby activating the Toll/interleukin-1 receptor (TIR)-domain-containing adapter-inducing interferon-β (TRIF) pathway, followed by increased expression of Interferon type I (IFN-I) and INF-γ-inducible protein 10 (IP-10) [25]. TLR7 receptors can recognize both ssRNA and dsRNA, while TLR8 only binds to ssRNA [26]. Thus, TLR7 stimulates the B cell response [27]. The injected mRNA in the vaccine is recognized by TLRs present on APCs, initiating the immune response through the activation of the Myeloid differentiation primary response 88 (MYD88) pathway and increasing the release of IFN-I response. The early shutdown of antigen expression following mRNA injection could be harmful due to the activation of PRRs [28]. The early shutdown of antigen expression following mRNA injection could be harmful due to the activation of PRRs.

The cytoplasm also contains different sets of PRRs, which specifically differentiate non-self mRNA from self mRNA. These include retinoic acid-inducible gene (RIG)-I and melanoma differentiation-associated protein 5 (MDA5), which sense dsRNA and activate a signaling cascade.

When activated, both RIG-I and MDA5 recruit mitochondrial antiviral signaling protein (MAVS), ultimately leading to increased expression of Nuclear Factor kappa-B (NF-κB) and IFN-I. Inflammasomes are biochemical complexes responsible for caspase-I activity and the activation of Interleukin-1 Beta (IL-1β). Inflammasome formation, induced by nucleotide-binding oligomerization domain (NOD)-like receptors (NLR), leads to IL-1β production, and triggers the process of apoptosis [29].

Subsequently, released IFN-I modulates the paracrine effect and signals an antineoplastic state to neighboring cells through IFN-stimulated gene [30].

5.2.2 INDUCTION OF ADAPTIVE IMMUNE RESPONSE BY MRNA VACCINE

For mRNA-based vaccines targeting cancer, the therapeutic effect is expected to come from the antitumor T-cells. DCs take up the mRNA from the vaccine, which is then translated, and the corresponding antigen-Major Histocompatibility Complex (MHC) I/II complex is presented on the cell surface [31] (Figure 5.1). These activated DCs then enter the lymph node, where they activate the T-cells by binding the MHC I/II complex with $CD8^+$/ $CD4^+$ (Cluster of Differentiation) receptors present on T-cells, leading to clonal expansion of T-cells [32] (Figure 5.2). Tumor cells release chemokines (CC) such as CC-ligand (CCL) 5 and CC-receptor 7 (CC-7) to attract cytokines, perforins, and granzymes. Chemokines also attract the activated T-cells to the tumor tissue [33]. Cytotoxic T-cells are activated by MHC Class I/II molecules, which act upon the target cells by directly destroying them. Neoepitopes generated by somatic mutation act as tumor-derived peptides. These neoepitopes are considered good candidates for cancer vaccines (CVs), as T-cells recognize them once they bind to MHC molecules [34].

5.3 PROGRESS OF MRNA-BASED VACCINE IN PRECLINICAL AND CLINICAL SETTINGS

An effective vaccine should deliver the chosen antigen into the host and initiate an immune response. This highlights the importance of selecting the appropriate antigen for incorporation into the vaccine [31]. The types of antigens are discussed in subsections.

5.3.1 TUMOR-ASSOCIATED ANTIGENS (TAAS)

In the case of CVs, it is crucial to target a specific antigen that can elicit the desired immunogenicity. It should be noted that this antigen is predominantly expressed in cancer cells and not in healthy cells. While many TAAs are common to both cancerous and normal cells, they are overexpressed in tumor cells. However, caution must be exercised as this antigen can induce tolerance during negative selection, and overcoming this tolerance could potentially lead to autoimmune diseases [35]. For example, an oncofetal antigen may have already induced tolerance and could hinder the immune response to TAAs. Ideally, a non-mutated TAA that elicits T-cell responses

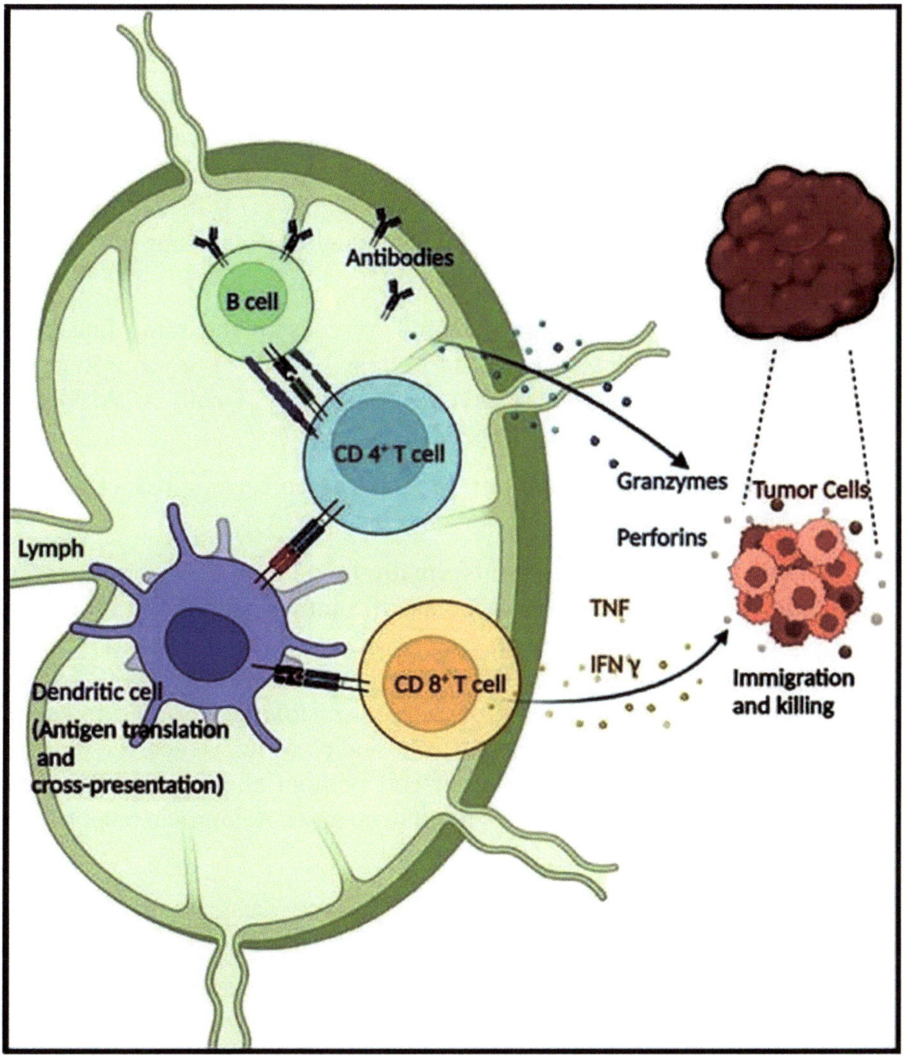

FIGURE 5.2 Principal mechanism of activation of adaptive immune responses by mRNA vaccine.

Note: To stimulate adaptive immunity during development or progression, a vaccine requires a pathogen-specific immunogen as well as an adjuvant. The adjuvant stimulates the adaptive immune response and provides the necessary second signal for T-cell and B-cell activation and proliferation. DCs containing mRNA from the vaccine undergo translation to produce MHC I or II. They then migrate into the lymph node where they present the mRNA molecule to T-cells through MHC-I or II molecules that complex with CD8+ or CD4+ on the surface of the T-lymphocytes. This results in the activation and proliferation of T-cells. Activated T-cells release various cytokines, granzymes, and perforins that are responsible for killing tumor cells, while B-cells produce antibodies against target cells.

would be an appropriate choice, but recent clinical trials have shown limited success with TAAs [36]. TAAs can be classified into the following categories [37]:

1. **Oncospermatogonal Antigen (Cancer/Testis Antigen (CTA)):** Typically found on spermatocytes [Melanoma Antigen Gene (MAGE), cancer/testis antigen G antigen (GAGE), B melanoma antigens (BAGE), New York Esophageal Squamous Cell Carcinoma-1 (NY-ESO-1)].
2. **Differentiation Antigen:** Found on cells of the same lineage as the tumor [Tyrosinase-related protein 1 (TRP-1), glycoprotein 100 (gp100), melanoma antigen recognized by T-cells 1 (MART-1), B-lymphocyte antigen CD20 (CD20)].
3. **Oncofetal Antigen:** Found in fetal and embryonic tissues [Alpha-fetoprotein, carcinoembryonic antigen (CEA)].
4. **Overexpressed Antigen:** Normally found proteins that are upregulated in cancer cells [prostate-specific antigen (PSA), wild-type p53, human epidermal growth factor receptor-2 (HER2/neu), epidermal growth factor receptor (EGFR)]. Cancer/testis antigens (CTA) are expressed in human tumors of various histological origins, but not in normal tissues except for the testis and placenta. Hence, the recognition only occurs at the tumor cell without altering non-cancerous cells and therefore, it does not induce any autoimmune responses or diseases.

5.3.2 TUMOR-SPECIFIC ANTIGENS (TSAS)

The cancer disease is characterized by mutations occurring in the cells. Therefore, each mutation can be considered a potential therapeutic target. Mutations are limited to non-normal or cancerous cells, known as tumor cells. Tumor-specific antigens (TSAs), also known as neoantigens, are specific to tumor cells and have lower tolerance [31]. The discovery of TSAs has allowed for the development of CVs that target specific antigens. This has also justified the use of tumor-infiltrating lymphocyte (TIL) infusion in cancer treatment.

It is estimated that only 10% of mutations result in mutant peptides that have a higher ability to bind to MHC. Furthermore, only 1% of these high-affinity peptides can be detected by T-cells in cancer patients [38]. In addition to their low tolerance, TSAs possess a significant level of mutation, which is crucial for tumor growth. Targeting these proteins would provide a dual benefit.

TSAs are also considered immunodominant epitopes. For example, a study showed that CD8$^+$ T-cells against mutated neoantigens dominated the immune response in melanoma patients, compared to CD8$^+$ T-cells recognizing shared antigens [37]. TSAs can be further classified as proposed in the study [38]:

i. Mutated TSA (mTSA): derived from DNA sequences in canonical gene.
ii. Aberrantly expressed TSA (aeTSA) is derived from the cancer-specific expression of unmutated non-canonical transcripts that are not observed in normal tissues. TSAs provide personalized therapy to patients, but this approach can be costly and time-consuming. However, continued development and advancements in omics technology will further reduce the time and cost associated with tailor-made therapy using TSAs.

5.4 CURRENT PROGRESS AND NEW PERSPECTIVES

At present, cancer is the most promising field for mRNA-based therapeutics in vaccines. Out of the total number of trials in cancer, nearly half of them are focused on melanomas, brain tumors, and prostate carcinomas. Most of these trials are in the early stages of study and are unable to evaluate the effectiveness of the vaccine due to the lack of a benchmark treatment regime for cancer [39]. The tabular view of several preclinical studies (Table 5.2) and clinical trials of mRNA encoding TAAs (Table 5.3), TSAs (Table 5.4), and mRNA-vaccines with DCs as carriers (Table 5.5) is shown below.

5.5 DELIVERY CARRIERS FOR MRNA CANCER VACCINE

Before initiating the delivery system, it is essential to discuss the fundamental structure of mRNA and its derived vaccine structure (Figure 5.3), with detailed information provided in the section on naked RNA. Although the mRNA vaccine does not require nuclear penetration, it must traverse the electronegative phospholipid bilayer membrane to reach the cytoplasm. Achieving efficient entry of the mRNA vaccine is a crucial and challenging

TABLE 5.2 Preclinical Studies Involving Different Types of Target Antigens and Their Results

Sl. No.	Target Antigen	Results	References
1.	Human carcinoembryonic antigen.	Immunized mice generated anti-CEA antibodies after 3 weeks of mRNA vaccination.	[9]
2.	Tumor-derived RNA	Immunized mice showed protection against OVA-expressing tumor cell.	[13]
3.	The human melanoma-associated antigen gp100 mRNA.	Immunized mice showed delayed tumor growth as compared to the control group.	[40]
4.	MART1 mRNA	Immunized mice showed specific and significant protection against B16F10 melanoma at the dose of 12.5 μg per mouse.	[41]
5.	Free and protamine-complexed mRNA.	Strong and sustained immune response against OVA-expressing tumor cells. Useful in prophylactic as well as treatment settings.	[42]
6.	TAA mRNA together with TriMix.	The promising strategy of intranodal delivery of the combination mRNA vaccine.	[43]
7.	Tumor antigen MUC1.	Efficacy of nanoparticles as a carrier.	[44]
8.	TRP2	Significantly showed T-cell activation and proliferation.	[45]
9.	Synthetic poly-neo-epitope messenger RNA vaccines.	Induce tumor protection in mice.	[18]
10.	Containing T-cell costimulator OX40L, IL-23, and IL-36g.	Administration produced effective tumor destruction and protection against tumor rechallenge.	[46]
11.	BisCCL2/5i and PD-1 ligand inhibitor.	Long-term survival of the mouse was found when the combination of BisCCL2/5i and PD-1 ligand inhibitor (PD-Li) was administered in the model of liver and pancreatic cancer.	[47]
12.	A tumor suppressor gene-phosphatase and tensin homolog deleted on chromosome 10 (PTEN).	The study shows inhibition of tumor growth and restoration of PTEN function in the model of prostate cancer.	[48]
13.	The p53 tumor suppressor gene.	Anti-tumor activity in the animal model of hepatocellular carcinomas (HCCs) and non-small cell lung cancers (NSCLCs).	[49]

TABLE 5.3 Current Progress in Clinical Trials of mRNA-based cancer vaccines Encoding TAAs, Including Phase, Conditions, Types of Vectors, and Route of Administration

Sl. No.	NCT Number	Phase	Condition	Vector	Route	Intervention
1.	NCT02709616	1	Glioblastoma	DCs	ID, IV	Personalized cellular vaccine, DC-based cellular vaccine, tumor antigen-pulsed DC vaccine.
2.	NCT02808364	1	Glioblastoma	DCs	ID, IV	Personalized cellular vaccine.
3.	NCT02808416	1	Brain cancer	DCs	ID, IV	Personalized cellular vaccine.
4.	NCT02688686	1/2	Non-small-cell lung cancer with bone metastases.	DCs	SC	Genetically modified Dendritic cells.
5.	NCT03164772	1/2	Non-small-cell lung cancer (metastatic).	Protamine	ID	Drug: Durvalumab Drug: Tremelimumab Biological: BI 1361849
6.	NCT04163094	1	Ovarian cancer	Lipo-MERIT	IV	W_ova1 vaccine
7.	NCT04382898	1/2	Prostate cancer	Lipo-MERIT	IV	Biological: W_pro1 (BNT112) Drug: Cemiplimab
8.	NCT04526899	2	Melanoma stage III/IV	Lipo-MERIT	IV	Biological: BNT111 Biological: Cemiplimab
9.	NCT04534205	2	Metastatic recurrent head and neck cancer.	Lipo-MERIT	IV	Biological: BNT113 Biological: Pembrolizumab
10.	NCT05142189	2	Non-small cell lung cancer.	Liposomes	IV	Biological: BNT116 Biological: Cemiplimab Drug: Docetaxel
11.	NCT02410733	1	Advanced melanoma	Lipo-MERIT	IV	Biological: Lipo-MERIT

TABLE 5.4 The Current Progress and Results of Clinical Trials Evaluating the Safety and Efficacy of mRNA-based Vaccines Encoding TSAs, Including the Phase, Conditions, Types of Vectors, and Route of Administration

Sl. No.	NCT Number	Phase	Condition	Vector	Route	Intervention
1.	NCT02529072	1	Malignant glioma, astrocytoma, glioblastoma	DCs	IV	Drug: nivolumab Biological: DC
2.	NCT02316457	1	Breast cancer	–	IV	Biological: IVAC_W_bre1_uID
3.	NCT03289962	1	Melanoma, bladder cancer, colorectal cancer, renal cancer.	Lipo-MERIT	IV	Drug: Autogene cevumeran (RO7198457) Drug: Atezolizumab
4.	NCT03313778	1	Solid tumors	LNP	IM	Biological: mRNA-4157 Biological: Pembrolizumab
5.	NCT03480152	1/2	Melanoma, colon/gastrointestinal/genitourinary/hepatocellular cancer.	LNP	IM	Biological: NCI4650
6.	NCT03468244	NA	Advanced esophageal squamous carcinoma, gastric adenocarcinoma, pancreatic adenocarcinoma, colorectal adenocarcinoma.	–	SC	Biological: Personalized mRNA tumor vaccine
7.	NCT03908671	NA	Esophageal cancer, non-small cell lung cancer	–	–	Biological: personalized mRNA tumor vaccine.
8.	NCT04161755	1	Pancreatic cancer	Lipo-MERIT	–	Drug: Atezolizumab Biological: RO7198457 Drug: mFOLFIRINOX
9.	NCT03815058	2	Advanced melanoma	Lipo-MERIT	IV	Biological: RO7198457 Drug: Pembrolizumab
10.	NCT04486378	2	Colorectal cancer stage II/III	Lipo-MERIT	IV	Drug: RO7198457
11.	NCT03897881	2	Melanoma	LNP	–	Biological: mRNA-4157 Biological: Pembrolizumab
12.	NCT03948763	1	Neoplasms, carcinoma, non-small-cell lung, pancreatic neoplasms, colorectal neoplasms.	LNP	IM	Biological: V941 Biological: Pembrolizumab
13.	NCT05198752	1	Solid tumor	–	–	–
14.	NCT05192460	NA	Gastric/esophageal/liver cancer	–	–	–
15.	NCT05227378	NA	Gastric cancer	–	–	–

TABLE 5.5 Clinical Trials of mRNA-based Cancer Vaccines (DCs as Carrier)

Sl. No.	Trial No.	Phase	mRNA Vaccine Target Antigen	Route of Administration	Vaccine Combined with Treatment
1.	NCT01326104	2	Total tumor RNA (TTRNA)-loaded Dendritic cells.	ID	TTRNA-xALT
2.	NCT05000801	Not applicable	WT1/hTERT/Survivin	Not reported	Follow-up care
3.	NCT03396575	1	Total tumor RNA (TTRNA)-pulsed Dendritic cells	ID	Cyclophosphamide, fludarabine, temozolomide, TTRNA-xALT, autologous hematopoietic stem cells, GM-CSF, and Td.
4.	NCT03548571	2 and 3	hTERT, survivin, and mRNA from primary prostate cancer tissue.	ID	Temozolomide
5.	NCT04335890	1	Autologous tumor RNA with gp100, tyrosinase, PRAME, MAGE-A3, IDO, and different RNA coding driver mutations ((GNAQ/GNA11Q209 or R183, or the SF3B1R625, CYSLTR2L129Q or PLCB4D630)).	IV infusion	Standard therapy
6.	NCT03688178	2	CMV pp65-LAMP mRNA.	ID	Temozolomide, varlilumab, and Td.
7.	NCT04963413	1	CMV matrix protein pp65-flLAMP plus GM-CSF.	Not mentioned	Temozolomide, GM-CSF and Td.
8.	NCT01983748	3	Autologous tumor RNA (20 mio DC per vaccination).	IV infusions	None
9.	NCT01995708	1	CT7, MAGE-A3, and WT1 (Langerhans-type Dendritic cells).	ID	Standard treatment
10.	NCT04911621	1 and 2	WT1	ID	Chemoradiation (with or without standard therapy).

TABLE 5.5 *(Continued)*

Sl. No.	Trial No.	Phase	mRNA Vaccine Target Antigen	Route of Administration	Vaccine Combined with Treatment
11.	NCT01686334	2	WT1	ID	Follow-up care
12.	NCT02649582	1 and 2	WT1	ID	Temozolomide
13.	NCT00639639	1	CMV pp65-LAMP mRNA-loaded dendritic cell (CMV-DC).	ID	Autologous lymphocyte transfer and Td.
14.	NCT02649829	1 and 2	WT1	ID	Standard therapy
15.	NCT01456104	1	mRNA encoding full-length murine tyrosinase-related peptide 2 (TRP2) (Langerhans-type dendritic cells).	Electroporated	None
16.	NCT03083054	1 and 2	WT1	Electroporated	None
17.	NCT02465268	2	HCMV pp65-shLAMP or pp65-flLAMP.	Shot under the skin.	Temozolomide, GM-CSF, and Td.
18.	NCT01197625	1 and 2	hTERT, survivin, and mRNA from primary prostate cancer tissue.	Not mentioned	None
19.	NCT04157127	1	Pancreatic adenocarcinoma lysate plus mRNA.	ID	Standard therapy

task. When administered, naked mRNA is promptly recognized by the host immunity as exogenous nucleic acid and is rapidly degraded by nucleases, ultimately reducing vaccine efficacy. Molecules with a size of around a 1,000 Daltons or less can easily cross the lipid membrane through passive transport. However, naked mRNA is considerably larger in size and therefore requires a carrier to cross and enter the membrane of target cells [50]. Consequently, the ratio of translation to degradation remains low. Injecting naked mRNA vaccines also poses the risk of inflammation and undesired immune responses [51]. Chemical modification and carrier-mediated delivery have significantly extended the shelf life and translation [52] potential, such as enhancing cellular uptake of RNAs through lipid-derived and polymer-derived materials [53]. mRNA in vaccines can be delivered through viral and non-viral vectors, as shown in Figure 5.4. In the following section of this chapter, we will discuss various approaches for mRNA delivery in vaccines.

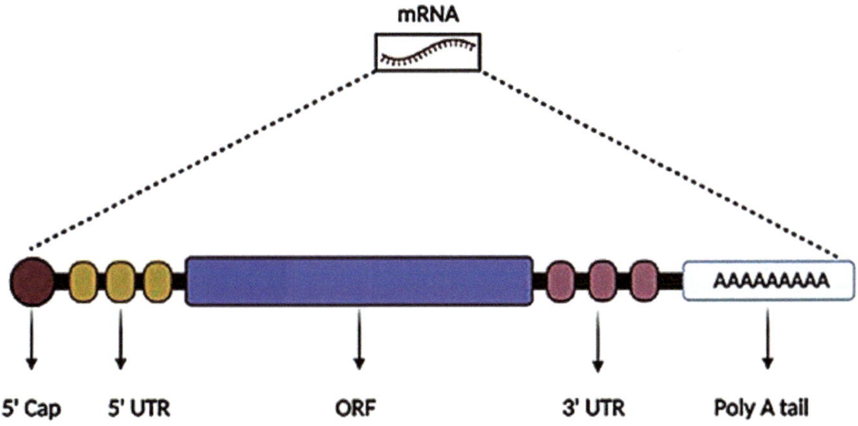

FIGURE 5.3 Schematic structure of mRNA.

Note: The structure starts with the 5' cap, which acts as a protectant against exonuclease cleavage. This is followed by a 5' UTR (untranslated region), which is involved in the initiation of the translation process. Next is an ORF (open reading frame), which is the segment being translated into the protein. After that, there is a 3' UTR (untranslated region), which is involved in the termination of the translation process. Finally, there is a 3' poly-A tail, which plays a role in the stability of the mRNA.

5.5.1 NAKED MRNA

Naked mRNA vaccines (Figure 5.3) are delivered by formulating them in a buffer and directly injecting them into the body. The mechanism of mRNA

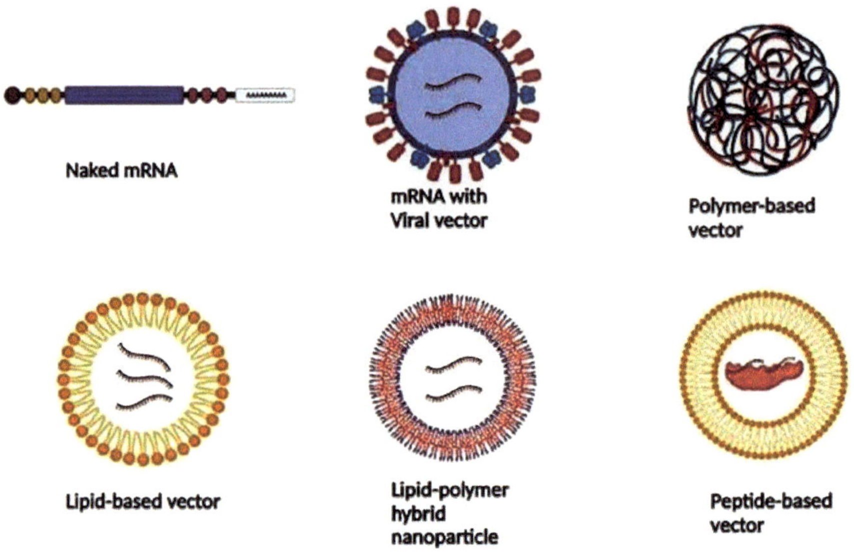

FIGURE 5.4 Various vectors used for the delivery of mRNA vaccine.

Note: The figure illustrates various vectors utilized for the delivery of mRNA vaccines. The first vector is naked mRNA, which consists of a 5' cap, 5' and 3' UTR, ORF segment, and poly-A tail. The second vector involves genetically modified viruses as carriers to transport mRNA into the cytoplasm. The third vector is a polymer-based delivery system, utilizing polymers such as polyethyleneimine (PEI) and poly-beta amino esters (PBAEs). The fourth vector, lipid-based vectors, is the most commonly employed method, where liposomal nanoparticles or liposome complexes are utilized to facilitate delivery. The fifth vector is a lipid-polymer hybrid nanoparticle, combining both lipids and polymers to form a polymeric matrix and lamellar lipid structure. The final vector is peptide-based, employing peptides like protamine as carriers for mRNA delivery.

diffusion remains unclear, as they cannot cross the intracellular membrane. Initially, it was proposed that internalization occurs through micropinocytosis [54]. Other studies have suggested that internalization is facilitated by mechanical force, such as hydrostatic pressure. Disruption of the membrane allows mRNA penetration into the cytoplasm [55]. Administration of naked mRNA through intramuscular (IM), subcutaneous (SC), and intradermal injection (ID) has proven to be effective [56]. However, efficient translation has been demonstrated through SC administration, which eliminates the need for a carrier and reduces costs. SC administration of mRNA induces both cellular and humoral immunity, providing an additional advantage [57]. Naked mRNA strongly stimulates TLRs and Protein kinase receptors (PKR) [30, 58]. The degradation of naked mRNA highlights the need for better engineering of the delivery system to protect mRNA and shield its charge. In a recent phase I clinical trial, individualized tumor mutation signatures were investigated, with 10 selected neoepitopes for each patient in melanoma stage III/IV. The trial observed a complete response in only 8% of patients and a partial response in another 8% (NCT02035956). Another phase I trial investigated TAAs, such as tyrosinase and gp100, in resected melanoma patients. The vaccine induced an immune response in 40% of patients at a low dose and in 33% of patients at a high dose (NCT03394937) [59].

5.5.2 VIRAL VECTORS

mRNA vaccine delivery utilizes genetically modified viruses as vectors. The genes are either partially or fully modified, such as model or therapeutic genes. The advantage lies in the fact that the replication of the RNA virus occurs in the cytoplasm, where the positive RNA strand can be used to translate the protein of interest. The viruses used for mRNA delivery include alphaviruses, picornaviruses, and flaviviruses [56]. However, alphaviruses may sometimes result in a cytopathogenic effect, which can be overcome through strategic modifications [60]. The occurrence of immunogenicity, cytotoxicity, and genome integration represents significant drawbacks of viral vectors, highlighting the need for the development of non-viral vectors for mRNA delivery [61].

5.5.3 POLYMER-BASED VECTORS

Polymers are functional materials that can deliver mRNA vaccines and provide protection against mRNA degradation [62]. Currently, polymers such as poly-amido-amine (PAA), poly-beta amino esters (PBAEs), and polyethyleneimine (PEI) are used as delivery systems. PEI was the first polymer demonstrated to deliver DNA into the mouse brain, highlighting its efficiency as a delivery vector [63]. Formulations using PEI were often prepared by directly mixing the RNA solution and PEI solution. For instance, intranasal vaccination with a PEI formulation containing mRNA encoding for HIV gp120 induced specific Abs against human immunodeficiency virus (HIV) infection in mice [64]. PEI has also been proven to be an mRNA carrier for *in-vitro* transfection and *in-vivo* efficacy. However, its potential toxicity and adverse effects have hindered its development [62]. An alternative is polyamidoamine (PAMAM), a positively charged polymer that has shown promise in carrying mRNA-based vaccines against Toxoplasma gondii, Ebola, and H1N1 influenza virus through intramuscular administration [54]. In addition to cationic polymers, anionic polymers like poly D, L-lactide-co-glycolide (PLGA) have also been used as vectors in several studies. However, negatively charged mRNA molecules do not efficiently fit into anionic polymers, so cationic lipid materials are added to achieve a lipid-polymer hybrid formulation [65]. While polymers have been found to be useful delivery carriers, the correlation between structural and biological activity is not yet satisfactory [66]. Research in this area continues to focus on strategies for novel candidates with improved biodegradability and efficient delivery.

5.5.4 LIPID-BASED VECTORS

The non-viral vectors, such as lipid-based vectors, are the most preferred vectors [62]. Both synthetic and naturally occurring lipids can be used to efficiently deliver mRNA, either by forming a liposomal complex or a liposomal nanoparticle (LNP). Lipids typically consist of three main components: a head, a linker, and a tail. The tail is hydrophobic, while the head is hydrophilic or polar in nature.

5.5.4.1 LIPOSOMAL COMPLEX

Liposomes are spherical structures composed of an outer phospholipid layer. The lipids have a head and a tail, and the interaction between these two components of opposite polarity promotes vesicle formation. This results in the formation of a vesicle with a core that contains the gene of interest in an aqueous environment. Initially, cationic liposomes were used as delivery materials in mRNA vaccines. Positively charged cationic lipids and negatively charged mRNA form aggregates through electrostatic interaction, and these multi-layered cystic complexes are called lipoplexes. Lipoplexes make it difficult for RNase to degrade the encapsulated mRNA. However, poor delivery effects can also be observed due to potential hindrance with negatively charged molecules in body fluids [50]. When used in a 1:1 ratio, polymers like 1,2-dioleoyl-3-trimethylammonium propane (DOTAP) and dioleoylphosphatidylethanolamine (DOPE) effectively transfer nucleic acid for the HIV-1 antigen Gag, successfully inducing an antigen-specific reaction in mice [67]. A Phase 1 clinical trial of a lipoplex formulation in patients with melanoma demonstrated an immune response in at least one TAA in 75% of subjects (NCT02410733). Another clinical study for Claudin-6 (CLDN6) showed engraftment of Chimeric Antigen Receptor (CAR)-T-cells in all patients (NCT04503278) [59].

5.5.4.2 LIPOSOME NANOPARTICLES (LNPS)

LNPs, initially used as a vector for siRNA, are well-known delivery systems for nucleic acid therapeutics or mRNA vaccines [68]. LNPs are formed using a bilayer shell of positively charged lipids, cholesterol, and polyethylene glycol (PEG), with a hydrophilic core where the target mRNA gene is placed [50]. The design of LNPs must ensure rapid metabolism and elimination to prevent tissue accumulation [69]. LNPs have the advantage of encapsulating multiple mRNA target genes and immunostimulants together to enhance overall immunogenicity [52]. Studies conducted in Phase 1/2 trials on patients with gastrointestinal cancer, investigating neoantigen-specific mRNA, showed that three out of four (75%) patients developed $CD4^+$ and $CD8^+$ T-cell responses against predicted mutation-specific neoepitopes (NCT03480152). Another study involving a personalized CV encoding several neoantigens formulated as LNPs for solid tumors demonstrated detectable neoantigen T-cell responses (NCT03313778) [59].

5.5.4.3 LIPID-POLYMER HYBRID NANOPARTICLES (LPNS)

Lipid-polymer hybrid nanoparticles (LPNs) have demonstrated effective therapeutic transfection *in vivo* and functional transfection of siRNA *in vitro* [56]. This hybrid formulation exhibits thermodynamic stability through hydrophobic, van der Waals, and electrostatic interactions [70]. Polycaprolactone, polylactic acid, and PLGA are commonly used polymers for LPNs formulation. DOTAP, lecithin, 1,2-Distearoyl-sn-glycero-3-phosphoethanolamine-(DSPE), and PEG are commonly used lipids for LPNs. The key components of LPNs are the lamellar lipid structure and the polymeric matrix core. Nucleic acid is embedded in both the core and the corona [71]. Lipid-based carriers overcome various barriers such as mRNA stability issues, benefiting from their high fusion capacity with cell membranes, long circulation time, endosome escape, and scalability in production. However, the challenge lies in the potential side effects due to their inflammatory nature. Further research will help overcome these shortcomings in the future.

5.5.5 *PEPTIDE-BASED VECTORS*

Several peptides have been utilized for the delivery of mRNA vaccines. Typically, positively charged peptides serve as the primary carriers for mRNA delivery. The presence of lysine and arginine residues imparts the positive charge and facilitates electrostatic interactions with negatively charged mRNA, forming a matrix [72]. Protamine, a cationic peptide, is widely employed as a vector for mRNA delivery in numerous experiments. Protamine offers two advantages: firstly, it protects mRNA from degradation by RNase, thereby enhancing its stability *in vitro*; secondly, it possesses adjuvant activity that activates TLR7 due to its similarity to the condensed viral genome [65]. Protamine also maintains stability under fluctuating storage conditions. Studies have demonstrated the ability of protamine to protect mRNA encoding glycoprotein for the rabies virus at elevated temperatures for extended durations [73]. However, the feasibility of the mRNA-protamine complex has been questioned, as a study showed poor translation of the complex containing β-galactosidase when injected into a glioblastoma tumor [74]. Currently, protamine remains the only peptide used as a carrier that has been evaluated in clinical trials of mRNA vaccines. Although well tolerated by patients, it fails to elicit the expected host immune response [65]. In a clinical trial of metastatic castration-resistant prostate cancer (PC), the use of protamine as a vector to deliver mRNA antigen did not show a significant difference in progression-free survival (PFS) (NCT01817738). Another Phase 1 study in non-small-cell lung cancer (NSCLC) demonstrated detectable antigen-specific immunity in 84% of patients (NCT01915524) [59].

5.6 CURRENT ROUTES OF ADMINISTRATION

The efficacy, stability, and safety of mRNA vaccines depend greatly on the route of administration, in addition to their composition. Vaccines can be administered systemically or through local application [54]. For certain diseases, a specific route of administration may be preferred. For example, liver disease can be treated with intravenous (IV), intramuscular, SC, or intraperitoneal administration [75], while pulmonary disease requires an intranasal or intratracheal route [76]. Ultimately, efficient delivery to the draining lymph node is crucial. This is because the presence of more immune cells allows the mRNA to be easily taken up by antigen-presenting cells

(APCs) through micropinocytosis, leading to an immune response [77]. Various routes of administration are described in subsections.

5.6.1 INTRAVENOUS (IV) INJECTIONS

mRNA is transported to the lymph node after IV injection. Once mRNA enters the lymph node, the immune system is activated, and this process is followed by the maturation of DCs through glycosylation-dependent interaction, which enhances antigen presentation [77]. The increased production of Tumor Necrosis Factor (TNF)-alpha also leads to the migration of DCs into the lymph nodes, ultimately resulting in the upregulation of CD69, an activation marker that supports the immune response [78].

5.6.2 INTRADERMAL (ID) INJECTION

The injection of mRNA via the ID route is preferable due to the presence of antigen-presenting cells (APCs) such as Langerhans cells in the epidermis and dendritic cells (DCs) in the dermis layer of the skin [79]. The mRNA is initially expressed locally after ID injection [77]. Following the ID injection of globin Untranslated region (UTR)-stabilized luciferase encoding mRNA into the pinna of the ear in mice, a peak of luminescence was observed approximately 17 hours later. This luminescence was still detectable after 3 days, indicating the appropriate activation of the immune response [80].

5.6.3 SUBCUTANEOUS (SC) INJECTION

The SC injection delivers the mRNA vaccine into the subcutis region of the skin, which is composed of loose adipose tissue and fewer immune cells compared to the dermis. The presence of loose adipose tissue allows for a larger injection volume than the ID layer, resulting in less pain but a reduced penetration rate [65].

5.6.4 INTRAMUSCULAR (IM) INJECTION

Due to the adjustable volume of injection and reduced risk of injection site reactions, the intramuscular (IM) route is the preferred administration route for vaccines [2]. A vaccine containing adjuvants induces inflammation at the injection site, eventually leading to the recruitment and activation of immune cells [81]. mRNA vaccines given through this route are first taken up by local myocytes and eventually generate an immune response via T-cells [77]. The volume of IM injection is higher than that of the ID route, and it has comparatively fewer adverse effects than the ID or SC route [65].

5.6.5 INTRANODAL (IN) INJECTION

When mRNA is injected directly into peripheral lymphoid organs, an immediate interaction occurs between antigen-presenting cells (APCs) and primed T-cells or B-cells. Therefore, intranasal (IN) injection is a more efficient route for the direct engulfment of mRNA by APCs [65]. However, IN injection is a complicated procedure that requires ultrasound guidance in humans, limiting its use [82]. Although SC or intramuscular (IM) administration are considered the two most preferred and safe routes for mRNA CVs, one must also consider the potential toxicity that may arise from IV administration. Nevertheless, IN or intra-nasal routes are now widely studied for the delivery of mRNA-based vaccines [2]. In comparison to ID injection, IN injection of ovalbumin (OVA) mRNA with TriMix as an adjuvant elicited significantly higher *in vivo* cytotoxicity by antigen-specific T-cells [83].

5.6.6 INTRATUMORAL INJECTION

Intratumoral administration of mRNA CV is an unconventional route. Intratumorally administered mRNA vaccine encoding IL-23, IL-36, and OX40L showed better antitumor activity than ID or subcutaneous (SC) administration, but the effect was also dependent on the frequency of administration [31]. A brief discussion of the advantages and disadvantages of various routes of administration is provided in Table 5.6.

5.7 CURRENT TRENDS AND CHALLENGES IN MRNA-BASED CANCER VACCINES

Clinical trials of numerous CVs have been ongoing to date. The US Food and Drug Administration (USFDA) has approved three CVs that have demonstrated therapeutic efficacy. These include a vaccine containing live

TABLE 5.6 Advantages and Disadvantages of Various Routes of Administration Used for mRNA-based Vaccines [65]

Route of Administration	Advantages	Disadvantages
IV injection	• Large injection volume. • Direct interaction with APCs and lymphoid organ.	• Rapid degradation of mRNA. • Systemic adverse effects.
ID injection	• Direct interaction with APCs.	• Less injection volume. • Local adverse effects.
SC injection	• Injection volume higher than ID. • Lesser local side effect.	• Degradation of mRNA.
IM injection	• Dense blood network. • Fewer local adverse effects.	• Less injection volume.
IN injection	High delivery efficiency.	• Skilled procedure required.

attenuated *Mycobacterium bovis* strain for non-muscle invasive bladder cancer, a DC vaccine for metastatic castration-resistant PC, and an oncolytic herpes virus vaccine for advanced melanoma [31]. The development of mRNA vaccines is influenced by intrinsic and extrinsic factors, tumor heterogeneity, HLA, and the tumor microenvironment (TME). Intrinsic factors such as stability and effectiveness can be improved by refining the mRNA structure and sequence, enhancing preparation and purification technologies, and improving delivery methods.

Another important aspect is the heterogeneity of HLA due to the polymorphism of HLA alleles among different geographical populations [84]. Therefore, the production of neoantigens remains a key focus in order to achieve a stronger and more specific effect with minimal adverse effects [31]. However, there are certain limitations that hinder the development of neoantigens. Overcoming these limitations could expedite the quest for mRNA vaccine fulfillment:

1. **Limited Amount of Antigen:** Tumors account for many mutations, but only a few meets the requirements of the antigen. Picking up specific mutations becomes important as they mainly occur in the non-coding region. Hence, research in this field becomes crucial.
2. **Screening Methods:** Another hurdle in vaccine production is the scarcity of methods to screen the neoantigen. Recent developments in the field of machine learning and bioinformatics may help solve this challenge.

3. **Preparations and Delivery:** Besides having different methods of delivery, it remains a challenge, especially for nucleic acid therapies.
4. **Heterogeneous Character of the Tumor:** The differences in mutation at every evolutionary step make the development of the vaccine difficult [37].

Tumor cells produce TGF-β, which can convert effector T-cells into regulatory T-cells (TRegs) that ultimately suppress other effector T-cells infiltrating the tumor. Tregs can express CD4 and CD25, thus inhibiting cytotoxic T lymphocytes (CTL) in the TME. This leads to the suppression of antitumor immunity. The presence of immunosuppressive cells such as $CD4^+$ Tregs and suppressor $CD8^+$ T-cells in the TME facilitates the release of a large number of soluble immunosuppressive factors such as TGF-β, IL-10, and vascular endothelial growth factor (VEGF) into the microenvironment [85]. The development of immune-based mRNA vaccines utilizes the concept of combination with adjuvants or immune checkpoint inhibitors (ICIs). Hence, the current trend in CVs aims to optimize the anticancer activity in a more specific way and simultaneously reduce toxicity by either combining mRNA vaccines and ICIs or adding adjuvants into the mRNA vaccines. Sometimes, the mRNA vaccine cannot produce the required adaptive immune response, leading to tolerance instead of an immune response. The addition of adjuvants in mRNA vaccines facilitates the draining of immune cells at the site of injection [86]. The strategy of combining adjuvants with mRNA vaccines particularly helps elderly cancer patients because aged DCs show reduced capacity to generate an inflammatory response to stimuli [87].

Chronic inflammation also leads to a decrease in the expression of the co-stimulatory receptor CD28, resulting in a reduced immune response [88]. One extensively studied adjuvant for mRNA vaccines is PAMPs, which stimulate the immune response through PRRs such as TLRs and NLRs. Ligands of TLR and NLR have been proven to be effective as vaccine adjuvants, ultimately enhancing the immune response [89]. In a murine tumor model, the use of TLR3, TLR7, and TLR9 agonists as adjuvants in vaccines demonstrated a robust immune response mediated by antigen specific $CD8^+$ T-cells [90]. The cGAS-STING pathway, which involves cyclic GMP-AMP synthase and stimulator of interferon genes (STING), plays a crucial and novel role in the immune system. Preclinical studies involving the co-administration of STING agonists with other cancer immunotherapies, such as CVs and ICIs like anti-programmed death 1 (PD-1) and cytotoxic T lymphocyte-associated antigen 4 (CTLA-4) Abs, as well as adoptive T-cell transfer therapies, have shown promising results [91, 92]. Currently, an early phase clinical trial is underway to evaluate the combination of ADU-S100 and MK-1454 with ICIs in patients with advanced/metastatic solid tumors or lymphoma [87]. On the other hand, cytokines can also be used as adjuvants. For example, granulocyte-macrophage colony-stimulating factor (GM-CSF) has proven to be effective in recruiting DCs to the site of injection in preclinical settings. However, in clinical settings, it has yielded disappointing results [86]. The administration of the mRNA vaccine through direct injection has been shown to be the most cost-effective method. However, the mRNA vaccine can also be successfully administered through various routes such as ID, intramuscular (IM), intranasal (IN), IV, and intranasal administration. The initial clinical trial of the mRNA vaccine in melanoma patients involved the administration of naked RNA, but it failed to demonstrate effectiveness. The self-adjuvanted RNActive® vaccine, which was successfully studied in phase I trials, utilized a novel technology developed by CureVac [39].

While the production method of the mRNA vaccine is simpler compared to conventional vaccine production, it lacks a manufacturing framework. To summarize the process, it can be divided into upstream processing, which involves enzymatic generation of mRNA, and downstream processing, which includes purification, LNP formation, and filling into the final product [93]. Once the mRNA product has been generated, it needs to be purified as the reaction mixture contains enzymes, residual Nucleoside Triphosphates (NTPs), and DNA templates [94]. The removal of impurities is crucial as they may hinder the efficiency of the mRNA vaccine. For example, the purification of modified mRNA using reverse phase High Performance Liquid Chromatography (HPLC) resulted in a 10–1,000-fold increase in protein production efficiency [95].

The manufacturing process of mRNA vaccines consists of three steps: synthesis, purification, and formulation. Specifically, the DNA template for *in vitro*-transcription (IVT) is stored at – 20°C. The mRNA is then produced, purified, and tested under the same temperature. It is subsequently formulated with LNPs and packed into vials. The entire process requires transportation under controlled conditions, which poses a challenge for large-scale production of mRNA vaccines. Continuous manufacturing could help achieve cost-effective production by reusing raw materials such as enzymes [96].

mRNA is conformationally more flexible than DNA. Intramolecular transesterification, a process in which the 2'-hydroxyl group of ribose breaks the phosphate backbone, eventually leads to strand cleavage. Therefore, most mRNA vaccine formulations need to be stored in freeze-dried conditions. According to study results, freeze-dried self-amplifying RNA remained stable for 10 months at 4°C [97]. The lyophilized LNP containing RNA retained its biochemical properties for 21 months at 4°C. Due to a lack of data, the thermostability data of LNPs containing Small interfering RNA (siRNA) can be used to estimate mRNA vaccine stability. *In vitro* studies have shown that the LNP complex remains stable for 5 months at 2°C [98]. A comparative study of freeze-dried LNP-mRNA with different lyoprotectants containing fresh LNPs without mRNA was conducted. The results showed mRNA expression in both *in vivo* and *in vitro* studies, but a mechanistic basis was not established [99]. Therefore, the process of lyophilization is an important consideration for achieving a stable product. Alternatively, spray drying, vacuum drying, or supercritical fluid drying could also be considered.

5.8 REGULATORY HURDLES IN MRNA CANCER VACCINES

The development of mRNA vaccines has been a long journey spanning almost 30 years. However, it was only two decades ago that these vaccines entered the clinical phase. The main reasons for the delay in making these vaccines available to the public include reactogenicity or immunogenic adverse reactions, large-scale production, and stringent regulatory requirements. Following the remarkable clinical outcomes of mRNA-based vaccines in preventing SARS-CoV-2 infection, efforts to develop new mRNA vaccines for other diseases have reached their peak. The proven effectiveness of mRNA COVID-19 vaccines has prompted international healthcare bodies to expedite the regulatory process by accelerating the development of guidance documents, with the World Health Organization (WHO) taking a leading role [100].

The regulatory hurdles for mRNA CVs are largely similar to those for other vaccines. These include ensuring the quality of excipients and adjuvants, maintaining a uniform manufacturing process, conducting preclinical evaluations for safety and efficacy, and carrying out clinical trials as well as long-term monitoring for adverse drug reactions after marketing approval. While the regulatory approval process for mRNA-based vaccines is expected to remain consistent in the near future, it is important to note that mRNA vaccines fall under the broad category of immunogens. Therefore, several stringent regulatory guidelines are specifically tailored to address the quality of starting materials, consistency, safety, and efficacy of the product. It is crucial to consider the regulatory aspects that may influence the transition from animal to human testing, particularly given the involvement of new technologies in mRNA vaccine development. Another challenge lies in the thermal stability of RNA, which affects vaccine development. The requirement for cold-chain transport and storage poses limitations on the use of mRNA vaccines, especially in third-world countries [101].

During the 74th meeting of the WHO Expert Committee on Biological Standardization held in October 2021, the committee published a document titled "Evaluation of the quality, safety, and efficacy of messenger RNA vaccines for the prevention of infectious diseases: regulatory considerations." The guidance document emphasizes the technical requirements for the manufacture and control of mRNA vaccines, including starting materials, raw materials, and excipients; process development and in-process controls; product characterization; consistency of manufacture; manufacture and control of both bulk purified mRNA (drug) and final formulated vaccine (product), and all related records. Therefore, all process and manufacturing records are also part of the regulatory considerations. If any changes are made to the process or composition, they need to be tested in preclinical studies again. Non-clinical evaluation is required for proof-of-concept and safety studies. Several other guiding documents need to be consulted, which are common regulations for any vaccines, namely the WHO Guidelines on the nonclinical evaluation of vaccines [102], Guidelines on the nonclinical evaluation of vaccine adjuvants and adjuvanted vaccines [103], and Guidelines on the clinical evaluation of vaccines: regulatory expectations [104]. The "Guidelines on the nonclinical evaluation of vaccine adjuvants and adjuvanted vaccines" provides very specific guidance on adjuvants used in vaccines, which are also applicable to mRNA vaccines.

These considerations primarily focus on the production, characterization, and quality assurance of lots to be used in nonclinical pharmacology studies, nonclinical toxicology studies, and first-in-human clinical trials. They also cover the rationale for the use of the adjuvant, selection of animal species for nonclinical evaluation of vaccine adjuvants and adjuvanted vaccines, nonclinical safety assessment in animals, and considerations for first-in-human clinical trials.

The nonclinical evaluation of mRNA vaccines includes pharmacology/immunology/proof-of-concept studies and safety/toxicity studies in animal models [105]. This is particularly important when approving mRNA CVs because unnatural nucleoside analogs with altered conformation have been shown to cause mitochondrial toxicities, resulting in polyneuropathy, myopathy, pancreatitis, liver steatosis, lactic acidosis, lipodystrophy, and even fatality [106, 107].

The complexity of document submission to regulatory authorities increases as the development process progresses. Therefore, maintaining proper records and documentation is crucial. According to the requirements set by the WHO for clinical evaluation of mRNA vaccines, these vaccines should undergo evaluation for safety, immunogenicity, and efficacy in humans. Clinical trials involving mRNA CVs should adhere to the WHO Guidelines for Good Clinical Practice (GCP) for trials on pharmaceutical products [108] and the WHO Guidelines on Clinical Evaluation of Vaccines: Regulatory Expectations [104]. The WHO guidelines for Good Manufacturing Practice (GMP) requirements for pharmaceuticals for human use should also be followed for vaccines.

In addition to the WHO considerations, manufacturers and marketing authorization holders of mRNA vaccines should follow the guidelines of the International Council for Harmonization (ICH) of Technical Requirements for Pharmaceuticals for Human Use (ICH) for quality and safety of vaccines. These guidelines include Q3C(R6) (impurities: guideline for residual solvents), Q3D(R1) (guideline for elemental impurities), and S2 (R1) (ICH guideline on genotoxicity testing and data interpretation for pharmaceuticals intended for human use) [110–112].

It is important to develop national and international guidelines for mRNA CVs to provide guidance to healthcare stakeholders. These guidelines should reflect the current knowledge and opinions. It should be noted that the regulatory requirements for marketing authorization of mRNA or DNA-based CVs differ from those for drugs. The mRNA CVs are considered as biologicals by the European Medicine Agency (EMA) and the USFDA. In 2011, the USFDA developed a guidance document for the industry on nucleic acid vaccines (NAVs) titled "Clinical Considerations for Therapeutic CVs." This document provides specific guidance to sponsors on submitting an Investigational New Drug application (IND) for a therapeutic CV. It focuses on recommendations for critical clinical considerations in investigational studies.

The guidance discusses considerations that are common to both early phase clinical trials (Phases 1 and 2) and late phase clinical trials (Phase 3). It also addresses considerations that are unique to specific stages of clinical development for therapeutic CVs. The guideline outlines specific requirements for Early Phase Clinical Trials, such as starting dose and dosing schedule, booster and maintenance therapy, dose escalation, and single-arm versus randomized Phase 2 trials in early development.

For Late Phase Clinical Trials, the guidance covers considerations such as safety profile data from early phase clinical trials, endpoints, statistical issues, control issues, delayed vaccine effect, autologous vaccine trials, and accelerated approval regulations.

Common considerations for both Early and Late Phase Clinical Trials include patient population, monitoring the immune response, biomarkers as evidence of efficacy, adjuvants used to stimulate immune response, multi-antigen vaccines, disease progression/recurrence immediately or shortly after the initial administration of CVs, and concomitant and subsequent therapies. This document exclusively applies to therapeutic CVs designed to elicit specific responses to tumor antigens for the purpose of treating established cancer patients. It does not apply to vaccines intended for the prevention or treatment of infectious diseases, products that induce or enhance non-specific immune responses, or products intended to prevent or reduce the incidence of cancer in individuals without cancer [113].

The European Union (EU) and the European Medicines Agency (EMA) have implemented more stringent regulatory guidelines for the use of CVs. Both the EU and EMA have established multiple sets of guidelines for the non-clinical and clinical development phases of CVs. During the non-clinical phase, the EU emphasizes the GMP of products, preclinical pharmacological and toxicological testing of vaccines, and non-clinical safety studies for conducting human clinical trials for pharmaceuticals. Separate guidelines for the quality, non-clinical, and clinical aspects of medicinal products containing genetically modified cells and gene therapy medicinal products are available in the EU and EMA. Specialized guidelines on "clinical evaluation of new vaccines" and "requirements for quality documentation concerning biological investigational medicinal products in clinical trials" must also be followed [114, 115].

Since the first mRNA CV trials in 2009, numerous mRNA-based CVs have been developed and their clinical trials have been conducted. Several databases, such as clinicalTrials.gov, and other resources, show that most of the CVs have reached Phases 1 or 2. However, there is limited data available for Phases 3 or 4 clinical evaluation of efficacy in humans. The major hurdles may be the stringent requirements for regulatory approval from national and international bodies.

5.9 ADVANCEMENT AND OPTIMIZATION IN THE STRATEGIES TO IMPROVE STABILITY AND DELIVERY OF VACCINES

Currently, there are two forms of mRNA vaccines based on their auto-replicative capacity: conventional/non-amplifying mRNA and self-amplifying mRNA.

5.9.1 CONVENTIONAL MRNA

The mRNA consists of various units, including a cap structure, 5'UTR, open reading frames (ORFs), the gene of interest, and a poly(A) tail. These sequences play a role in cis-regulatory functions, controlling gene expression, and managing degradation rates and translation. Conventional mRNA offers advantages such as a simplified construction process, small size, and the absence of additional encoded proteins [116]. However, it also has disadvantages, such as RNA instability and a short shelf life. To address these shortcomings and enhance mRNA stability and expression, structural optimization can be pursued, as discussed in the preceding segment.

5.9.2 SELF-AMPLIFYING MRNA

Self-amplifying mRNA or replicon was initially utilized in the development of a rationally designed West Nile virus vaccine. This approach resulted in a high titer of the virus in patients after a single immunization. Another study demonstrated conditional control over gene expression levels, achieving a 47-fold change by introducing viral-packed self-amplifying mRNA [117]. The objective is to enhance the stability and magnitude of gene expression of interest. Self-amplifying mRNA is relatively larger than conventional mRNA due to the additional ORF segment encoding nonstructural proteins. The ability to auto-replicate allows for increased expression of the gene of interest. Furthermore, *in-vivo* duration can be extended up to 2 months [116]. Therefore, self-replicating mRNA vaccines maintain the advantages of rapid development, convenient modular design, and entirely cell-free synthesis, even in small doses, thanks to their self-replicating properties [36].

Structural modifications can also contribute to the stability of mRNA vaccines. The basic structure of mRNA resembles that of eukaryotic mRNA, consisting of a protein-encoding ORF, 5' and 3' UTR, a 5' Cap, and a 3' poly(A) tail, as depicted in Figure 5.3. All these structural components can be modified to achieve increased stability and improved translational efficacy [118].

5.9.2.1 5'-CAP

The RNA undergoes post-translational changes, such as 5'-capping, which acts as an initiation factor for translation. Natural mRNA contains a 7-methylguanosine (m7G) moiety at the 5'-cap, which also serves as a protector against exonuclease cleavage and plays a key role in distinguishing between self and non-self mRNA molecules [118]. While capping of IVT mRNA can be achieved, the resulting RNA often fails to be translated in a significant number of mRNA transcripts. To address this issue, Antireverse cap analogs (ARCA) have been designed. When incorporated into the mRNA, ARCA significantly increases translational efficiency compared to conventional mRNA, more than doubling it [57]. Additionally, modifications within the triphosphate linkage can hinder decapping of the corresponding mRNA and enhance binding to initiation factor 4E, which is involved in ribosome recruitment [119]. However, ARCA caps have some limitations. They exhibit a relatively low capping efficiency of 60%–80%, the cap contains an additional O'methyl group at the C3 position, acting as an exogenous motif, and transcription must begin with guanine [2]. Another method to add a 5' cap is through an enzymatic approach, with the Vaccinia capping system being the most widely used. This system provides 100% capping efficiency with proper orientation, but for large-scale production, purification of the Vaccinia virus Capping Enzyme (VCE) is required [120].

5.9.2.2 5'-3' UNTRANSLATED REGIONS (UTRS)

The UTRs are not translated but contain important regulatory elements and protein expressions. The 5'-UTR is primarily involved in the initiation of the translational process and recruits ribosomes, whereas the 3'-UTR is responsible for the stability and half-life of the mRNA [121]. To optimize the 5'-UTR, one can avoid the presence of the Start codon (AUG) and non-canonical start codon (CUG), as these may disrupt the normal translation process. Additionally, highly stable structures that may reduce ribosome recruitment can be eliminated, and a shorter 5'-UTR can be incorporated [2]. The stability of the mRNA can be enhanced by adding two consecutive β-globin arranged head-to-tail at the 3'-UTR. For the better design of UTRs in therapeutic mRNA vaccines, understanding cell pharmacology is important, as the effects depend on the species and cell types [122].

5.9.2.3 THE OPEN READING FRAMES (ORFS)

Codon optimization is a controversial approach for improving translation. Replacing codons may offer greater stability but could also result in the formation of novel peptides with unknown biological activity *in vivo* [121]. The sequence can be optimized and designed to maintain the same ratio of each codon or to use the best pair of codons commonly found in highly expressed proteins of the target cell. Additionally, it is important to avoid hairpin loops in the ORF [123]. Nucleoside-modified mRNAs are created by incorporating modified nucleosides into mRNA transcripts, which have lower immunostimulatory activity and enhance translational efficiency. The most commonly used modifications for achieving low immunostimulatory activity and improved translational efficiency are 5-methylcytidine (m5C) and pseudouridine nucleosides [124]. Furthermore, purification through HPLC can eliminate dsRNA contamination, reducing the production of pro-inflammatory cytokines and prolonging the action of the mRNA [57].

5.9.2.4 THE POLY(A) TAIL

Poly(A) tail consists of approximately 250 nucleotides, although the number varies among different cells. The length of the tail plays a critical role in translational efficiency, degradation, and stability [123]. The formation of the "Closed-loop structure" occurs when the 3'-end poly(A) tail combines with the PABPs and interacts with the 5' cap through the initiation factor [121]. A study has suggested that shorter poly(A) sequences may enhance the close-loop structure and translation efficiency. Therefore, evaluating the role of the poly(A) tail length in the expression of IVT-mRNA is necessary [2]. There are two methods of adding a poly(A) tail: one involves encoding the tail on the DNA template, and the other involves using recombinant poly(A) polymerase during the transcription process [57].

5.9.3 MRNA-BASED DENDRITIC CELL CANCER VACCINES

For the delivery of the mRNA CV, various delivery routes are available for the injectable mRNA vaccine which has been discussed earlier (Table 5.6). This section will discuss the use of DCs in the development and delivery of mRNA vaccines. DCs have been in the limelight to be used as a vehicle for the CV as they have the ability to control and regulate the immune response [59]. Hence, the focus remains on the generation of *ex-vivo* antigen-loaded DCs that will elicit $CD8^+$ and $CD4^+$ T-cell responses in cancer patients [125]. The clinical trial conducted on various types of cancer shows that DC-based immunotherapy is well tolerated and produces anti-tumor immunity (Table 5.5). These studies have also reported that 8.5% of the patients who received DC-based therapy show similar results to the patients receiving dacarbazine treatment. Some of the parameters that need to be considered are antigen loading, DC maturation, and route of administration [125].

Antigen loading or the delivery of mRNA into the DCs involves various strategies, such as electroporation, lipid-derived carriers, and sonoporation. Electroporation is commonly used due to its high mRNA delivery efficiency. This method applies a high-voltage shock to disrupt the cell membrane, allowing mRNA to enter the cell [65]. The rapid delivery of mRNA protects it from degradation by extracellular ribonucleases. On the other hand, lipid-based carriers are taken up by endocytosis, where the lipid compound fuses and releases the mRNA content. Sonoporation involves the use of ultrasound-responsive microbubbles, which enter the cell and implode to release the mRNA directly into the cytosol [125].

The route of administration for mRNA-loaded DCs is mainly through ID, SC, IV, and IN injections. Various studies on mice have shown that the ID and SC routes of vaccine administration are superior to IV injection [126]. However, a comparison study did not reveal any significant difference between IN and ID delivery of DCs in patients with melanoma [127]. Since there is a lack of data suggesting a clear winner, a combined approach of administration routes could be used. For example, a study observed a broad immune response when mRNA-modified Trimix-DCs were administered via IV and ID routes, compared to only ID delivery [109]. This suggests that, in addition to the antigen, the route of administration also affects vaccine efficiency.

5.10 CONCLUSION AND FUTURE PROSPECTS

The mRNA-based immunotherapeutic follows proteins as an alternative treatment. After the foremost research related to the administration of mRNA into preclinical models, the concerned area has rapidly evolved and revealed potential as a descendant's biopharmaceutical for vaccine advancement. Due to their swift progress capability, immense effectiveness, post-safety status, and lower cost of production, mRNA vaccines are considered highly important and confirm the next-generation vaccines in the field of immunotherapy. In recent years, these products have significantly increased the thrust in research progression, resulting in remarkable achievements in the field. The introduction of recent COVID-19 vaccines in the market has paved the way for extensive recognition in cancer and other modalities. Clinical trial data of mRNA vaccines related to cancer and infectious diseases have shown promising results. Therefore, the next decade will be tremendously important for the expansion of innovative mRNA-based therapy. It is likely that mRNA vaccines will be developed for other modalities beyond infectious diseases and cancer, as the technologies have rapidly advanced and become refined and optimized. However, certain areas in mRNA vaccine technologies still need to be advanced and evaluated.

It has been observed that mRNA-based vaccines are an encouraging means of innovative therapy and preventive vaccines alongside cancer and infectious diseases, as evidenced by the numerous preclinical and clinical trial applications (CTAs) submitted to regulatory bodies. However, diverse difficulties related to mRNA's charge, bulky size, key instability, and vulnerability to enzymatic destruction hinder the development of vaccines from ground research to market availability. These obstacles restrict the wider application of these vaccines and justify the need for better-quality vectors, applicable vaccine delivery systems, and materials.

Recently, mRNA COVID-19 vaccines have been introduced to the market with remarkable efficacy. As a result, researchers are now expressing their hopes to develop mRNA vaccines for a wide range of applications. Over the past few years, clinical trials for cancer have been conducted with a small number of patients, showing promising outcomes. Numerous clinical trials are currently underway to evaluate mRNA vaccines for various types of cancer. However, no mRNA CV has been approved by the USFDA for use as a standalone treatment or as an adjunct to cancer therapy. Through ongoing CV trials, preliminary results will help determine the safety profile and efficacy of mRNA CVs. Ultimately, additional extensive and detailed clinical trials are necessary to enhance our understanding of the immunogenicity and safety of these vaccines.

mRNA is naturally produced by our body's cells on a daily basis for protein production. Physiological cells have a regulatory mechanism to control the amount of protein produced, ensuring that only the required amount is generated and preventing excessive protein production. The breakdown of mRNA vaccines by the body's cells is efficiently confirmed and controlled by the poly(A) tail. Currently, researchers are primarily focused on designing vaccines that have a stay period in the cell no longer than necessary to induce immunity, while avoiding short- and long-term adverse drug reactions.

The medical community places great emphasis on the long-term safety of vaccines. Past data on vaccines has shown that delayed effects can occur after vaccination. However, these effects are typically observed within six weeks after vaccine administration. Regulatory authorities in different countries require a minimum of eight weeks of post-vaccine safety profile data before granting marketing approval. mRNA vaccines have been under development for many years and have undergone extensive evaluation. Decades of reviewing mRNA vaccines have not revealed any long-term post-vaccination side effects. mRNA COVID-19 vaccines have been used for over two years, benefiting more than 9.5 billion people, with no significant toxicity observed. Nevertheless, the long-term effects of these vaccines still need to be assessed. Physiological cells have a regulatory mechanism to control the production of proteins, ensuring that only the required amount is produced and preventing excessive protein generation. The timely breakdown of mRNA vaccines by body cells is confirmed and controlled by the poly(A) tail in a highly efficient manner. Currently, researchers are primarily focused on designing vaccines that have a stay period in the cell no longer than necessary to induce immunity. This is to avoid any short- or long-term adverse drug reactions. The medical community is particularly cautious about the long-term safety of vaccines, as past data has shown that delayed effects can occur after vaccination. However, these effects are typically observed within six weeks after vaccine administration. Therefore, regulatory authorities in different countries require a minimum of eight weeks of post-vaccine safety profile data before granting marketing approval. mRNA vaccines have been under development for many years and have undergone extensive review. Decades of reviewing mRNA vaccines have not revealed any long-term post-vaccination side effects. mRNA COVID-19 vaccines have been in use for over two years, benefiting more than 9.5 billion people, with no significant observed toxicity. Nevertheless, the long-term effects of these vaccines still need to be assessed.

The comparatively simple and straightforward manufacturing process of the mRNA vaccine is the most significant benefit over traditional products. The cost per dose and manufacturing price of the vaccine can be influenced by the specific unit operations used during the process. Ultimately, the quantity of RNA per dose, product titer, and scale of production has a major impact on the cost of the vaccine. Therefore, it is crucial to adopt sustainable and cost-effective manufacturing processes. The recovery and purification of the biosynthetic vaccine product currently fall short in terms of establishment, relying on approaches that lack commercial scalability and economic viability. One potential solution to these problems is to transition towards continuous manufacturing. By implementing innovative production methods that allow for material recycling and recirculation, along with high-throughput purification methods and distinct analytical processes, adaptable, sustainable, and profitable mRNA vaccine production can be achieved. In conclusion, further scalability and optimization are necessary in the manufacturing processes of mRNA vaccines.

Insufficient thermostability and cryogenic stowage pose significant challenges for mRNA vaccines in the market. Maintaining the proper temperature is crucial for clinical efficacy, highlighting the need for a thermostable vaccine with high therapeutic efficacy. This remains a challenge for researchers and scientists. By gaining a deeper understanding of the underlying reasons and mechanisms of instability, along with the selection of suitable expert stabilization processes, we can make progress towards developing highly stable mRNA vaccines. These advanced mRNA vaccines, with enhanced stability, will facilitate easy shipping and efficient storage throughout the entire vaccine supply chain. They will also enable manufacturers to distribute vaccines worldwide more quickly in the future.

In the biopharmaceutical industry, mRNA vaccines are leading the way. Investors, sponsors, and research organizations must align themselves for the market availability of mRNA vaccines for new diseases. Increased community and private collaborations can create a more favorable environment for the expansion of mRNA-based vaccines. Future clinical trials should focus on stratified patient populations. In conclusion, the future of vaccine development looks promising. mRNA vaccines have the potential to overcome these challenges and pave the way for more mRNA vaccines in clinical settings. The rapid development of mRNA CVs will bring new possibilities to cancer therapeutics and establish mRNA CVs as a proud member of the immunotherapy family.

KEYWORDS

- **antigen presenting cells**
- **biological response modifier**
- **cancer**
- **chemokines**
- **chimeric antigen**
- **dendritic cells**
- **gene therapy**
- **immunotherapy**
- **mRNA vaccines**
- **tumor-associated antigen**

REFERENCES

1. Pardi, N., Hogan, M. J., Porter, F. W., & Weissman, D., (2018). mRNA vaccines — A new era in vaccinology. *Nature Reviews Drug Discovery, 17*(4), 261–279.
2. Miao, L., Zhang, Y., & Huang, L., (2021). mRNA vaccine for cancer immunotherapy. *Mol. Cancer, 20*(1), 41.
3. Fiedler, K., Lazzaro, S., Lutz, J., Rauch, S., & Heidenreich, R., (2016). mRNA cancer vaccines. *Recent Results Cancer Res., 209*, 61–85.
4. Chung, J. Y., Thone, M. N., & Kwon, Y. J., (2021). COVID-19 vaccines: The status and perspectives in delivery points of view. *Adv. Drug. Deliv. Rev., 170*, 1–25.
5. Brenner, S., Jacob, F., & Meselson, M., (1961). An unstable intermediate carrying information from genes to ribosomes for protein synthesis. *Nature, 190*(4776), 576–581.
6. Wolff, J. A., Malone, R. W., Williams, P., Chong, W., Acsadi, G., Jani, A., et al., (1990). Direct gene transfer into mouse muscle *in vivo*. *Science, 247*(4949), 1465–1468.
7. Verbeke, R., Lentacker, I., De Smedt, S. C., & Dewitte, H., (2019). Three decades of messenger RNA vaccine development. *Nano Today, 28*, 100766.
8. Martinon, F., Krishnan, S., Lenzen, G., Magné, R., Gomard, E., Guillet, J. G., et al., (1993). Induction of virus-specific cytotoxic T lymphocytes *in vivo* by liposome-entrapped mRNA. *European Journal of Immunology, 23*(7), 1719–1722.
9. Conry, R. M., LoBuglio, A. F., Wright, M., Sumerel, L., Pike, M. J., Johanning, F., et al., (1995). Characterization of a messenger RNA polynucleotide vaccine vector. *Cancer Research, 55*(7), 1397–1400.
10. Furuichi, Y., & Miura, K. I., (1975). A blocked structure at the 5′ terminus of mRNA from cytoplasmic polyhedrosis virus. *Nature, 253*(5490), 374–375.
11. Krieg, P., & Melton, D., (1984). Functional messenger RNAs are produced by SP6 *in vitro* transcription of cloned cDNAs. *Nucleic Acids Research, 12*(18), 7057–7070.
12. Malone, R. W., Felgner, P. L., & Verma, I. M., (1989). Cationic liposome-mediated RNA transfection. *Proceedings of the National Academy of Sciences, 86*(16), 6077–6081.
13. Boczkowski, D., Nair, S. K., Snyder, D., & Gilboa, E., (1996). Dendritic cells pulsed with RNA are potent antigen-presenting cells *in vitro* and *in vivo*. *The Journal of Experimental Medicine, 184*(2), 465–472.
14. Heiser, A., Coleman, D., Dannull, J., Yancey, D., Maurice, M. A., Lallas, C. D., et al., (2002). Autologous dendritic cells transfected with prostate-specific antigen RNA stimulate CTL responses against metastatic prostate tumors. *The Journal of Clinical Investigation, 109*(3), 409–417.
15. Karikó, K., Buckstein, M., Ni, H., & Weissman, D., (2005). Suppression of RNA recognition by toll-like receptors: The impact of nucleoside modification and the evolutionary origin of RNA. *Immunity, 23*(2), 165–175.
16. Weide, B., Pascolo, S., Scheel, B., Derhovanessian, E., Pflugfelder, A., Eigentler, T. K., et al., (2009). Direct injection of protamine-protected mRNA: Results of a phase 1/2 vaccination trial in metastatic melanoma patients. *Journal of Immunotherapy, 32*(5), 498–507.
17. Kreiter, S., Selmi, A., Diken, M., Koslowski, M., Britten, C. M., Huber, C., et al., (2010). Intranodal vaccination with naked antigen-encoding RNA elicits potent prophylactic and therapeutic antitumoral immunity antitumoral immunity by intranodal vaccination with RNA. *Cancer Research, 70*(22), 9031–9040.

18. Kreiter, S., Vormehr, M., Van, De. R. N., Diken, M., Löwer, M., Diekmann, J., et al., (2015). Mutant MHC class II epitopes drive therapeutic immune responses to cancer. *Nature, 520*(7549), 692–696.
19. Sahin, U., Derhovanessian, E., Miller, M., Kloke, B. P., Simon, P., Löwer, M., et al., (2017). Personalized RNA mutanome vaccines mobilize poly-specific therapeutic immunity against cancer. *Nature, 547*(7662), 222–226.
20. BioNTech Receives, (2021). Fast track designation for Its FixVac Candidate BNT111 in Advanced Melanoma. 2021.
21. Maruggi, G., Chiarot, E., Giovani, C., Buccato, S., Bonacci, S., Frigimelica, E., et al., (2017). Immunogenicity and protective efficacy induced by self-amplifying mRNA vaccines encoding bacterial antigens. *Vaccine, 35*(2), 361–368.
22. Richner, J. M., Himansu, S., Dowd, K. A., Butler, S. L., Salazar, V., Fox, J. M., et al., (2017). Modified mRNA vaccines protect against Zika virus infection. *Cell, 168*(6), 1114–1125. e10.
23. Heil, F., Hemmi, H., Hochrein, H., Ampenberger, F., Kirschning, C., Akira, S., et al., (2004). Species-specific recognition of single-stranded RNA via toll-like receptor 7 and 8. *Science, 303*(5663), 1526–1529.
24. Deng, Z., Tian, Y., Song, J., An, G., & Yang, P., (2022). mRNA Vaccines: The dawn of a new era of cancer immunotherapy. *Front. Immunol., 13*, 887125.
25. Patinote, C., Karroum, N. B., Moarbess, G., Cirnat, N., Kassab, I., Bonnet, P. A., et al., (2020). Agonist and antagonist ligands of toll-like receptors 7 and 8: Ingenious tools for therapeutic purposes. *European Journal of Medicinal Chemistry, 193*, 112238.
26. Ablasser, A., Poeck, H., Anz, D., Berger, M., Schlee, M., Kim, S., et al., (2009). Selection of molecular structure and delivery of RNA oligonucleotides to activate TLR7 versus TLR8 and to induce high amounts of IL-12p70 in primary human monocytes. *The Journal of Immunology, 182*(11), 6824–6833.
27. Hua, Z., & Hou, B., (2013). TLR signaling in B-cell development and activation. *Cellular & Molecular Immunology, 10*(2), 103–106.
28. Zhang, C., Maruggi, G., Shan, H., & Li, J., (2019). Advances in mRNA vaccines for infectious diseases. *Front. Immunol., 10*, 594.
29. Gergen, J., & Petsch, B., (2021). mRNA-based vaccines and mode of action. *Curr. Top. Microbiol. Immunol., 2021.*
30. Anderson, B. R., Muramatsu, H., Nallagatla, S. R., Bevilacqua, P. C., Sansing, L. H., Weissman, D., et al., (2010). Incorporation of pseudouridine into mRNA enhances translation by diminishing PKR activation. *Nucleic Acids Research, 38*(17), 5884–5892.
31. He, Q., Gao, H., Tan, D., Zhang, H., & Wang, J. Z., (2022). mRNA cancer vaccines: Advances, trends, and challenges. *Acta Pharmaceutica Sinica B., 2022.*
32. Hollingsworth, R. E., & Jansen, K., (2019). Turning the corner on therapeutic cancer vaccines. *NPJ Vaccines, 4*(1), 1–10.
33. Melief, C. J., Van, H. T., Arens, R., Ossendorp, F., & Van, D. B. S. H., (2015). Therapeutic cancer vaccines. *The Journal of Clinical Investigation, 125*(9), 3401–3412.
34. Türeci, Ö., Vormehr, M., Diken, M., Kreiter, S., Huber, C., & Sahin, U., (2016). Targeting the heterogeneity of cancer with individualized neoepitope vaccines. *Clinical Cancer Research, 22*(8), 1885–1896.
35. Bringmann, A., Held, S. A., Heine, A., & Brossart, P., (2010). RNA vaccines in cancer treatment. *J. Biomed. Biotechnol., 2010*, 623687.

36. Tsao, S. Y., (2022). Potential of mRNA vaccines to become versatile cancer vaccines. *World J. Clin. Oncol., 13*(8), 663–674.
37. Schietinger, A., Philip, M., & Schreiber, H., (2008). Specificity in cancer immunotherapy. *Seminars in Immunology.* Elsevier.
38. Janelle, V., Rulleau, C., Del Testa, S., Carli, C., & Delisle, J. S., (2020). T-cell immunotherapies targeting histocompatibility and tumor antigens in hematological malignancies. *Frontiers in Immunology, 11*, 276.
39. Rosa, S. S., Prazeres, D. M. F., Azevedo, A. M., & Marques, M. P. C., (2021). mRNA vaccines manufacturing: Challenges and bottlenecks. *Vaccine, 39*(16), 2190–2200.
40. Zhou, W. Z., Hoon, D., Huang, S., Fujii, S., Hashimoto, K., Morishita, R., et al., (1999). RNA melanoma vaccine: Induction of antitumor immunity by human glycoprotein 100 mRNA immunization. *Human Gene Therapy, 10*(16), 2719–2724.
41. Mockey, M., Bourseau, E., Chandrashekhar, V., Chaudhuri, A., Lafosse, S., Le Cam, E., et al., (2007). mRNA-based cancer vaccine: Prevention of B16 melanoma progression and metastasis by systemic injection of MART1 mRNA histidylated lipopolyplexes. *Cancer Gene Therapy, 14*(9), 802–814.
42. Fotin-Mleczek, M., Duchardt, K. M., Lorenz, C., Pfeiffer, R., Ojkic-Zrna, S., Probst, J., et al., (2011). Messenger RNA-based vaccines with dual activity induce balanced TLR-7 dependent adaptive immune responses and provide antitumor activity. *Journal of Immunotherapy, 34*(1), 1–15.
43. Van, L. S., Goyvaerts, C., Maenhout, S., Goethals, L., Disy, A., Benteyn, D., et al., (2012). Preclinical evaluation of TriMix and antigen mRNA-based antitumor therapy mRNA-based antitumor vaccination. *Cancer Research, 72*(7), 1661–1671.
44. Liu, L., Wang, Y., Miao, L., Liu, Q., Musetti, S., Li, J., et al., (2018). Combination immunotherapy of MUC1 mRNA nano-vaccine and CTLA-4 blockade effectively inhibits growth of triple negative breast cancer. *Molecular Therapy, 26*(1), 45–55.
45. Wang, Y., Zhang, L., Xu, Z., Miao, L., & Huang, L., (2018). mRNA vaccine with antigen-specific checkpoint blockade induces an enhanced immune response against established melanoma. *Molecular Therapy, 26*(2), 420–434.
46. Hewitt, S. L., Bai, A., Bailey, D., Ichikawa, K., Zielinski, J., Karp, R., et al., (2019). Durable anticancer immunity from intratumoral administration of IL-23, IL-36γ, and OX40L mRNAs. *Science Translational Medicine, 11*(477), eaat9143.
47. Wang, Y., Tiruthani, K., Li, S., Hu, M., Zhong, G., Tang, Y., et al., (2021). mRNA delivery of a bispecific single-domain antibody to polarize tumor-associated macrophages and synergize immunotherapy against liver malignancies. *Advanced Materials, 33*(23), 2007603.
48. Islam, M. A., Xu, Y., Tao, W., Ubellacker, J. M., Lim, M., Aum, D., et al., (2018). Restoration of tumor-growth suppression *in vivo* via systemic nanoparticle-mediated delivery of PTEN mRNA. *Nature Biomedical Engineering, 2*(11), 850–864.
49. Kong, N., Tao, W., Ling, X., Wang, J., Xiao, Y., Shi, S., et al., (2019). Synthetic mRNA nanoparticle-mediated restoration of p53 tumor suppressor sensitizes p53-deficient cancers to mTOR inhibition. *Science Translational Medicine, 11*(523), eaaw1565.
50. Liu, T., Liang, Y., & Huang, L., (2021). Development and delivery systems of mRNA vaccines. *Front. Bioeng. Biotechnol., 9*, 718753.
51. Broderick, K. E., & Humeau, L. M., (2017). *Enhanced Delivery of DNA or RNA Vaccines by Electroporation* (pp. 193–200). RNA Vaccines: Springer.

52. Kowalzik, F., Schreiner, D., Jensen, C., Teschner, D., Gehring, S., & Zepp, F. (2021). mRNA-Based Vaccines. *Vaccines. 9*(4), 390. https://doi.org/10.3390/vaccines9040390.
53. Riley, R. S., June, C. H., Langer, R., & Mitchell, M. J., (2019). Delivery technologies for cancer immunotherapy. *Nature Reviews Drug Discovery, 18*(3), 175–196.
54. Nitika, Wei, J., & Hui, A. M., (2022). The delivery of mRNA vaccines for therapeutics. *Life (Basel), 12*(8).
55. Stewart, M. P., Langer, R., & Jensen, K. F., (2018). Intracellular delivery by membrane disruption: Mechanisms, strategies, and concepts. *Chemical Reviews, 118*(16), 7409–7531.
56. Wadhwa, A., Aljabbari, A., Lokras, A., Foged, C., & Thakur, A., (2020). Opportunities and challenges in the delivery of mRNA-based vaccines. *Pharmaceutics, 12*(2).
57. McNamara, M. A., Nair, S. K., & Holl, E. K., (2015). RNA-based vaccines in cancer immunotherapy. *J. Immunol. Res., 2015*, 794528.
58. Edwards, D. K., Jasny, E., Yoon, H., Horscroft, N., Schanen, B., Geter, T., et al., (2017). Adjuvant effects of a sequence-engineered mRNA vaccine: Translational profiling demonstrates similar human and murine innate response. *Journal of Translational Medicine, 15*(1), 1–18.
59. Lorentzen, C. L., Haanen, J. B., Met, Ö., & Svane, I. M., (2022). Clinical advances and ongoing trials on mRNA vaccines for cancer treatment. *The Lancet Oncology, 23*(10), e450-e8.
60. Lundstrom, K., (2019). RNA viruses as tools in gene therapy and vaccine development. *Genes, 10*(3), 189.
61. Ramamoorth, M., & Narvekar, A., (2015). Non viral vectors in gene therapy: An overview. *Journal of Clinical and Diagnostic Research: JCDR, 9*(1), GE01.
62. Kowalski, P. S., Rudra, A., Miao, L., & Anderson, D. G., (2019). Delivering the messenger: Advances in technologies for therapeutic mRNA delivery. *Molecular Therapy, 27*(4), 710–728.
63. Boussif, O., Lezoualc'h, F., Zanta, M. A., Mergny, M. D., Scherman, D., Demeneix, B., et al., (1995). A versatile vector for gene and oligonucleotide transfer into cells in culture and *in vivo*: Polyethylenimine. *Proceedings of the National Academy of Sciences, 92*(16), 7297–7301.
64. Li, W., Ma, L., Guo, L. P., Wang, X. L., Zhang, J. W., Bu, Z. G., et al., (2017). West Nile virus infectious replicon particles generated using a packaging-restricted cell line is a safe reporter system. *Scientific Reports, 7*(1), 1–10.
65. Zeng, C., Zhang, C., Walker, P. G., & Dong, Y., (2020). Formulation and delivery technologies for mRNA vaccines. *Curr. Top. Microbiol. Immunol.*, 2020.
66. Gary, D. J., Puri, N., & Won, Y. Y., (2007). Polymer-based siRNA delivery: Perspectives on the fundamental and phenomenological distinctions from polymer-based DNA delivery. *Journal of Controlled Release, 121*(1, 2), 64–73.
67. Pollard, C., Rejman, J., De Haes, W., Verrier, B., Van, G. E., Naessens, T., et al., (2013). Type I IFN counteracts the induction of antigen-specific immune responses by lipid-based delivery of mRNA vaccines. *Molecular Therapy, 21*(1), 251–259.
68. Islam, M. A., Reesor, E. K., Xu, Y., Zope, H. R., Zetter, B. R., & Shi, J., (2015). Biomaterials for mRNA delivery. *Biomaterials Science, 3*(12), 1519–1533.
69. Stanton, M. G., (2018). Current status of messenger RNA delivery systems. *Nucleic Acid Therapeutics, 28*(3), 158–165.

70. Mukherjee, A., Waters, A. K., Kalyan, P., Achrol, A. S., Kesari, S., & Yenugonda, V. M., (2019). Lipid–polymer hybrid nanoparticles as a next-generation drug delivery platform: State of the art, emerging technologies, and perspectives. *International Journal of Nanomedicine, 14*, 1937.
71. Colombo, S., Cun, D., Remaut, K., Bunker, M., Zhang, J., Martin-Bertelsen, B., et al., (2015). Mechanistic profiling of the siRNA delivery dynamics of lipid–polymer hybrid nanoparticles. *Journal of Controlled Release, 201*, 22–31.
72. Grau, M., Walker, P. R., & Derouazi, M., (2018). Mechanistic insights into the efficacy of cell penetrating peptide-based cancer vaccines. *Cellular and Molecular Life Sciences, 75*(16), 2887–2896.
73. Stitz, L., Vogel, A., Schnee, M., Voss, D., Rauch, S., Mutzke, T., et al., (2017). A thermostable messenger RNA based vaccine against rabies. *PLoS Neglected Tropical Diseases, 11*(12), e0006108.
74. Scheel, B., Aulwurm, S., Probst, J., Stitz, L., Hoerr, I., Rammensee, H. G., et al., (2006). Therapeutic anti-tumor immunity triggered by injections of immunostimulating single-stranded RNA. *European Journal of Immunology, 36*(10), 2807–2816.
75. Hajj, K. A., Melamed, J. R., Chaudhary, N., Lamson, N. G., Ball, R. L., Yerneni, S. S., et al., (2020). A potent branched-tail lipid nanoparticle enables multiplexed mRNA delivery and gene editing *in vivo*. *Nano Letters, 20*(7), 5167–5175.
76. Kim, H. J., Ogura, S., Otabe, T., Kamegawa, R., Sato, M., Kataoka, K., et al., (2019). Fine-tuning of hydrophobicity in amphiphilic polyaspartamide derivatives for rapid and transient expression of messenger RNA directed toward genome engineering in brain. *ACS Central Science, 5*(11), 1866–1875.
77. Heine, A., Juranek, S., & Brossart, P., (2021). Clinical and immunological effects of mRNA vaccines in malignant diseases. *Mol. Cancer, 20*(1), 52.
78. Duewell, P., Kisser, U., Heckelsmiller, K., Hoves, S., Stoitzner, P., Koernig, S., et al., (2011). ISCOMATRIX adjuvant combines immune activation with antigen delivery to dendritic cells *in vivo* leading to effective cross-priming of CD8+ T cells. *The Journal of Immunology, 187*(1), 55–63.
79. Klechevsky, E., (2013). Human dendritic cells—Stars in the skin. *European Journal of Immunology, 43*(12), 3147–3155.
80. Probst, J., Weide, B., Scheel, B., Pichler, B., Hoerr, I., Rammensee, H., et al., (2007). Spontaneous cellular uptake of exogenous messenger RNA *in vivo* is nucleic acid-specific, saturable and ion dependent. *Gene Therapy, 14*(15), 1175–1180.
81. Liang, S. L., Quirk, D., & Zhou, A., (2006). RNase L: Its biological roles and regulation. *IUBMB Life, 58*(9), 508–514.
82. Senti, G., & Kündig, T. M., (2015). Intralymphatic immunotherapy. *World Allergy Organization Journal, 8*, 9.
83. Le Moignic, A., Malard, V., Benvegnu, T., Lemiègre, L., Berchel, M., Jaffrès, P. A., et al., (2018). Preclinical evaluation of mRNA trimannosylated lipopolyplexes as therapeutic cancer vaccines targeting dendritic cells. *Journal of Controlled Release, 278*, 110–121.
84. Krishna, S., & Anderson, K. S., (2016). *T-Cell Epitope Discovery for Therapeutic Cancer Vaccines* (pp. 779–796). Vaccine Design: Springer.
85. Vesely, M. D., Kershaw, M. H., Schreiber, R. D., & Smyth, M. J., (2011). Natural innate and adaptive immunity to cancer. *Annual Review of Immunology, 29*(1), 235–271.
86. Paston, S. J., Brentville, V. A., Symonds, P., & Durrant, L. G., (2021). Cancer vaccines, adjuvants, and delivery systems. *Front. Immunol., 12*, 627932.

87. Cuzzubbo, S., Mangsbo, S., Nagarajan, D., Habra, K., Pockley, A. G., & McArdle, S. E. B., (2020). Cancer vaccines: Adjuvant potency, importance of age, lifestyle, and treatments. *Front. Immunol., 11*, 615240.
88. Moro-García, M. A., Alonso-Arias, R., & López-Larrea, C., (2013). When aging reaches CD4+ T-cells: Phenotypic and functional changes. *Frontiers in Immunology, 4*, 107.
89. Maisonneuve, C., Bertholet, S., Philpott, D. J., & De Gregorio, E., (2014). Unleashing the potential of NOD-and toll-like agonists as vaccine adjuvants. *Proceedings of the National Academy of Sciences, 111*(34), 12294–12299.
90. Maynard, S. K., Marshall, J. D., MacGill, R. S., Yu, L., Cann, J. A., Cheng, L. I., et al., (2019). Vaccination with synthetic long peptide formulated with CpG in an oil-in-water emulsion induces robust E7-specific CD8 T cell responses and TC-1 tumor eradication. *BMC Cancer, 19*(1), 1–12.
91. Jiang, M., Chen, P., Wang, L., Li, W., Chen, B., Liu, Y., et al., (2020). cGAS-STING, an important pathway in cancer immunotherapy. *Journal of Hematology & Oncology, 13*(1), 1–11.
92. Gulen, M. F., Koch, U., Haag, S. M., Schuler, F., Apetoh, L., Villunger, A., et al., (2017). Signaling strength determines proapoptotic functions of STING. *Nature Communications, 8*(1), 1–10.
93. Kis, Z., Kontoravdi, C., Shattock, R., & Shah, N., (2020). Resources, production scales and time required for producing RNA vaccines for the global pandemic demand. *Vaccines, 9*(1), 3.
94. Kwon, H., Kim, M., Seo, Y., Moon, Y. S., Lee, H. J., Lee, K., et al., (2018). Emergence of synthetic mRNA: *In vitro* synthesis of mRNA and its applications in regenerative medicine. *Biomaterials, 156*, 172–193.
95. Karikó, K., Muramatsu, H., Ludwig, J., & Weissman, D., (2011). Generating the optimal mRNA for therapy: HPLC purification eliminates immune activation and improves translation of nucleoside-modified, protein-encoding mRNA. *Nucleic Acids Res. 39*(21), e142. doi:10.1093/nar/gkr695.
96. Chen, J., Chen, J., & Xu, Q., (2022). Current developments and challenges of mRNA vaccines. *Annual Review of Biomedical Engineering, 24*, 85–109.
97. Geall, A. J., Mandl, C. W., & Ulmer, J. B., (2013). RNA: The new revolution in nucleic acid vaccines. *Semin. Immunol., 25*(2), 152–159.
98. Uddin, M. N., & Roni, M. A., (2021). Challenges of storage and stability of mRNA-based COVID-19 vaccines. *Vaccines (Basel), 9*(9).
99. Crommelin, D. J. A., Anchordoquy, T. J., Volkin, D. B., Jiskoot, W., & Mastrobattista, E., (2021). Addressing the cold reality of mRNA vaccine stability. *J. Pharm. Sci., 110*(3), 997–1001.
100. Liu, M. A., Zhou, T., Sheets, R. L., Meyer, H., & Knezevic, I., (2022). WHO informal consultation on regulatory considerations for evaluation of the quality, safety, and efficacy of RNA-based prophylactic vaccines for infectious diseases. *Emerging Microbes & Infections, 11*(1), 384–391.
101. Knezevic, I., Liu, M. A., Peden, K., Zhou, T., & Kang, H. N., (2021). Development of mRNA vaccines: Scientific and regulatory issues. V*accines (Basel), 9*(2).
102. *WHO Guidelines on Nonclinical Evaluation of Vaccines*. In: WHO Expert Committee on Biological Standardization: Fifty-fourth report. Geneva: World Health Organization; 2005: Annex 1 (WHO Technical Report Series, No. 927; https://cdn.who.int/media/

docs/default-source/biologicals/annex1nonclinical.p31-63.pdf (accessed on 11 March 2024).
103. *Guidelines on the Nonclinical Evaluation of Vaccine Adjuvants and Adjuvanted Vaccines.* In: WHO Expert Committee on Biological Standardization: Sixty-fourth report. Geneva: World Health Organization; 2014: Annex 2 (WHO Technical Report Series, No. 987; https://cdn.who.int/media/docs/default-source/biologicals/vaccine-standardization/trs_987_annex2.pdf?sfvrsn=ea91caca_3 (accessed on 11 March 2024).
104. *Guidelines on the Clinical Evaluation of Vaccines: Regulatory Expectations.* In: WHO Expert Committee on Biological Standardization: Sixty-seventh report. Geneva: World Health Organization; 2017: Annex 9 (WHO Technical Report Series, No, https://cdn.who.int/media/docs/default-source/prequal/vaccines/who-trs-1004-web-annex-9.pdf (accessed on 11 March 2024).
105. *Evaluation of the Quality, Safety and Efficacy of Messenger RNA Vaccines for the Prevention of Infectious Diseases: Regulatory Considerations.* https://cdn.who.int/media/docs/default-source/biologicals/ecbs/post-ecbs-who-regulatory-considerations-document-for-mrna-vaccines---final-version---29-nov-2021_tz.pdf (accessed on 11 March 2024).
106. Johnson, A. A., Ray, A. S., Hanes, J., Suo, Z., Colacino, J. M., Anderson, K. S., et al., (2001). Toxicity of antiviral nucleoside analogs and the human mitochondrial DNA polymerase. *Journal of Biological Chemistry, 276*(44), 40847–40857.
107. Moyle, G., (2000). Toxicity of antiretroviral nucleoside and nucleotide analogues. *Drug Safety, 23*(6), 467–481.
108. *Guidelines for Good Clinical Practice (GCP) for Trials on Pharmaceutical Products.* In: WHO Expert Committee on the Use of Essential Drugs: Sixth report. Geneva: World Health Organization; 1995: Annex 3 (WHO Technical Report Series, No. 850; https://apps.who.int/iris/bitstream/handle/10665/37340/WHO_TRS_850.pdf (accessed on 11 March 2024).
109. Van, N. A. M., Benteyn, D., Wilgenhof, S., Corthals, J., Heirman, C., Neyns, B., et al., (2012). Intravenous and intradermal TriMix-dendritic cell therapy results in a broad T-cell response and durable tumor response in a chemorefractory stage IV-M1c melanoma patient. *Cancer Immunology, Immunotherapy, 61*(7), 1033–1043.
110. *Impurities: Guideline for Residual Solvents – Q3C(R6).* ICH Harmonized Guideline. Final version: 20 October 2016. International Council for Harmonization of Technical Requirements for Pharmaceuticals for Human Use; 2016 https://database.ich.org/sites/default/files/Q3C-R6_Guideline_ErrorCorrection_2019_0410_0.pdf (accessed on 11 March 2024).
111. *Guideline for Elemental Impurities – Q3D(R1).* ICH Harmonized Guideline. Final version: 22 March 2019. International Conference for Harmonization of Technical Requirements for Pharmaceuticals for Human Use; 2019. https://database.ich.org/sites/default/files/Q3D-R1EWG_Document_Step4_Guideline_2019_0322.pdf (accessed on 11 March 2024).
112. *ICH Guideline S2 (R1) on Genotoxicity Testing and Data Interpretation for Pharmaceuticals Intended for Human Use.* Step 5. London: European Medicines Agency; June 2012. Document EMA/CHMP/ICH/126642/2008; https://www.ema.europa.eu/en/documents/scientific-guideline/ich-guideline-s2-r1-genotoxicity-testing-data-interpretation-pharmaceuticals-intended-human-use-step_en.pdf (accessed on 11 March 2024).

113. *Guidance for Industry: Clinical Considerations for Therapeutic Cancer Vaccines*, (2011). Available at: https://www.fda.gov/media/82312/download (accessed on 11 March 2024).
114. *Guideline on Clinical Evaluation of New Vaccines*, (2006). European Medicines Agency; Document EMEA/CHMP/VWP/164653/2005; https://www.ema.europa.eu/en/documents/scientific-guideline/guideline-clinical-evaluation-new-vaccines_en.pdf (accessed on 11 March 2024).
115. *Guideline on the Requirements for Quality Documentation Concerning Biological Investigational Medicinal Products in Clinical Trials*, (2018). Document EMA/CHMP/BWP/534898/2008 rev. 1 corrigendum; https://www.ema.europa.eu/en/documents/scientific-guideline/guideline-requirements-quality-documentation-concerning-biological-investigational-medicinal_en-0.pdf (accessed on 11 March 2024).
116. Deering, R. P., Kommareddy, S., Ulmer, J. B., Brito, L. A., & Geall, A. J., (2014). Nucleic acid vaccines: Prospects for non-viral delivery of mRNA vaccines. *Expert. Opin. Drug Deliv., 11*(6), 885–899.
117. Bell, C. L., Yu, D., Smolke, C. D., Geall, A. J., Beard, C. W., & Mason, P. W., (2015). Control of alphavirus-based gene expression using engineered riboswitches. *Virology, 483*, 302–311.
118. Kim, S. C., Sekhon, S. S., Shin, W. R., Ahn, G., Cho, B. K., Ahn, J. Y., et al., (2022). Modifications of mRNA vaccine structural elements for improving mRNA stability and translation efficiency. *Mol. Cell Toxicol., 18*(1), 1–8.
119. Schlake, T., Thess, A., Fotin-Mleczek, M., & Kallen, K. J., (2012). Developing mRNA-vaccine technologies. *RNA Biol., 9*(11), 1319–1330.
120. Fuchs, A. L., Neu, A., & Sprangers, R., (2016). A general method for rapid and cost-efficient large-scale production of 5′ capped RNA. *RNA, 22*(9), 1454–1466.
121. Qin, S., Tang, X., Chen, Y., Chen, K., Fan, N., Xiao, W., et al., (2022). mRNA-based therapeutics: Powerful and versatile tools to combat diseases. *Signal Transduct. Target Ther., 7*(1), 166.
122. Von, N. A. G. O., Poleganov, M. A., Rechner, C., Plaschke, A., Kranz, L. M., Fesser, S., et al., (2019). Improving mRNA-based therapeutic gene delivery by expression-augmenting 3′ UTRs identified by cellular library screening. *Molecular Therapy, 27*(4), 824–836.
123. Linares-Fernández, S., Lacroix, C., Exposito, J. Y., & Verrier, B., (2020). Tailoring mRNA vaccine to balance innate/adaptive immune response. *Trends in Molecular Medicine, 26*(3), 311–323.
124. Weng, Y., Li, C., Yang, T., Hu, B., Zhang, M., Guo, S., et al., (2020). The challenge and prospect of mRNA therapeutics landscape. *Biotechnol. Adv., 40*, 107534.
125. Benteyn, D., Heirman, C., Bonehill, A., Thielemans, K., & Breckpot, K., (2015). mRNA-based dendritic cell vaccines. *Expert Rev. Vaccines, 14*(2), 161–176.
126. Okada, N., Tsujino, M., Hagiwara, Y., Tada, A., Tamura, Y., Mori, K., et al., (2001). Administration route-dependent vaccine efficiency of murine dendritic cells pulsed with antigens. *British Journal of Cancer, 84*(11), 1564–1570.
127. Lesterhuis, W. J., De Vries, I. J. M., Schreibelt, G., Lambeck, A. J., Aarntzen, E. H., Jacobs, J. F., et al., (2011). Route of administration modulates the induction of dendritic cell vaccine–induced antigen-specific T cells in advanced melanoma patients intradermal versus intranodal DC vaccination of melanoma. *Clinical Cancer Research, 17*(17), 5725–5735.

CHAPTER 6

Utilization of Microbial Genes as a Vaccine Against Cancer

GARIMA VARSHNEY,[1] RISHABHA MALVIYA,[1] SONALI SUNDRAM,[1] and DHANALEKSHMI UNNIKRISHNAN MEENAKSHI[2]

[1]*Department of Pharmacy, School of Medical and Allied Science, Galgotias University, Greater Noida, Uttar Pradesh, India*

[2]*College of Pharmacy, National University of Science and Technology, Muscat, Oman*

ABSTRACT

Cancer is a worldwide cause of death. Its causes include physiological, chemical, and organic toxins. Carcinoma, leukemia, sarcoma, and CNS cancers are specific types of cancers that grow uncontrollably in a particular organ (benign) or throughout the entire body (malignant). The successful treatment of cancer remains a challenge due to its malignant properties and resistance to therapy. Traditional treatments such as radiation therapy, chemotherapy, and immunotherapy have limitations in controlling cancer to a certain extent. As a result, the utilization of microbial genes has emerged as a promising approach in cancer treatment. The use of microbial genes in treating cancer was first identified in the 1800s. Various approaches, including enzyme-based therapy, bacteriocins, bacterial spores, and bacterial toxins (B.T.), are known for treating cancer using microorganisms as vaccines. Genes from *Clostridium, Salmonella,* and *Listeria* are employed in the treatment of cancer. While microbial genes cannot completely cure cancer, their high specificity, replication, and tumor susceptibility to microorganisms offer a successful means of preventing the disease from spreading worldwide to a greater extent. However, there are certain limitations to using

Cancer Vaccination and Challenges, Volume 1: Strategies for Therapeutic Cancer Vaccine Development. Rishabha Malviya, Bhupendra Prajapati, and Sonali Sundram (Eds.)

microbes against cancer, such as their high reactivity, uncontrollable infection, and limited therapeutic range. Research on cancer will continue until a successful treatment is discovered.

6.1 INTRODUCTION TO CANCER

Any part of the body can be affected by the wide category of diseases known as cancer [1], which occurs when a number of living organisms cells develop uncontrollably and migrate to other body areas [2]. As depicted in Figure 6.1, the rapid development of aberrant cells that outgrow their normal bounds can subsequently infect other body parts and spread to other organs [1]. Eventually, a lump or tumor will form from these aberrant cells [3].

These tumors can be cancerous or benign. Malignant cancer can spread to various regions to form new tumors and invade nearby tissues of the body through a process known as metastasis. Non-cancerous tumors do not invade or disseminate neighboring tissues. Malignant tumors can regrow after removal, but benign ones normally do not. The three main gene types that are commonly influenced by the genetic changes that lead to cancer are proto-oncogenes, tumor suppressor genes, and DNA repair genes. These alterations are considered as the "drivers" of cancer [2].

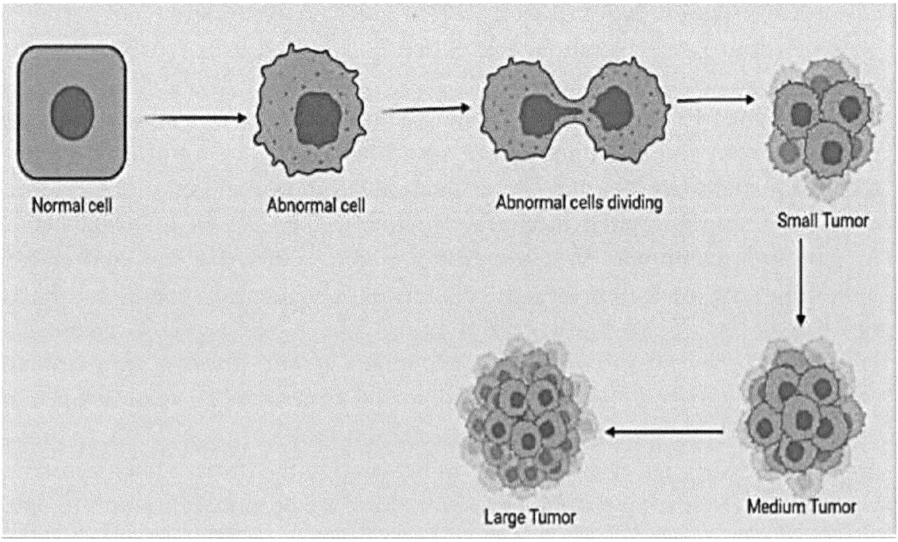

FIGURE 6.1 Representation of cancer cells growth.

6.1.1 *FOUR PHASES ARE TYPICAL FOR MALIGNANCIES*

- **Level I:** No tumor has progressed to the tissues such as lymph glands or others and is localized to a small site.
- **Level II:** The tumor has grown but is not widely spread.
- **Level III:** The tumor has increased in size and may have invaded lymph glands and other tissues.
- **Level IV:** The tumor has metastasized to distant body parts. Also referred to as metastatic or advanced cancer [3].

6.1.2 *CAUSES OF TUMOR*

Physiological toxins, such as UV and electromagnetic radiation, synthetic toxins like alcohol, chrysotile, tobacco, smoking, food adulterants, and arsenic (contaminated drinking water), as well as organic toxins like specific microbes such as viruses, bacteria, or parasites, can all contribute to its development [1].

6.1.3 *CERTAIN TYPES OF CANCERS*

1. **Carcinoma:** A malignant growth, typically originates in the epithelium or in the group of cells that surround internal organs. Carcinoma has the following subcategories: Transitional cell carcinoma, basal cell cancer, cutaneous squamous cell cancer, and adenocarcinoma.
2. **Sarcoma:** It is a type of cancer that begins in the soft or supportive group of cells, such as chondrocytes, fat, musculoskeletal system, or blood vessels.
3. **Leukemia:** A malignancy of white blood cells, originates in the bone marrow, which produces blood cells.
4. **Both Non-Hodgkin Lymphoma and Myeloma (Kahler Disease):** These are malignancies that start in immune system cells.
5. **CNS Cancers:** These are those that affect the anatomy and physiology of the brain and spinal cord [4]. Cancer remains the leading cause of death globally [5–8]. In the year 2018, there were 9.6 million deaths and 18.1 million new cases of the disease reported [5,

9]. It is projected that there will be 17 million deaths by 2030 [5, 6]. These figures emphasize the urgency of developing new effective therapies. Conventional cancer treatments, such as surgery, radiation, chemotherapy, and immunotherapy, all have their limitations. Treating multiple tumors can be challenging due to the complex nature of cancer, including issues with measurement, location, stage, and malignancy [5, 10]. Moreover, resistance often develops, leading to reduced efficacy of radiation, chemotherapy, and immunotherapy, resulting in poor cancer control and various adverse effects during or after treatment. Additional therapies, such as cell therapy, meal therapy, laser therapy, insulin modulatory therapy, HAMLET, telomerase therapy, thermal therapy, dichloroacetate, and non-invasive radiofrequency treatment for cancer, have been suggested to enhance and improve the effectiveness of conventional cancer treatment [5, 3, 18].

6.2 CANCER VACCINES (CV)

A vaccine is a live product that provides active acquired immunity against specific infections or malignancies [11, 12]. Vaccination typically involves the administration of a weakened or inactivated form of the microorganism, its toxin, or one of its surface proteins. This stimulates the body's immune system to recognize the substance as a threat, eliminate it, and then identify and eliminate any subsequent exposure to the same substance. Vaccines can be either preventive, reducing the signs and symptoms of an impending disease caused by a natural or "wild" virus, or therapeutic, targeting a disease that has already developed, such as cancer [12–15]. Some vaccines provide complete sterilizing immunity [11, 16].

CVs are designed to elicit an immune response from the body's immune system against tumor antigens. Despite decades of research and development, only a small number of CVs have been approved for human use, while many more are currently undergoing various stages of clinical testing. The effectiveness of these CVs depends on factors such as the type of antigen and toxin, the tumor microenvironment (TME), the tumor's defense mechanisms, and the vaccine formulation. As shown in Figure 6.2, tumor antigens are transported to the lymphatic system after immunization, where they stimulate specific lymphocytes (T and B lymphocytes) that recognize the antigen. B lymphocytes directly identify antigens, while T lymphocytes are activated by dendritic cells (DC), which present the vaccine antigen on major histocompatibility complex (MHC) molecules. Activated B lymphocytes and effective T lymphocytes produce antibodies (Abs) that penetrate tumors and induce tumor cell death [17].

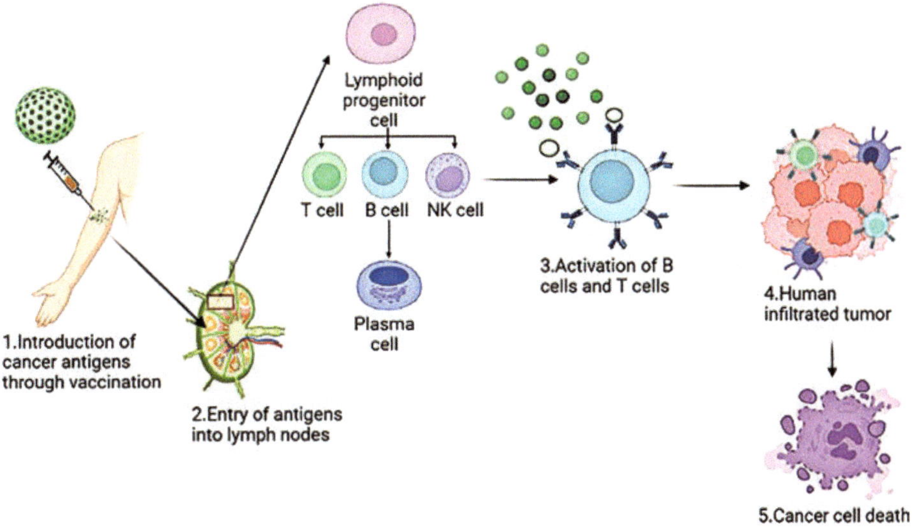

FIGURE 6.2 Representation of cancer vaccination.

6.3 MICROBIAL-GENES-BASED CANCER THERAPY

Worldwide, the prevalence of cancer-related mortality is increasing, necessitating the development of efficient measures to control this threat. The pathophysiology of solid tumors imposes strict restrictions on the entry and accumulation of anti-tumor chemotherapeutic medicines in the body [24]. Although conventional chemotherapy drugs are still commonly used, bacteriotherapy has shown impressive results due to its high specificity, controllability post-execution, and oncogenic effectiveness in numerous studies on living organisms. However, there are still several issues with using bacteria as antitumor agents in clinical practice. Genes, which are present on DNA segments on chromosomes, are the basic units of heredity. Genetically modified microorganisms will become increasingly useful for both diagnostic and therapeutic purposes [36]. It is possible to modify and take advantage of the intrinsic characteristics of microorganisms [37].

Various applications of microorganisms as cancer therapeutic options have been explored so far. Potential approaches include using live, weakened microbes as anticancer agents and as gene-controlled enzyme prodrug treatment vectors [18]. Tumor-targeting bacteria are exceptional delivery systems for therapeutic payloads that target cancer, thanks to their tumor selectivity and limitless gene packaging capacity. This infinite gene packaging capability enables the expression of large or multiple target genes, facilitating the construction of signaling networks that allow bacteria to perform complex tasks in cancer treatment [39]. With advancements in genetic engineering, medicines can now be enhanced with molecules such as cytokines, drug-metabolizing enzymes, tumor triggering agents, necrosis receptors, and even RNA intervention to provide improved therapeutic effects. Clinical trials of these medicines have shown promising outcomes. As an alternative to conventional cancer treatment, experimentation has now begun to explore the potential of microbes as a primary or complementary therapy [40]. The history of microbial gene-based therapy is depicted in Figure 6.3.

In their respective studies, German doctors Wilhelm Busch and Friedrich Fehleisen found that some tumors shrank after unintentional infections with a superficial form of cellulitis, erysipelas caused by Streptococcus pyogenes, which occurred while patients were in hospitals [20]. Independently, American doctor William Coley observed how one of his neck cancer patients started to improve after contracting erysipelas. This patient had been infected with the bacteria. He initiated the first extensively recorded practice of using bacteria and their antigens to treat advanced cancer. In the late 1800s, he developed a less dangerous vaccination using two dead bacterial species, Serratia marcescens and *S. pyogenes*, to simulate an illness and its associated fever without the risk of a real infection [21, 22]. Furthermore, the vaccine was successfully used to treat a variety of tumors, including sarcomas, carcinomas, lymphatic system cancers, melanomas, and myelomas [19].

The mechanism underlying the activity of Coley's toxin has long been a subject of interest for cancer researchers. Recent advancements in immunology and biotechnology have rekindled this interest [5].

Through the use of genetic engineering, numerous attenuated bacteria have been developed that can specifically target tumors and exhibit various anticancer effects. This technique, known as bacteria-mediated cancer therapy (BMCT), has gained recent attention. Modern medical technology [24] has allowed for the examination of diverse bacterial species, including Streptococcus [25], Bifidobacterium [26], Clostridium [27], and Salmonella species [28, 29].

Live, weakened, or genetically altered bacterial species [30] that target tumors [31] possess the inherent ability to colonize [30] and infiltrate [31] tumors. They can reproduce within the tumors with specificity, preventing the formation of malignant cells [30].

Studies conducted on animal tumor models have shown that live cancer-targeting bacteria can selectively colonize tumors or cancerous lymph nodes, suppress tumor growth, and increase survival time following systemic infection. For example, the most well-known attenuated strain of *Salmonella Typhimurium*, VNP20009, exhibits a tumor-to-liver colonization ratio greater than 1000:1 and is more than 10,000 times more attenuated than the wild-type strain [32, 33]. It also demonstrates strong inhibitory effects on tumor growth in models. Tumor-targeting bacteria can enhance the effects of chemotherapy medications while minimizing systemic harm to the patient and overcoming penetration barriers [5].

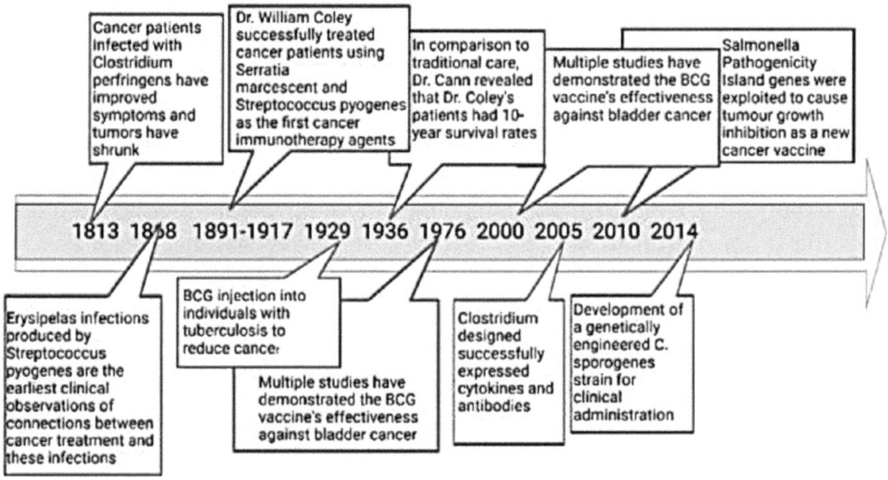

FIGURE 6.3 Representation of the history of microbial-genes-based therapy [36, 42, 43].

6.4 DIFFERENT APPROACHES TO BACTERIAL DELIVERY

6.4.1 GENETICALLY MODIFIED BACTERIA AS A VACCINE AGAINST CANCER

An innovative alternative to conventional cancer treatment is gene therapy. The main benefits of gene therapy are its ability to selectively target and eliminate tumor cells [54, 59–61]. Genetically altered bacteria may also be more effective in fighting tumors and less pathogenic to the host [5, 60, 62]. The use of biogenetically modified bacteria to demonstrate anticancer drugs, as well as tumor-specific antigens, has recently been studied in several research studies to develop a novel approach to cancer therapy [5, 62–64]. Table 6.1 depicts different microorganisms and their treatment strategies used as anticancer agents. In malignancies, genetically modified bacteria have been observed to multiply more significantly compared to normal tissues [5, 65, 66]. The use of biogenetically modified microbes that carry specific therapeutic genes has been one strategy to overcome these limitations. These microbial vectors can serve as effective adjuvant therapies for certain cancer treatments by selectively generating the desired protein in the TME. Therefore, bacteria act as vectors or carriers to transport therapeutic proteins, pro-drug converting enzymes, cytotoxic peptides, or anticancer drugs to solid tumors [19]. With the aid of such a system, *in vivo* drug manufacturing can be controlled in terms of timing and location. The methods for regulating gene expression in this manner fall into three main categories: internal triggering, self-triggering (quorum sensing-QS), and external triggering [5]. The processing of specific antigens can be modified to enhance the immune response. For example, it has been shown that gene sequences generating fusion proteins connected to intracellular trafficking sites can enhance and optimize antigen presentation [49, 52, 53]. Salmonellas er.VNP20009 and C.butyM55 were utilized as agents in animal tumor models to induce specific colonization of tumors, resulting in strong defense reactions and hazardous secondary responses. However, some of these research findings did not yield the expected outcomes [54–58]. Nevertheless, it is possible to genetically engineer *C. ace* and *C. bei* strains to express genes for specific microbial enzymes such as cytosine deaminase (CD), nitroreductase, or m-TNF, leading to more favorable anticancer effects. Several studies have demonstrated that bacteria can produce immunoglobulins that bind to HIF1. Clinical studies have shown that modified *S. typhi* and *Clostridium novyi* NT, expressing HlyE or Stx2 (an acidic pH-responsive promoter) or recA (a 38 kDa protein necessary for DNA maintenance and repair), can stimulate the host defense system to produce cytokines IL2, IL4, IL18, and CC 21, ultimately resulting in tumor regression and termination. Combining bacterial therapy with radiotherapy, chemotherapy, immunotherapy, or other forms of cancer treatment may be an innovative and effective strategy [5]. *In situ* vaccination (ISV) is another approach to cancer prevention. Vaccines use adjuvants and antigens to stimulate the immune system, but identifying the precise antigens for cancer has proven challenging. ISV involves injecting the adjuvant directly into the tumor, using it as the source of the antigen. The main objective is to induce a strong local antitumor response, reversing local tumor-mediated immune suppression, and subsequently spreading this response throughout the body to combat known or suspected metastatic disease [49–51].

TABLE 6.1 Representation of Microorganisms Used in Cancer Therapy

Treatment Strategy	Microorganism	Cancer	Implementation	References
Vectors to carry tumoricidal agents	*Clostridium novyi*	Solid tumors	Clinical trials	[5, 44, 45]
Complementary therapy	*Mycobacterium bovis*	Bladder cancer	Commonly used	[44]
Bacterial toxins/enzymes	*Salmonella enterica, Serovar typhimurium*	Melanoma	Clinical trial	[5, 44, 46]
Additional therapy supporting chemotherapy	*Magnetococcus marinus*	Solid tumors and some metabolic tumors.	Experimental research (animal studies)	[44]
Immunotherapeutic agents	*Streptococcus pyogenes*	Lymphangioma	Commonly used	[5, 44, 47]
Bacterial toxins	*Pseudomonas aeruginosa*	Anti-tumor	No data	[5, 48]

6.4.2 ENZYME PRODRUG (INACTIVE DRUG) TREATMENT

Cancer-based selective release of inactive drugs is an exceptional approach to circumvent the unfavorable adverse effects of bacterial therapy. In this technique, anaerobic bacteria are used to deliver an enzyme that can convert a non-harmful prodrug into a lethal one. The delivery is specifically targeted to tumor areas, where bacteria thrive in necrotic and hypoxic regions. To enhance effectiveness, both the inactive and activated drugs must be capable of crossing cell membranes, as the inactive drug is activated in microbes before the activated drug enters cancerous cells. This is because the inactive drug is only converted to the active and harmful drug within the tumor region [5, 19, 67, 68].

Tests have been conducted with *C. sporogenes* using the enzymes NR, which converts the inactive drug CB1954 into a DNA cross-coupling agent, and CD, which converts the non-toxic 5FC into the anticancer drug 5FU. However, these combinations have yielded poor *in vivo* results so far, despite their ability to eradicate cancer cells *in vitro* and deliver large amounts of enzymes to malignancies. An exogenous enzyme is precisely delivered to tumors using CD developed in *Clostridium acetobutylicum*. This type of vector system is valuable for cancer gene therapy, as the majority of human solid tumors contain hypoxic and necrotic regions [19, 67–70].

The bacterial enzyme arginine deiminase-I has a molecular weight of roughly 46 kDa. This enzyme metabolizes the amino acid arginine by cleaving and converting it to citrulline, which ultimately leads to the release of ammonium. When coupled to polyethylene glycol (PEG) polymers, arginine deaminase is known as ADI-PEG20. It has an anticancer impact by triggering autophagy, limiting tumor cell growth and reproduction, and inducing caspase-independent apoptosis [34, 71]. As a result, arginine deaminase has been used as an anticancer drug, and phase II clinical trials are currently underway. This enzyme shows promising potential as a cancer treatment for several malignant cell lines, including liver carcinoma cell lines [34, 72].

6.4.3 BACTERIAL TOXINS (B.T.)

To carry out specific tasks, bacteria develop and release a group of highly toxic proteins called B.T. These toxins have been proven to be an effective strategy for treating tumors [73]. B.T have the ability to destroy cells or, at lower concentrations, disrupt the cellular processes that regulate differentiation, cell death, and proliferation. These disruptions are associated with the development of cancer and can either promote abnormal cellular behavior or suppress it. Cell-cycle inhibitors, including CDTs and CIF, prevent cell division and are expected to weaken the body's immune system by inhibiting the growth of lymphocyte colonies. On the other hand, cell-cycle stimulators like CNF encourage cell reproduction while interfering with cell multiplication [19, 74]. B.T, such as DT, CPE, and PE, that specifically target these toxins could be beneficial for targeting and killing cancer cells because tumor cells express a variety of distinct antigens [75]. These DTs have shown promise in cancer treatment in both mice and humans due to their optimized and desirable anti-cancer effects [73].

6.4.4 BACTERIAL SPORES

Spores (anaerobic bacteria) may germinate and flourish in the dead tissue found inside tumors, making them the ideal type of bacteria to utilize against cancer [19]. Numerous *Bifidobacteria*, *Salmonella* sp. Strains [76], *Escherichia coli* [77, 78], and *Clostridium* [27] have been discovered to specifically target and eradicate cancerous cells in regions with low levels of oxygen, offering a targeted and specified tumor therapy [5]. The follicles of the *C. novyi*-NT strain, an engineered edited variety free of fatal antigens, have demonstrated localized activity with no general adverse effects. When mice were injected with *C. histolyticum* spores intratumorally, noticeable lysis of tumor tissues was observed. The same effect was seen when mice were given intravenously administered *C. sporogenes* spores. Furthermore, animals undergoing an intravenous (IV) drug test had *Clostridium* only found in tumors and not in normal organs [19, 73].

6.4.5 BACTERIOCINS

They are positively charged peptides that originate from various bacterial species through their ribosomes. Bacteriocins exhibit toxicities that are specific to cancer cells, are biodegradable, and are not immunogenic. These bacteriocins may work in conjunction with other cancer treatments to combat cancer [80]. Conversely, they bind to tumor cell membranes instead of normal cell membranes, which lack charge and are therefore less favorable for bacterial attachment [81]. This is because the majority of tumor cell membranes have anionic properties. Colicins, a type of bacteriocin secreted by Enterobacteriaceae such as *E. coli*, have demonstrated anti-tumor effects against numerous tumor cell lines *in vitro,* including breast, colon, bone, and HeLa cancer cells [81, 79].

6.5 DIFFERENT MICROBES USED AS ANTICANCER

6.5.1 CLOSTRIDIUM

Research on *Clostridium* began due to its naturally anaerobic characteristics. The presence of low levels of oxygen and cell death in cancer tissues make them resistant to conventional therapies. Among the most studied strains is *Clostridium novyi*, a Gram-positive, obligate anaerobe, and spore-forming microbe. It is being explored for its potential in developing cancer therapeutic treatments that induce cell death, particularly in cases with poor prognosis. *C. novyi*-NT, with its toxin gene deleted, can selectively colonize cancer tissues and exhibits fewer negative effects due to reduced exotoxin production. It has shown promise in research involving sarcomas, renal cancer, CNS cancer, and colorectal cancer (CRC), where it has been used to study selective colonization, immune cell infiltration, and cytokine release leading to cancer tissue death.

A patient with retroperitoneal leiomyosarcoma, who is resistant to conventional therapy and has multiple metastases, experienced shrinkage of the tumor in their right shoulder following intra-tumoral treatment. This localized treatment approach is recommended over systemic injection to minimize side effects. Patient recruitment for further clinical trials is currently ongoing. The immune system cells exhibit aggression towards cancer cells, inducing their death without causing harm to normal cells. This process occurs after the cells germinate, triggering an inflammatory response. These derivatives have been utilized in genetic engineering as delivery systems for cytokines such as TNF, IL-12, and IL-2. The aim is to achieve maximum concentrations of these cytokines within the tumor, while minimizing systemic harm. Additionally, *C. novyi*-NT and *C. sporogenes* stimulate the release of specific immunoglobulins (antibodies) against HIF-1, the primary element involved in the response to low oxygen conditions (hypoxia) within cancers. The oncolytic properties of this strain were enhanced through genetic therapy, yielding promising results. The tumor-targeting characteristics of *Clostridium sporogenes* make it a valuable tool. Nitroreductase (NR) and CD codification are facilitated by *E. coli*-derived genes. *In vivo* antitumoral effects are achieved through the metabolism of potent drugs by these enzymes within tumors. Although they are not highly immunogenic, *Clostridium* spores can invade multiple organs when administered systemically [82].

6.5.2 LISTERIA MONOCYTOGENES (LM)

Gram-positive, facultative intracellular bacteria known as Lm are present in the body. Numerous studies conducted in past decades have demonstrated its potential as a therapeutic agent for tumors, exhibiting multiple effects and mechanisms [83]. In a favorable immune microenvironment, it can be utilized for both primary and metastatic tumors. Its selective colonization is facilitated by the latter, while the production of reactive oxygen species (ROS) promotes their eradication [84]. Lm also affects the TME by reducing T-regulatory cells and immunomodulatory substances such as TGF and IL10 [85]. However, the primary characteristic of Lm lies in its ability to selectively infect antigen-presenting cells (APCs), facilitating the processing and presentation of foreign and self-antigens [86]. These qualities make Lm a valuable immune stimulant agent.

Genetic engineering has been employed to investigate these characteristics of Lm and identify recombinant strains capable of releasing cancer toxins [87]. These strains can serve as crucial vectors in vaccinations, enhancing cellular responses and overcoming immune tolerance to certain cancers [23]. This can be achieved by introducing plasmids encoding tumor antigens or integrating them into the bacterial chromosome. Additionally, these toxins, such as ActA or LLO, can be demonstrated as chimeric proteins. Lm's utilization of ActA for intercellular proliferation and motility enhances the immune response to cancer toxins (antigens) with low immunogenicity. The objective of this experimental research was to evaluate the efficacy of recently developed vaccinations, specifically examining cervical cancer models using Lm-LLO-E7 [82].

6.5.3 *SALMONELLA*

Due to its adaptability, *Salmonella ser*. has been extensively studied as a bacterium [35]. These newly engineered strains have demonstrated anti-tumor activity. Phase I clinical trials were conducted at the beginning of the 21st century to evaluate the efficacy of gene modifications involving msb-B and purI gene deletions. The deletion of the msbB gene aimed to reduce TNF-related toxicity, which could lead to septic shock [76]. The msb-B gene is responsible for lipid A synthesis. On the other hand, the deletion of the purI gene allowed the bacterium to selectively colonize tumors. As a result, these strains became dependent on external purine sources and their growth was limited to regions with high rates of cell renewal [32].

These strains would specifically colonize purine-rich tumor tissues, which provide ideal environments [32]. The primary strain used in experimental studies based on this thesis is *Salmonella typhimurium* VNP20009. The findings of this study determined the maximum tolerable dose (MTD) of this bacteria, the maximum toxic dose, and adverse drug reactions by maximizing the formulation of pro-inflammatory cytokines. Negative effects observed included thrombocytopenia, anemia, persistent bacteria in the blood, elevated bilirubin levels, nausea and vomiting, increased alkaline phosphatase enzyme, and decreased phosphate levels in the serum. Only 3 patients showed evidence of tumor colonization, and there was no indication of tumor regression [56]. Although the results of this study were not encouraging, it served as a starting point for future research to identify modified doses that could enhance effectiveness, tumor restriction, and other beneficial outcomes.

A study was initiated to investigate whether using them as vectors for tumor gene therapy could enhance the therapeutic effect of bacteria [38]. Initial research was conducted using a modified and weakened strain of Salmonella that expressed the suicide prodrug-activating enzyme from *E. coli* [41]. The TAPET-CD strain was created by inserting these genes into the VNP20009 chromosome using the Donnenberg and Karper method [82]. The mechanism of action of this enzyme involves converting 5-fluorocytosine (5-FC), a limited cytotoxic antifungal drug, into 5-fluorouracil (5-FU), a cytotoxic antimetabolite that can induce cellular apoptosis [41]. Unfortunately, the results were not promising. Two out of three treated patients did not exhibit tumor reduction, although their condition did seem to be improving [82].

6.6 ADVANTAGES OF MICROBIAL GENES-BASED CANCER THERAPY

- Tumor-targeting anaerobes with high specificity.
- Increased potential for customization based on specific geographical regions.
- Ability to migrate through the body, accumulate in tumor areas, and penetrate cancer cells.
- Enhanced vulnerability of tumor tissues to metabolites and bacteria.
- Capability to replicate and disseminate throughout the body, eliminating the need for continuous treatment, and the ability to be inhibited and controlled by antibiotics within the therapeutic window.

6.7 DISADVANTAGES OF MICROBIAL GENES-BASED CANCER THERAPY

- Due to the frequent use of difficulty, a narrow therapeutic window was required for direct intra-tumor injection.
- The high reactivity and inability to lyse tumors and cancer cells make septic shock a possibility. Furthermore, the total combination therapy carries risks due to the bacteria's unpredictable pathogenicity and its metabolites.
- There is a risk of uncontrolled infection caused by the bacteria and its metabolites [34].

6.8 CONCLUSION

Cancer-related deaths are becoming more commonplace globally, necessitating the development of effective defenses against this menace. The body's ability to absorb anti-tumor chemotherapeutic drugs is strictly constrained by the physiology of solid tumors. While traditional chemotherapy medications remain the mainstream, bacteriotherapy has shown promising results due to its high specificity and oncogenic efficacy in various live organism investigations. However, the use of microbes as anticancer medicines in clinical settings still faces challenges. Several microorganisms, including *Salmonella, Listeria,* and *Clostridium,* have been explored as alternative cancer treatments. In the future, both diagnostic and therapeutic applications of genetically modified microbes are expected to increase. The intrinsic properties of microbes can be modified for this purpose and utilized to our advantage.

KEYWORDS

- **adverse drug reaction**
- **bacteria**
- **cancer**
- **clostridium strains**
- **genes**
- **malignancy**
- **microorganisms**
- **Salmonella typhimurium**
- **vaccine**

REFERENCES

1. Wolfgang Grisold, Riccardo Soffietti, Stefan Oberndorfer, & Guido Cavaletti (2021). *Effects of Cancer Treatment on the Nervous System*, 2153–2153.
2. Martin, E., (2019). Defining cancer: Causes and treatments. *Microreviews in Cell and Molecular Biology, 5*(2).
3. Jayasri, J., (2018). Herbal therapies in cancer prevention and treatment. *International Journal of Oncological Nursing, 4*(1), 24–28.

4. Tasnim, H., (2021). *Immunotherapy in the Management of Lung Cancer* Doctoral dissertation, Brac University.
5. Ivanenko, N., (2021). Biofilm and tumor: Interpretation of interaction and treatment strategies: Review. *Medical Science of Ukraine (MSU), 17*(1), 104–120.
6. Melo, B. R., Diniz, M. A., Casemiro, F. G., Figueiredo, L. C., Santos-Orlandi, A. A., Haas, V. J., Orlandi, F. D., & Gratão, A. C., (2017). Cognitive and functional assessment about elderly people users of health public service. *Escola Anna Nery, 21.*
7. Ramos, R. A., García, D. M., & Cruz, E. R., (2019). Inulin-type fructans: Effect on gut microbiota, obesity and satiety. *Gaceta Médica Espirituana, 21*(2), 134–145.
8. Salari, A., Saeedi, E., Hadji, M., Jalaeefar, A., Zendehdel, K., & Shirkhoda, M., (2021). The role of lymph node ratio in predicting survival in patients with esophageal squamous cell carcinoma. *Journal of Iranian Medical Council, 4*(4), 296–302.
9. Nandini, C., Roopakala, N., & Andanigoudar, K. B., (2022). Barriers to early diagnosis and treatment among breast cancer survivors: Qualitative study in tertiary care Hospital, North Karnataka. *National Journal of Community Medicine, 13*(3), 190–194.
10. Datta, N. R., Ordóñez, S. G., Gaipl, U. S., Paulides, M. M., Crezee, H., Gellermann, J., Marder, D., et al., (2015). Local hyperthermia combined with radiotherapy and-/or chemotherapy: Recent advances and promises for the future. *Cancer Treatment Reviews, 41*(9), 742–753.
11. Dai, H., Han, J., & Lichtfouse, E., (2021). Smarter cures to combat COVID-19 and future pathogens: A review. *Environmental Chemistry Letters, 19*(4), 2759–2771.
12. https://www.legis.iowa.gov/docs/iac/rule/657.39.11.pdf (accessed on 11 March 2024).
13. Bol, K. F., Aarntzen, E. H., Pots, J. M., Olde, N. M. A., Van De, R., M. W., Scharenborg, N. M., De Boer, A. J., et al., (2016). Prophylactic vaccines are potent activators of monocyte-derived dendritic cells and drive effective anti-tumor responses in melanoma patients at the cost of toxicity. *Cancer Immunology, Immunotherapy, 65*(3), 327–339.
14. Raulet, D., Coscoy, L., & Portnoy, D., (2017). In: Danielle, K., (ed.), *Berkeley Scientific Journal.*
15. https://doi.org/10.2217/fvl.15.60.
16. *doi:* https://doi.org/10.1038/d41586-020-02400-7.
17. Grimmett, E., Al-Share, B., Alkassab, M. B., et al., (2022). Cancer vaccines: Past, present, and future; a review article. *Discov. Onc., 13*, 31. https://doi.org/10.1007/s12672-022-00491-4.
18. Patyar, S., Joshi, R., Byrav, D. S., Prakash, A., Medhi, B., & Das, B. K., (2010). Bacteria in cancer therapy: A novel experimental strategy. *Journal of Biomedical Science, 17*(1), 1–9.
19. Patyar, S., Joshi, R., Byrav, D. P., et al., (2010). Bacteria in cancer therapy: A novel experimental strategy. *J. Biomed. Sci., 17*, 21. https://doi.org/10.1186/1423-0127-17-21.
20. Nauts, H. C., (1980). *The Beneficial Effects of Bacterial Infections on Host Resistance to Cancer End Results in 449 Cases: A Study and Abstracts of Reports in the World Med.* Literature (1775–1980) and Personal Communications. Cancer Research Inst.
21. Richardson, M. A., Ramirez, T., Russell, N. C., & Moye, L. A., (1999). Coley toxins immunotherapy: A retrospective review. *Alternative Therapies in Health and Medicine, 5*(3), 42.
22. Zacharski, L. R., & Sukhatme, V. P., (2005). Coley's toxin revisited: Immunotherapy or plasminogen activator therapy of cancer? *Journal of Thrombosis and Haemostasis, 3*(3), 424–427.

23. Hosseinidoust, Z., Mostaghaci, B., Yasa, O., Park, B. W., Singh, A. V., & Sitti, M., (2016). Bioengineered and biohybrid bacteria-based systems for drug delivery. *Advanced Drug Delivery Reviews, 106*, 27–44.
24. Badie, F., Ghandali, M., Tabatabaei, S. A., Safari, M., Khorshidi, A., Shayestehpour, M., Mahjoubin-Tehran, M., et al., (2021). Use of *Salmonella* bacteria in cancer therapy: Direct, drug delivery and combination approaches. *Frontiers in Oncology, 11*, 624759.
25. Maletzki, C., Linnebacher, M., Kreikemeyer, B., & Emmrich, J., (2008). Pancreatic cancer regression by intratumoural injection of live *Streptococcus* pyogenes in a syngeneic mouse model. *Gut., 57*(4), 483–491.
26. Zhu, H., Li, Z., Mao, S., Ma, B., Zhou, S., Deng, L., Liu, T., et al., (2011). Antitumor effect of sFlt-1 gene therapy system mediated by bifidobacterium infantis on Lewis lung cancer in mice. *Cancer Gene Therapy, 18*(12), 884–896.
27. Agrawal, N., Bettegowda, C., Cheong, I., Geschwind, J. F., Drake, C. G., Hipkiss, E. L., Tatsumi, M., et al., (2004). Bacteriolytic therapy can generate a potent immune response against experimental tumors. *Proceedings of the National Academy of Sciences, 101*(42), 15172–15177.
28. Wang, Y., Chen, J., Tang, B., Zhang, X., & Hua, Z. C., (2013). Systemic administration of attenuated *Salmonella* typhimurium in combination with interleukin-21 for cancer therapy. *Molecular and Clinical Oncology, 1*(3), 461–465.
29. Yoon, W., Yoo, Y., Chae, Y. S., Kee, S. H., & Kim, B. M., (2018). Therapeutic advantage of genetically engineered *Salmonella* typhimurium carrying short hairpin RNA against inhibin alpha subunit in cancer treatment. *Annals of Oncology, 29*(9), 2010–2017.
30. Sarotra, P., & Medhi, B., (2016). Use of bacteria in cancer therapy. *Current Strategies in Cancer Gene Therapy, 2016*, 111–121.
31. Sieow, B. F., Wun, K. S., Yong, W. P., Hwang, I. Y., & Chang, M. W., (2021). Tweak to treat: Reprograming bacteria for cancer treatment. *Trends in Cancer, 7*(5), 447–464.
32. Clairmont, C., Lee, K. C., Pike, J., Ittensohn, M., Low, K. B., Pawelek, J., Bermudes, D., et al., (2000). Biodistribution and genetic stability of the novel antitumor agent VNP20009, a genetically modified strain of *Salmonella typhimuvium. Journal of Infectious Diseases, 181*(6), 1996–2002.
33. Luo, X., Li, Z., Lin, S., Le, T., Ittensohn, M., Bermudes, D., Runyab, J. D., et al., (2001). Antitumor effect of VNP20009, an attenuated *Salmonella*, in murine tumor models. *Oncology Research Featuring Preclinical and Clinical Cancer Therapeutics, 12*(11, 12), 501–508.
34. Rommasi, F., (2021). Bacterial-based methods for cancer treatment: What we know and where we are. *Oncology and Therapy, 15*, 1–32.
35. Zhang, L., Gao, L., Zhao, L., Guo, B., Ji, K., Tian, Y., Wang, J., et al., (2007). Intratumoral delivery and suppression of prostate tumor growth by attenuated *Salmonella* enterica serovar typhimurium carrying plasmid-based small interfering RNAs. *Cancer Research, 67*(12), 5859–5864.
36. Sedighi, M., Zahedi, B. A., Hamblin, M. R., Ohadi, E., Asadi, A., Halajzadeh, M., Lohrasbi, V., et al., (2019). Therapeutic bacteria to combat cancer; current advances, challenges, and opportunities. *Cancer Medicine, 8*(6), 3167–3181.
37. Nair, N., Kasai, T., & Seno, M., (2014). Bacteria: Prospective savior in battle against cancer. *Anticancer Research, 34*(11), 6289–6296.
38. Forbes, N. S., (2010). Engineering the perfect (bacterial) cancer therapy. *Nature Reviews Cancer, 10*(11), 785–794.

39. Duong, M. T. Q., Qin, Y., You, S. H., et al., (2019). Bacteria-cancer interactions: Bacteria-based cancer therapy. *Exp Mol. Med., 51*, 1–15. https://doi.org/10.1038/s12276-019-0297-0.
40. Torres, W., Lameda, V., Olivar, L. C., Navarro, C., Fuenmayor, J., Pérez, A., Mindiola, A., et al., (2018). Bacteria in cancer therapy: Beyond immunostimulation. *Journal of Cancer Metastasis and Treatment, 4*, 4.
41. Aghi, M., Hochberg, F., & Breakefield, X. O., (2000). Prodrug activation enzymes in cancer gene therapy. *The Journal of Gene Medicine, 2*(3), 148–164.
42. Rommasi, F., (2022). Bacterial-based methods for cancer treatment: What we know and where we are. *Oncol. Ther., 10*, 23–54. https://doi.org/10.1007/s40487-021-00177-x.
43. https://doi.org/10.1016/j.ctarc.2021.100436.
44. Łukasiewicz, K., & Fol, M., (2018). Microorganisms in the treatment of cancer: Advantages and limitations. *J. Immunol. Res., 2018*, 2397808. doi: 10.1155/2018/2397808. PMID: 29682586; PMCID: PMC5848056.
45. Cheong, I., Huang, X., Bettegowda, C., Diaz, Jr. L. A., Kinzler, K. W., Zhou, S., & Vogelstein, B., (2006). A bacterial protein enhances the release and efficacy of liposomal cancer drugs. *Science, 314*(5803), 1308–1311.
46. Xiong, G., Husseiny, M. I., Song, L., Erdreich-Epstein, A., Shackleford, G. M., Seeger, R. C., Jäckel, D., et al., (2010). Novel cancer vaccine based on genes of *Salmonella* pathogenicity island 2. *International Journal of Cancer, 126*(11), 2622–2634.
47. Fehleisen, F., (1882). Ueber die züchtung der erysipelkokken auf künstlichem nährboden und ihre übertragbarkeit auf den menschen. *Dtsch Med Wochenschr, 8*(31), 553–554.
48. Puri, R. K., (1999). Development of a recombinant interleukin-4-Pseudomonas exotoxin for therapy of glioblastoma. *Toxicologic Pathology, 27*(1), 53–57.
49. Forbes, N. S., Coffin, R. S., Deng, L., Evgin, L., Fiering, S., Giacalone, M., Gravekamp, C., et al., (2018). White paper on microbial anti-cancer therapy and prevention. *Journal for Immunotherapy of Cancer, 6*(1), 1–24.
50. Hammerich, L., Binder, A., & Brody, J. D., (2015). *In situ* vaccination: Cancer immunotherapy both personalized and off-the-shelf. *Molecular Oncology, 9*(10), 1966–1981.
51. Pierce, R. H., Campbell, J. S., Pai, S. I., Brody, J. D., & Kohrt, H. E., (2015). In-situ tumor vaccination: Bringing the fight to the tumor. *Human Vaccines & Immunotherapeutics, 11*(8), 1901–1909.
52. Hartman, Z. C., Wei, J., Glass, O. K., Guo, H., Lei, G., Yang, X. Y., Osada, T., et al., (2011). Increasing vaccine potency through exosome antigen targeting. *Vaccine, 29*(50), 9361–9367.
53. Nair, S. K., Boczkowski, D., Morse, M., Cumming, R. I., Lyerly, H. K., & Gilboa, E., (1998). Induction of primary carcinoembryonic antigen (CEA)-specific cytotoxic T lymphocytes *in vitro* using human dendritic cells transfected with RNA. *Nature Biotechnology, 16*(4), 364–369.
54. Nallar, S. C., Xu, D. Q., & Kalvakolanu, D. V., (2017). Bacteria and genetically modified bacteria as cancer therapeutics: Current advances and challenges. *Cytokine, 89*, 160–172.
55. Thamm, D. H., Kurzman, I. D., King, I., Li, Z., Sznol, M., Dubielzig, R. R., Vail, D. M., & MacEwen, E. G., (2005). Systemic administration of an attenuated, tumor-targeting *Salmonella* typhimurium to dogs with spontaneous neoplasia: Phase I evaluation. *Clinical Cancer Research, 11*(13), 4827–4834.

56. Toso, J. F., Gill, V. J., Hwu, P., Marincola, F. M., Restifo, N. P., Schwartzentruber, D. J., Sherry, R. M., et al., (2002). Phase I study of the intravenous administration of attenuated *Salmonella* typhimurium to patients with metastatic melanoma. *Journal of Clinical Oncology: Official Journal of the American Society of Clinical Oncology, 20*(1), 142.
57. Schlechte, H., & Elbe, B., (1988). Recombinant plasmid DNA variation of *Clostridium* oncolyticum—Model experiments of cancerostatic gene transfer. Zentralblatt für bakteriologie, mikrobiologie und hygiene. *Series A: Medical Microbiology, Infectious Diseases, Virology, Parasitology, 268*(3), 347–356.
58. Felgner, S., Pawar, V., Kocijancic, D., Erhardt, M., & Weiss, S., (2017). Tumor-targeting bacteria-based cancer therapies for increased specificity and improved outcome. *Microbial Biotechnology, 10*(5), 1074–1078.
59. Mengesha, A., Dubois, L., Chiu, R. K., Paesmans, K., Wouters, B. G., Lambin, P., & Theys, J., (2007). Potential and limitations of bacterial-mediated cancer therapy. *Frontiers in Bioscience-Landmark, 12*(10), 3880–3891.
60. Nuyts, S., Van, M. L., Theys, J., Landuyt, W., Lambin, P., & Anné, J., (2001). The use of radiation-induced bacterial promoters in anaerobic conditions: A means to control gene expression in clostridium-mediated therapy for cancer. *Radiation Research, 155*(5), 716–723.
61. Ginn, S. L., Amaya, A. K., Alexander, I. E., Edelstein, M., & Abedi, M. R., (2018). Gene therapy clinical trials worldwide to 2017: An update. *The Journal of Gene Medicine, 20*(5), e3015.
62. Zhang, S., Chen, Y., Wang, J., Tang, F., Miao, T., & Li, M., (2014). Role of nontoxigenic *Clostridium novyi* in solid tumor therapy. *Reviews in Medical Microbiology, 25*(3), 71–76.
63. Zhou, S., Gravekamp, C., Bermudes, D., & Liu, K., (2018). Tumor-targeting bacteria engineered to fight cancer. *Nature Reviews Cancer, 18*(12), 727–743.
64. Jiang, S. N., Phan, T. X., Nam, T. K., Nguyen, V. H., Kim, H. S., Bom, H. S., Choy, H. E., et al., (2010). Inhibition of tumor growth and metastasis by a combination of *Escherichia coli*–mediated cytolytic therapy and radiotherapy. *Molecular Therapy, 18*(3), 635–642.
65. Dang, L. H., Bettegowda, C., Huso, D. L., Kinzler, K. W., & Vogelstein, B., (2001). Combination bacteriolytic therapy for the treatment of experimental tumors. *Proceedings of the National Academy of Sciences, 98*(26), 15155–5160.
66. Liu, X., Jiang, S., Piao, L., & Yuan, F., (2016). Radiotherapy combined with an engineered of *Salmonella typhimurium* inhibits tumor growth in a mouse model of colon cancer. *Experimental Animals, 2016*, 16–0033.
67. Jabbari, P., (2020). Using bacterial vectors for probable vaccines: From molecular mechanism to cancer therapy. *World Journal of Vaccines, 10*(2), 33–42.
68. Brüggemann, H., & Gottschalk, G., (2009). *Clostridia: Molecular Biology in the Post-Genomic Era*. Caister Academic Press.
69. Theys, J., Landuyt, W., Nuyts, S., et al., (2001). Specific targeting of cytosine deaminase to solid tumors by engineered *Clostridium acetobutylicum*. *Cancer Gene Ther., 8*, 294–297 https://doi.org/10.1038/sj.cgt.7700303.
70. Liu, S. C., Minton, N., Giaccia, A., *et al., (*2002*)*. Anticancer efficacy of systemically delivered anaerobic bacteria as gene therapy vectors targeting tumor hypoxia/necrosis. *Gene Ther., 9*, 291–296. https://doi.org/10.1038/sj.gt.3301659.

71. Holtsberg, F. W., Ensor, C. M., Steiner, M. R., Bomalaski, J. S., & Clark, M. A., (2002). Poly (ethylene glycol)(PEG) conjugated arginine deiminase: Effects of PEG formulations on its pharmacological properties. *Journal of Controlled Release, 80*(1–3), 259–271.
72. https://doi.org/10.1158/0008-5472.CAN-08-3157.
73. Huang, X., Pan, J., Xu, F., Shao, B., Wang, Y., Guo, X., & Zhou, S., (2021). Bacteria-based cancer immunotherapy. *Advanced Science, 8*(7), 2003572.
74. Nougayrède, J. P., Taieb, F., De Rycke, J., & Oswald, E., (2005). Cyclomodulins: Bacterial effectors that modulate the eukaryotic cell cycle. *Trends in Microbiology, 13*(3), 103–110.
75. Shafiee, F., Aucoin, M. G., & Jahanian-Najafabadi, A., (2019). Targeted diphtheria toxin-based therapy: A review article. *Frontiers in Microbiology, 10*, 2340.
76. Low, K. B., Ittensohn, M., Le, T., Platt, J., Sodi, S., Amoss, M., Ash, O., et al., (1999). Lipid A mutant *Salmonella* with suppressed virulence and TNFα induction retain tumor-targeting *in vivo*. *Nature Biotechnology, 17*(1), 37–41.
77. Min, J. J., Kim, H. J., Park, J. H., Moon, S., Jeong, J. H., Hong, Y. J., Cho, K. O., et al., (2008). Noninvasive real-time imaging of tumors and metastases using tumor-targeting light-emitting *Escherichia coli*. *Molecular Imaging and Biology, 10*(1), 54–61.
78. Min, J. J., Nguyen, V. H., Kim, H. J., Hong, Y., & Choy, H. E., (2008). Quantitative bioluminescence imaging of tumor-targeting bacteria in living animals. *Nature Protocols, 3*(4), 629–636.
79. Song, S., Vuai, M. S., & Zhong, M., (2018). The role of bacteria in cancer therapy–enemies in the past, but allies at present. *Infectious Agents and Cancer, 13*(1), 1–7.
80. Kaur, S., & Kaur, S., (2015). Bacteriocins as potential anticancer agents. *Frontiers in Pharmacology, 6*, 272.
81. Chumchalova, J., & Šmarda, J., (2003). Human tumor cells are selectively inhibited by colicins. *Folia Microbiologica, 48*(1), 111–115.
82. Nemunaitis, J., Cunningham, C., Senzer, N., Kuhn, J., Cramm, J., Litz, C., Cavagnolo, R., et al., (2003). Pilot trial of genetically modified, attenuated *Salmonella* expressing the *E. coli* cytosine deaminase gene in refractory cancer patients. *Cancer Gene Therapy, 10*(10), 737–744.
83. Weiskirch, L. M., Pan, Z. K., & Paterson, Y., (2001). The tumor recall response of antitumor immunity primed by a live, recombinant *Listeria monocytogenes* vaccine comprises multiple effector mechanisms. *Clinical Immunology, 98*(3), 346–357.
84. Jahangir, A., Chandra, D., Quispe-Tintaya, W., Singh, M., Selvanesan, B. C., & Gravekamp, C., (2017). Immunotherapy with Listeria reduces metastatic breast cancer in young and old mice through different mechanisms. *Oncoimmunology, 6*(9), e1342025.
85. Pan, Z. K., Ikonomidis, G., Pardoll, D., & Paterson, Y., (1995). Regression of established tumors in mice mediated by the oral administration of a recombinant *Listeria monocytogenes* vaccine. *Cancer Research, 55*(21), 4776–4779.
86. Wallecha, A., Singh, R., & Malinina, I., (2013). *Listeria monocytogenes* (Lm)-LLO immunotherapies reduce the immunosuppressive activity of myeloid-derived suppressor cells and regulatory T cells in the tumor microenvironment. *Journal of Immunotherapy, 36*(9), 468–476.
87. Ikonomidis, G., Paterson, Y., Kos, F. J., & Portnoy, D. A., (1994). Delivery of a viral antigen to the class I processing and presentation pathway by *Listeria monocytogenes*. *The Journal of Experimental Medicine, 180*(6), 2209–2218.

CHAPTER 7

Utilization of Protein and Peptides as Cancer Vaccines: Advancement and Challenges Assessment of Ethnomedicinal Plants: Shiwalik Hills, Himachal Pradesh, India

PATEL GAYATRI and PATEL PRIYA

Ramanbhai Patel College of Pharmacy, Charotar University of Science and Technology, CHARUSAT Campus, Changa, Anand, Gujarat, India

ABSTRACT

Vaccines have significantly improved the detection and management of chronic diseases over the past 20 years. According to WHO data, cancer is projected to become the leading cause of death worldwide in 2020, accounting for nearly 10 million deaths, or approximately one in every six deaths. Consequently, there has been a growing interest in the development of cancer vaccines (CV) as a means of disease prevention rather than just treatment. The use of proteins and peptides for vaccine delivery has garnered continued interest due to advancements in administration, stability, and design. This review focuses on recent advancements in the field and explores the potential of protein- and peptide-based vaccines for treating various types of cancer.

7.1 INTRODUCTION

Despite decades of scientific and clinical research, as well as clinical trials of promising new treatments, cancer continues to be a leading cause of morbidity and mortality [1]. The uncontrolled cell division is a characteristic

Cancer Vaccination and Challenges, Volume 1: Strategies for Therapeutic Cancer Vaccine Development. Rishabha Malviya, Bhupendra Prajapati, and Sonali Sundram (Eds.)

feature of cancer, and the ability of these cells to invade other tissues leads to the formation of tumor masses, vascularization, and metastasis (spread of cancer throughout the body) [2]. Angiogenesis, the process by which new blood vessels are formed from pre-existing ones, is essential for normal growth and development, but it also plays a significant role in the transformation of benign tumors into malignant ones. Chemotherapy, which involves the administration of cytotoxic substances to cancer cells, is one of the main methods used to treat cancer [3].

Despite significant advancements in recent years, the majority of cancer treatments still involve surgery or relatively unchanged chemotherapy, radiation, and hormone therapies [4]. For example, traditional chemopreventive agents (such as DNA alkylating agents) that target actively proliferating polyps may also harm healthy growing cells and may not completely eliminate quiescent or non-proliferating cancer cells. The development of modern drugs is also associated with the emergence of drug resistance, which can be caused by defects in drug transporters or detoxifying enzymes that affect the drug's interaction with its target [5]. Drug resistance in cancer can also result from defects in DNA repair processes and programmed cell death pathways. Even after successful initial therapy, patients still face the risk of cancer recurrence. The development of small molecules as medications has driven efforts to discover new cancer treatments, but unexpected side effects have necessitated the exploration of alternative strategies.

7.2 KEY CONCEPTS OF VACCINE IMMUNOLOGY

Vaccines utilize the body's ability to develop defense mechanisms against microbes. The immune system can quickly recognize and eliminate pathogenic germs that it has previously encountered [6]. Vaccines contain destroyed or inactivated germs, which trigger an immune response that closely resembles the body's natural response to bacteria. Since the germs in vaccines are killed or inactivated, they do not cause the disease, making vaccination a less risky method of building immunity [7].

Vaccines, along with good hygiene practices, are the most effective ways to prevent the spread of infectious diseases. Uncontrolled, some infectious diseases have proven to be deadly.

7.2.1 THE IMMUNE SYSTEM

The immune system is capable of distinguishing between self-molecules and structures and those that are not. By identifying and eliminating them, this skill aims to defend the organism from the spread of diseases [8]. The immune system consists of two major parts: the innate (inborn) immune system and the adaptive (specific) immune system.

7.2.1.1 INNATE IMMUNITY SYSTEM

The innate immune system provides a rapid response to invading pathogens. It is triggered when specific molecular structures of germs are recognized by cells and molecules. The innate immune system works to prevent or limit the replication and spread of these pathogens. Inflammation is one of the earliest responses of the innate immune system to infection in vertebrates. Infected and injured cells produce specific chemicals, such as histamine and prostaglandins, which initiate inflammation. This response increases sensitivity to pain, widens local blood vessels, and attracts white blood cells called neutrophils, which can phagocytose pathogens as an initial defense. Neutrophils can release chemokines and cytokines, which attract other immune cells like macrophages and natural killer cells (NK-cell). Macrophages reside in tissues and can both ingest and kill infections. NK-cells use a process called apoptosis to recognize and destroy infected cells and some cancer cells. Apoptosis is characterized by the breakdown of DNA and proteins and the disintegration of the cell [9].

The innate immune system responds to pathogens in a variety of ways in the early stages of illness, but it does not provide long-lasting immunity. When viruses are able to bypass this initial line of defense, the innate immune systems of invertebrates activate the adaptive immune system [10].

7.2.1.2 ADAPTIVE IMMUNITY SYSTEM

The adaptive immune system possesses the remarkable ability to enhance pathogen recognition, tailor a response specific to the actual structure of the pathogen and remember that response to be prepared for future encounters with the same or closely related pathogens. The adaptive immune system activates the bone marrow and the thymus, which produce B cells and T cells, respectively, to generate humoral immunity and cellular immunity. Generally, T cells attack body cells that have been infected by germs or have developed cancer, while B cells produce antibodies (Abs) that directly combat pathogens. In response to antigens (molecules recognized as non-self), B cells produce Abs when they are activated. Antigens can include bacterial toxins (B.T), cancerous structures, or bacteria themselves. The Abs produced are specific to that particular antigen. The primary role of Abs is to either neutralize (render inactive) invaders to prevent their reproduction, or to mark them for elimination (performed by other immune cells). Similar to B cells, T cells have antigen-specific surface receptors [11]. T cells can specialize in various roles, such as attracting and activating macrophages, stimulating B cells to produce Abs, or directly eliminating infected cells (cytotoxic T cells, also known as killer cells). Even after the infection has been cleared, this heightened response is maintained through immunological memory. Whenever the virus re-enters the body, the immune system is able to respond more rapidly. This ability is sustained by memory cells, which retain certain characteristics of the previously encountered pathogen and can trigger a robust reaction upon recognition of the pathogen again [12].

7.2.1.3 TUMOR IMMUNOLOGY

There is a significant amount of evidence suggesting that the immune system often eliminates tumors before they become visible, a process known as immune surveillance [13]. The immune response to a tumor is highly complex, not yet fully understood, and influenced by various factors, particularly the type of tumor. Effective tumor control requires the involvement of both cytotoxic T cells and NK-cell s. Another important characteristic is the high rate of tumor cell apoptosis. Apoptotic cells can release protein fragments that activate NK cells and other cells, inducing interferon production if the synthesis rate is high and there are more apoptotic cells than macrophages can phagocytize. Interferon plays a role in immune surveillance by enhancing the ability of anti-tumor T cells to recognize tumor cells. Tumor cells have the ability to produce tumor-associated antigens (TAAs), which are proteins that are more abundant on the surface of tumor cells compared to normal cells [14]. In response, the immune system employs T cells to eliminate tumor cells after identifying these antigens as foreign.

7.3 A VACCINE'S IMMUNOLOGICAL REQUIREMENTS

7.3.1 SELECTING THE BEST ANTIGEN AFTER BEING IDENTIFIED

Vaccines are designed to stop the disease indicator that results from pathogenic infection. Most of the time, this happens by minimizing the illness's effects rather than totally preventing it. When choosing the right antigens to utilize as the basis for a vaccine, it is therefore very helpful to have a grasp of the disease's etiology and natural immune control [15]. The complexity of pathogen-derived material contained in vaccines generated from pathogens can vary. Our improved understanding of basic immunology and selection techniques has resulted in more thoroughly characterized vaccinations than ever before, inspiring a more logical approach to vaccine creation [16].

7.3.2 *INDUCTION OF INNATE IMMUNE RESPONSES*

As previously described, a combination of factors and stimuli can activate the immune system. Therefore, when developing a vaccine, especially when using highly pure or refined antigens, it is important to consider more than just the presence of a "foreign" antigen to elicit an immune response. Some inactivated whole viruses and highly refined component antigen formulations may lack certain molecular characteristics and defensive triggers that are necessary for the innate immune system to function. These antigens may not be as immunogenic as desired, but they are designed to reduce excessive inflammatory reactions. In such cases, the use of adjuvants can replace the missing innate triggers by simulating them, restoring the proper balance between the need for protective responses and their tolerability [17].

7.3.3 *HELPFUL CD4D T-CELL INDUCTION*

The activation of CD4 T cells is largely determined by the nature of the early inflammatory response. Adjuvants used in vaccines can therefore influence the activation and subsequent development of CD4 T cells, which in turn affects the magnitude and nature of the adaptive immune system's response [18].

7.3.4 *EFFECTOR CELL SELECTION AND TARGETING*

It is crucial to realize that not all immune responses to a given pathogen or antigen are optimal for evoking protection; some infections have evolved strategies to evade or manipulate the immune response. Although antibody titers are frequently thought of as adequate indicators of immune protection, they may not actually be the means by which the best protection is attained [19]. It is possible that there are no immunological correlates of immunity and protection, or that they are not well understood. Clinical trials are still used in modern vaccine development to learn about clinical effectiveness and, if possible, the immunological profiles of those who are protected [20].

7.4 COMPARISON OF CONVENTIONAL VACCINES TO PROTEIN AND PEPTIDE-BASED VACCINES

Comparison of conventional to protein and peptide-based vaccines are explained in Table 7.1.

TABLE 7.1 Comparison of Conventional Vaccines to Protein and Peptide-based Vaccines

Conventional Vaccines	Protein and Peptide-based Vaccines
According to Jenner's discovery, the use of cowpox to treat human patients could effectively stop the spread of smallpox. This strategy, aimed at eradicating fatal diseases, led to the development of scalable vaccination systems that are constantly being improved [21].	Researchers from the Netherlands' Institute for Animal Science and Health announced the development of the first synthetic peptide vaccination in 1994 [22].
Historically, pathogens or their component parts have been used in attenuated or inactivated forms in infection-targeting vaccines.	With the help of peptides, vaccines can target very precise epitopes while lowering the risk of allergic or autoimmune reactions.
Only a small fraction is responsible for inducing a proper protective response.	Specificity and safety [23].
These vaccines work because they include the complete disease-causing agent and provide the recipient with lifelong immunity to it.	Reduced production costs and lead times for vaccines are other intriguing benefits of using peptides as immunogens.
The main drawback is that common contaminants present in these formulations can lead to allergic or autoimmune reactions. This happens because infections can develop in regular culture media that contain biological byproducts and antibiotics. When administered to patients, these residues often trigger a severe allergic reaction because they are not completely eliminated when the pathogen becomes inactive [25].	A disadvantage is that peptides usually lack immunogenicity, which prevents them from successfully inducing an immune response [24].
The potential for the pathogen to revert to a virulent form after being attenuated or deactivated is another disadvantage, which puts patients, medical personnel, and vaccine production employees at risk.	Another drawback is that these molecules are extremely unstable in our body and can be quickly destroyed by common proteases before they can successfully trigger an immune response.
Vaccines do not target specific domains.	By targeting incredibly specific regions, the use of peptides in vaccine development can modify the immune response.
Difficult to modify.	To enhance stability, immunogenicity, and solubility, peptides can be readily modified, for instance, by incorporating lipids, carbohydrates, phosphate, acetyl, and amide groups.

7.5 CONSIDERATIONS FOR PROTEIN AND PEPTIDE VACCINES

Proteins, mAbs, and peptides are utilized as part of the "biologics" approach to cancer treatment. In contrast to peptides, mAbs and large protein ligands can be toxic to the bone marrow and liver at certain concentrations, which can cause damage to these organs due to their non-selective uptake into the reticuloendothelial system. Additionally, their large size makes it challenging to deliver them to tumors. As a result, these biomolecules are primarily used for hematological malignancies, and they are delivered to arterial targets on the mucosal surface of the tumor. On the other hand, peptides offer several advantages, including their small size, ease of synthesis and modification, ability to penetrate tumors, and high biocompatibility. Chemical modifications, such as the incorporation of D-amino acids or cyclization, can prevent peptide degradation by proteolysis [26].

There are numerous peptide candidates being developed in preclinical and clinical settings. Peptides have been most frequently studied in clinical trials for cancer symptoms (18%) and metabolic disease symptoms (17%) since 2000. Peptides can be used in various ways to treat cancer, including as medications (such as angiogenesis inhibitors), hormones, and vaccinations. They can also be used as agents to target tumors and deliver radioactive substances and cytotoxic drugs (targeted chemotherapy and radiation therapy). Peptides can serve as diagnostic tools and biomarkers in cancer research due to their ability to bind to different receptors and participate in various metabolic pathways [27]. The concept of cancer vaccinations has evolved over the past decade, with clinical research focusing on delivering vaccines based on specific antigens to enhance anticancer immunity. This approach involves using vaccines made of peptides derived from the protein sequence of potential tumor tissue or specialized antigens. TAAs are produced by tumor cells and can be recognized by the host's immune system, specifically T cells. These TAAs can be administered intravenously to cancer patients with the aim of triggering a broad immune response that may lead to the elimination of cancer cells in various organ tissues. Active immunotherapy, commonly known as vaccination, involves inducing an effective, tumor-specific immune response in the host, which can ultimately result in tumor regression [28]. Any protein or peptide produced by a tumor cell with a mutated structure can serve as a tumor antigen. The mutation in the relevant gene leads to the production of such abnormal proteins. Numerous clinical trials have investigated the therapeutic potential of TAA peptide vaccination or active immunization in individuals with metastatic cancer.

CTL ($CD8^+$ T-cells or killer T cells) epitopes are the most prevalent TAAs [29]. Typically, 8 to 10 amino acids in length, peptide antigens have 2–3 major anchor residues that bind to the MHC class I molecule and 2–3 residues that connect to T-cell receptors (TCRs). Due to their focus on peptides produced by MHC class I molecules, CTLs are effective immune system effectors against tumor cells. The T-cell antigen receptor (TCR) on T cells can identify a small peptide located in the epitope pocket of an MHC molecule. The two subtypes of MHC are class I and class II, with class I molecules present on all nucleated cells and class II molecules found on specific antigen-presenting cells. MHC molecules are also known as human leukocyte antigens [30].

As previously mentioned, MHC class II molecules are expressed on antigen-presenting cells (APCs) that have the ability to phagocytose antigens or bind them to surface Abs. $CD4^+$ T cells can recognize antigens associated with these molecules.

Carcinoembryonic antigen (CEA) (breast, colon, gastric, pancreatic, and non-small-cell lung cancers (NSCLCs)), HER-2/neu immunodominant peptide (lung, breast, or ovarian cancer), prostate-specific membrane antigen (PSMA) (melanoma), and Mucin-1 are several prominent peptide vaccines that have undergone phase I, phase II, and phase III studies. GV-1001, an alternative vaccine manufactured from the catalytic component of telomerase reverse transcriptase, is currently under research. It is a promiscuous MHC class II peptide in injectable form. GV-1001 is now being used in phase II clinical trials to treat pancreatic cancer, NSCLC, and liver cancer [31].

7.6 CHALLENGES FACING ALL VACCINES

All immunizations intended to prime the immune system for encounters with cancer, pathogenic microorganisms, or autoimmunity mediators share some of these problems.

7.6.1 PICKING THE APPROPRIATE ANTIGEN

Effective vaccinations have historically utilized live, attenuated organisms. These vaccinations carry a moderate yet significant risk of activation, which may lead to illness or other unfavorable side effects, despite their success at the population level. Early CVs were developed using intact tumor cells that had undergone radiation therapy or other forms of inactivation, as there were no tumor antigens available [32]. This vaccination approach resulted in tumor rejection in mouse models and elicited immune responses specific to the tumor. These early vaccines either did not employ tumor tissue lines that had undergone multiple *in vivo* or *in vitro* passages and accumulated numerous mutations, or they did. Whole tumor-cell injections pose significant health risks, similar to vaccines based on whole viruses that raise concerns about reactivation and disease progression. The main concern is the potential development of autoimmunity. Immature dendritic cells (DCs) process dying cells and self-antigens in tissues without strong activation signals, such as those provided by pathogens, and the immune system does not respond to these antigens. To generate potent immunity, the tumor-cell vaccine requires chemicals that activate DCs. However, in the case of whole tumor cells, it is expected that activated DCs will not only prime immunity but also present tumor-specific antigens and other antigens (autoantigens) that are often subject to peripheral tolerance. Studies using animal models and clinical trials have accumulated evidence of autoimmune reactions occurring after vaccination, confirming the reality of this situation.

The primary advantage of immunotherapy over other forms of therapy is the specificity it offers. However, this advantage can be undermined when using whole cancer cells or complex mixtures of tumor-derived material. The immune system has the ability to identify tumor-expressed epitopes and target tumor cells while sparing healthy ones. Extensive efforts have been made in the past two decades to develop tumor antigens that can benefit from this specificity [33].

Numerous tumor antigens have been identified, and pre-clinical investigations have shown that cancer vaccinations based on these antigens can promote long-term memory and tumor-specific immunity without inducing autoimmune diseases. In fact, extensive research has demonstrated that vaccines containing epitopes derived from MAGE3, HER2/NEU17, mucin 1, or other MAGE gene family members, mammaglobin, or CEA are immunogenic without causing autoimmune illnesses.

Despite the availability of well-established tumor antigens, recent developments in the field of CVs have once again focused on using whole tumor cells or whole-cell lysates as antigens. These complex mixtures may be more immunogenic and elicit a stronger antitumor immune response compared to separate antigens found in mice carcinogen-induced tumors, as they are likely to include specific tumor antigens that are unique to each tumor. Studies conducted on mice transgenic for common tumor antigens have shown that these antigens can elicit an equally robust anticancer response and lead to tumor rejection [34].

Discontent with the clinic's current level of success is a concerning issue that may discourage the use of CVs based on recognized antigens. Before underestimating the potential of defined tumor antigen-based vaccines and reverting to undefined tumor mixtures that may lead to autoimmunity, it is crucial to remember that antigen-based vaccines have shown success in animal models, primarily in tumor prevention. Currently, these vaccines are only being tested for severe illnesses, often in cases where conventional therapy has failed. Therefore, they have not yet had the opportunity to replicate the same level of success in humans [35].

7.6.2 STIMULATING THE PROPER TYPE OF IMMUNOLOGICAL REACTION

Systemic immunity is designed to monitor the systemic condition known as metastatic cancer. However, many primary tumors originate in the mucosa, where the mucosal immune system first encounters them. More attention is being given to antigens, adjuvants, and delivery systems for vaccines that can effectively enhance mucosal and systemic immunity. The mucosal immune system has evolved to maintain a healthy balance between a rapid response to viruses, insensitivity to food or other environmental antigens, and non-pathogenic bacterial flora. Mucosal vaccines must carefully maintain this regulated balance while also enhancing the protective response [36].

There is currently a question of whether an immune response can effectively treat organ malignancies such as lung, liver, or brain tissue. Most research conducted with existing animal models, especially with transplantable tumors originating in subcutaneous (SC) sites, does not contribute new knowledge to this issue. Therapeutic CVs are expected to strengthen an already present, active, but immature immune response instead of eliciting entirely new immune responses. This provides further justification to consider whether a specific vaccination should be administered to promote mucosal immunity rather than systemic immunity [37]. When targeting a tumor with a mucosal origin, such as bladder cancer, colon cancer (CC), cervical cancer, squamous cell carcinoma of the head and neck (SCCHN), lung adenocarcinoma, or cervical cancer, a mucosal route of immunization may be more effective in boosting the response. Future development of CVs will require a deeper understanding of mucosal immunity [38].

7.6.3 ELICITATION OF LONG-TERM MEMORY

Some vaccines have the ability to stimulate immune memory, which is a crucial defense mechanism, while others do not. Understanding what immunological memory is and what is necessary for its development and maintenance has recently become a focus of research. The lack of distinct markers that can differentiate memory T cells from other T cells has been a major obstacle in this field. However, recent advancements in the use of chemokine receptors have successfully allowed the differentiation of memory cells and other functional T cell subsets. Additionally, indicators such as glycoproteins resembling mucins are starting to be reported. These markers will be valuable in determining how tumor antigens, adjuvants, and injection routes impact the immune response they elicit, not only in terms of complexity and intensity, but also in terms of the type of memory response they trigger.

There is a general consensus that a strong primary immune response is necessary to generate a significant number of memory cells. However, there is ongoing debate regarding the importance of antigens in this process, and it remains unclear what factors influence the lifespan of memory T cells [39]. These concerns are particularly relevant for therapeutic CVs. Vaccines based on specific antigens are expected to enhance immunity in the presence of these antigens, as cancer patients' immune systems are exposed to tumor antigens over an extended period of time. Prior to vaccination, the patient already had tumor-specific T cells, but it is unknown whether these cells are effector cells or a combination of effector and memory cells. According to several studies, memory cells are unique to cancers and can be identified there. However, it is uncertain what percentage of T cells in cancer patients are tumor-specific and how vaccination affects them, as distinguishing effector cells from memory cells and the persistence of tumor antigens pose challenges [40].

It is also unknown whether long-term memory would ever be recovered in conditions such as cancer. Some criteria are being developed in the context of chronic viral diseases and to a lesser extent in cancer, particularly the requirement for TH cell activation and innate immunity. Recent studies indicate that the development of T-cell memory occurs in stages. It begins with naive cells that develop into effector cells when the antigen is present, then effector memory cells when the antigen is restricted, and finally central memory cells when the antigen is completely eliminated. Even as a prophylactic measure, it is assumed that cancer immunizations would achieve this in healthy young adults. However, it is less clear how a therapeutic vaccination would trigger this entire differentiation process in the context of chronic antigen [41].

7.6.4 ADDITIONAL DIFFICULTIES FOR CANCER VACCINES

7.6.4.1 IMMUNE SYSTEM AGING

A significant amount of time has passed since the thymus ceased production of naive T cells, and CVs are currently undergoing testing on elderly cancer patients aged 65 to 80 years old. Consequently, in order for an effector-cell population to expand in response to a vaccine, at least one or more memory T cells within the patient's T-cell repertoire must be capable of recognizing the vaccine antigen. The "best match" T cells that would have been selected from a diverse pool of naive clones earlier in life may or may not be present in the T cells that respond to the vaccine. Animal models demonstrate that young mice exhibit stronger primary reactions compared to old mice, which is empirically true. As individuals age, their ability to generate the initial response and convert it into memory is impaired. This is attributed to alterations in immune system function caused by aging [42]. Preclinical research utilizing mouse models and clinical trials for CVs currently exhibit notable disparities. Most studies do not employ aged mice, while those that do report challenges in inducing effective antitumor immune responses, as well as an age-related increase in cancer susceptibility due to shifting patterns of T-cell subsets. Given that these vaccinations will primarily be administered to older patients, it is crucial to place greater emphasis on designing cancer therapeutic vaccines that can at least partially address age-related concerns. For example, it has been demonstrated that activating the co-stimulatory molecule 4-1BB (CD137) enhances T-cell responses in aged mice. It is possible that inactivating harmful regulators such as cytotoxic T lymphocyte antigen 4 (CTLA4) or activating additional co-stimulatory molecules that have not yet been tested will have a similar effect. Additionally, while a large number of adjuvants may be beneficial in young mice, a small number may enhance immune responses in the elderly. In older rats, CpGDNA appears to be more effective in boosting cellular and humoral immunity [43].

The decline in immune function associated with aging suggests that children with cancer may be more receptive to therapeutic cancer vaccinations compared to adults. However, such experiments have been conducted infrequently. In contrast to clinical research on older patients, the results of a DC-based vaccine trial on children aged 3 to 17 with recurrent neuroblastomas, sarcomas, and kidney malignancies are unfortunately not very promising [44]. This indicates that immunosuppressive standard therapy and/or prior treatment negatively impact the immune system even in young patients. It implies that effective vaccination methods would require vaccination not only in the early stages of the disease but also in the absence of immunosuppressive conventional medication [45].

7.6.4.2 IMMUNE EVASION AND TUMOR-INDUCED IMMUNOSUPPRESSION

When a tumor is discovered, the immune system and the tumor have already engaged in several interactions. The immune system might not have recognized a tumor if it was developing slowly and not significantly destroying the nearby healthy tissue. Tumor cells undergo additional mutations over that time, some of which promote proliferation and invasion [46]. In addition to defense mechanisms like wound healing and clotting, because DCs have been activated, the adaptive immune system is notified when the tumor expands and starts to destroy tissue. These cells “ferry” tumor and tissue fragments to the lymph nodes that drain where T cells may expose it. Cellular and humoral responses in cancer patients that are unique to the tumor show that the immune system has recognized the tumor. The immune system has tried to eliminate the tumor when the tumor cells stop expressing the specific tumor antigens or MHC molecules. However, persistent tumor growth suggests that the immune system may have eventually been unable to identify the tumors. The term “cancer immunoediting” refers to this immunosurveillance procedure that modifies the tumor but does not completely eradicate it [47]. Tumors can affect the immune system in a variety of ways, and numerous immunological effector pathways have been found to have functional flaws. Cancer patients experience inhibition of DC development and activity. T-cell activation and function are also significantly abnormal, as was originally observed in tumor-bearing mice and later discovered in humans with a variety of tumor forms. These effects might be mediated by the cytokines generated by tumor cells, such as transforming growth factor (TGF), interleukin 10 (IL-10), and other cytokines, as well as other, less well-known soluble chemicals or cell-surface molecules.

Furthermore, the inappropriate activation of innate immunity can hinder adaptive antitumor immunity. Studies have demonstrated that the activation of macrophages and polymorphonuclear cells induced by tumors leads to oxidative stress in cancer patients, resulting in a significant reduction in T-cell function. Additionally, it has been observed that the activation of NKT cells, which can produce high levels of IL-13, may suppress the immune response to cancer [48].

To facilitate therapeutic intervention, efforts are underway to gain a better understanding of these immunosuppressive pathways at the molecular level. There are encouraging reports that, in a limited number of cases, immunization has been able to partially reverse these issues. Further research is needed to comprehend the role of regulatory T cells (TRegs) in cancer immunity. A subset of $CD4^{+}CD25^{+}$ T cells has shown the ability to suppress autoimmunity, suggesting that these lymphocytes may become highly potent immune cells in response to increased presentation of autoantigens during tumor progression. Limited animal studies suggest that removing these cells may yield some benefits [49]. Prior to attempting to modulate a patient's immune system using therapeutic CVs, it is crucial to thoroughly understand the patient's immune system. Since many immunosuppressive processes are shared among various tumor types, it may be more effective to develop a treatment plan that includes lifting immunosuppression prior to therapeutic vaccination [50].

7.7 ADJUVANTS AND FORMULATIONS FOR PROTEIN AND PEPTIDE VACCINES

Whether they are composed of complete cells, certain proteins, or peptides, all CVs must contain adjuvants. Currently, only two adjuvants, squalene-oil-water emulsion (MF59) and aluminum-based salts (alum), are authorized for use in clinical trials. However, numerous additional compounds that enhance vaccination immunogenicity have been studied and proven effective in both animal models and humans [51]. The mechanisms underlying the adjuvant effect of many novel adjuvants are better understood as they are recognized molecules. Adjuvants can activate T cells, NK-cells, and other innate system cells to release cytokines, in order to promote the survival of antigen-specific T cells. Interleukin-2 (IL-2), granulocyte-macrophage colony-stimulating factor (GM-CSF), IL-12, IL-4, and several other cytokines have been used as adjuvants in CVs. Additionally, bacterial products have long been used as effective adjuvants [52]. Monophosphoryl lipid A (MPL) from Salmonella and lipopolysaccharide (LPS) from gram-negative bacteria are the two most well-known examples. More recently, it has been discovered that bacterial DNA has significant immunostimulatory activity due to the presence of unmethylated CpG dinucleotides. These and other bacterial chemicals are recognized by various unique receptors produced by DCs, macrophages, and possibly NK cells and other innate system cells. Consequently, these cells develop, become activated, and start producing pro-inflammatory cytokines. Many of these receptors, which belong to the Toll-like receptor (TLR) family, can recognize invading pathogens and are present within or on the surface of certain cells. Bacterial compounds have attracted the attention of tumor immunologists due to their exceptional ability to activate cytotoxic T lymphocytes (CTLs) [53].

The discovery that different antigen-processing pathways control how antigenic peptides are presented by either MHC class I molecules to $CD8^+$ T cells (endogenous pathway) or MHC class II molecules to $CD4^+$ T cells led to the development of a class of adjuvants that could deliver antigens to the desired processing pathway (exogenous pathway). Vaccines that contain all antigen types besides nucleic acids primarily deliver antigens to antigen-presenting cells (APCs) via the exogenous pathway. This then promotes the production of Abs and the activation of $CD4^+$ T cells. Before the cytoplasmic antigen is transported to the endoplasmic reticulum (ER) and attached to MHC class I molecules, it must first be broken down by the proteasome. Adjuvants, such as poly(D, L-lactic-co-glycolic acid) (PLGA) microspheres, virus-like particles (VLPs), and immunostimulatory complexes (ISCOMs), which are micelles comprised of cholesterol and Quil A, are small particles that efficiently transmit antigens to the cytoplasm. To successfully transfer antigen to APCs for presentation by both MHC class I and class II molecules, adjuvants in vaccination formulations must be particulate. Heat-shock proteins (HSPs) might also be one of these adjuvants. Antigens are successfully delivered to the MHC class I pathway, activating APCs [54].

7.7.1 MICRORESERVOIR DELIVERY SYSTEMS

Microreservoir (microparticulate) carriers can often serve as a viable alternative to soluble polymeric carriers. Liposomes, micelles, polymer microparticles, and cell ghosts are examples of such systems [55]. Compared to "direct" molecular conjugates, the use of these carriers results in a significantly higher ratio of active moiety to carrier. Moreover, due to the complete separation of drug molecules from their surroundings by the carrier wall during parenteral administration, they provide a higher level of protection against enzymatic degradation and other harmful chemicals. Another advantage of these carriers is that a single carrier can transport multiple drug species. However, it should be noted that not all microparticles can be eliminated through renal filtration due to their size. The main drawback of microreservoir carriers is their tendency to be absorbed by RES cells, particularly in the liver and spleen [56].

7.7.2 MICELLES AND LIPOSOMES

The liposomes are the most extensively investigated and well-suited particulate drug carriers for encapsulating peptides and proteins. Liposomes consist of concentric spherical phospholipid bilayers that enclose an aqueous space, forming vesicles. These particles have minimal toxic or antigenic reactions and are completely biocompatible and biologically inactive. Peptides and proteins can be encapsulated within the aqueous compartment of liposomes. Various methods for liposome preparation only require adjustments that do not compromise the integrity of the peptide or protein. Conventional liposomes are rapidly eliminated by cells of the RES, similar to other microparticulate delivery techniques [57]. Efforts have been made to prolong the circulation lifespan of liposomes, allowing the development of liposomes that can deliver medication to targets other than the RES. This has been achieved by incorporating flexible hydrophilic polymers with modified surfaces, typically PEG, onto long-circulating liposomes. This prevents plasma proteins from adhering to the liposome surface, thereby hindering recognition and absorption by the RES. Similar to macromolecules, liposomes can accumulate in tumors of various origins due to the EPR effect.

Modifying long-circulating liposomes to deliver peptide- and protein-based medications to the tumor is a simple process. However, using different small-sized delivery methods, such as micelles with added medication or conjugates of peptides and proteins with polymers, may prove to be more effective. Among the various pharmacological micelles, polymeric micelles made from amphiphilic PEG-phospholipid conjugates are particularly interesting due to their durability [58]. Unlike liposomes, these tiny particles lack internal aqueous space. By chemically attaching hydrophobic "anchors" to the surface of these particles, pharmaceutical substances like peptides or proteins can be incorporated into micelles [59].

7.8 WHO RECOMMENDED DIRECTIVES FOR THE PRODUCTION AND QUALITY CONTROL OF SYNTHETIC PEPTIDE VACCINES?

To ensure the consistency of safety and efficacy of peptide vaccines used in humans, this guideline serves as a foundation for their development, manufacturing, and control [60]. These guidelines include:

- Control of starting materials, including background data on the synthesis of the peptide of interest;
- Control of the manufacturing process;
- Control of the final product.

7.8.1 GUIDELINE FOR CONTROL OF PRODUCTION

7.8.1.1 PEPTIDE SYNTHESIS AND CHARACTERIZATION

The peptide synthesis process should be fully described, including any in-process controls. The description should encompass the sources and specifics of the raw materials used, the methods employed, the coupling and deprotection conditions, the requirements for progressing to the next step of the synthesis, and any integrated groups (such as glycosyl or lipid). If the synthesis is conducted in solution, a flowchart and details of the intermediate peptides should be provided.

Evidence should be offered to support the following:

- The correct amino acids have been added in the required order, and the appropriate amount and kind of modifications (such as the incorporation of lipid or glycosyl moieties) have been made, resulting in the primary peptide sequence having the intended structure.
- The primary and, whenever feasible, minor impurities of the peptide monomer have been detected and characterized.
- Over a number of subsequent batches, the peptide maintains a consistent level of quality.

7.8.1.2 CONJUGATES

The following requirements should be met by a conjugated product:

- Conjugation proof needs to be established;
- Reduced side effects (such as carrier cross-linking) should be the goal;
- It is important to gather proof that conjugation does not change the antigenic sequences;
- Residual reagent and by-product levels ought to be kept to a minimum.

7.8.1.3 CARRIERS

A carrier may have peptides conjugated to it or integrated into it. Since an immune reaction to the carrier may outweigh responses to the peptide, care should be given when selecting a carrier. It is best to avoid carriers that could trigger autoimmunity as well as those that could cause pre-existing hypersensitivity in the target population.

7.8.1.4 POLYMERIZED, CYCLIZED, OR CARRIER-SYNTHESIZED PEPTIDES

Particular attention should be paid to the following:

- The relative molecular mass distribution of the product remains consistent from batch to batch.
- The relative ratios of various peptides are consistent when appropriate.
- The nature of the product(s) is clearly identified with the aid of a suitable selection of complementary analytical procedures.
- The nature of the product(s) is clearly identified with the aid of a suitable selection of complementary analytical procedures.

7.8.1.5 PRESERVATIVES

If a preservative is added, the content of the preservative should be determined and the amount used should be shown to have no deleterious effect on individual vaccine components, nor cause any unexpected adverse reactions in humans. Preservatives should not be added to single-dose preparations.

7.8.2 GUIDELINE FOR THE FORMULATED FINAL PRODUCT

7.8.2.1 INTERACTION WITH ADJUVANTS

It is important to specify how the various vaccine components interact with the adjuvant and demonstrate that this interaction is consistent across different batches. For instance, the extent to which the different species in a polymeric preparation bind to the adjuvant may be influenced by factors such as molecular mass, hydrophobicity, or pH. Consequently, the formulation process may unintentionally cause fractionation of the preparation.

7.8.2.2 STABILITY STUDIES

Stability studies that are adequate are a crucial component of vaccine development. The outcomes are used to determine the product's maximum shelf life under ideal storage circumstances. Real-time stability tests should be conducted for each component for which an effectiveness claim is made in order to verify the product's immunogenicity and lack of toxicity up until its expiration date. Accelerated stability studies may offer preliminary proof that the product is stable, but they cannot replace real-time tests for licensing purposes.

7.8.2.3 POTENCY

The functional activity of the vaccine or its mode of action in humans need not necessarily be reflected in the potency assay. Where appropriate, a suitable test of vaccine efficacy should be included. The major goal of the potency assay is to demonstrate batch-to-batch consistency using techniques that partially rely on biological activity. Therefore, it is important to consider an appropriate *in vitro* or *in vivo* assay for immunogenicity or antigenicity. A comparison with a reference material should be conducted in such an assay, and the results should be statistically validated. The specifics of these assays must be determined on a case-by-case basis due to potential variations in the mechanism of action of vaccines.

7.9 ANTIGENICITY AND IMMUNOGENICITY OF CANCER

The ability of a tumor to elicit an immune response that can halt its growth is referred to as tumor immunogenicity. However, there is still much to be learned about the functioning of this intricate process. The concept that the immune system may impede the spread of cancer was initially proposed by Paul Ehrlich in 1909. Burnet later formulated a theory to account for this, suggesting that tumors release neo-antigens that trigger immune responses capable of eradicating new cancers [61]. Lewis Thomas adopted an evolutionary perspective, postulating that organisms possess highly evolved defense mechanisms similar to graft rejection processes that prevent the development of cancer. The concepts of cancer immunosurveillance and immunoediting were inspired by these hypotheses. The immune system distinguishes between self-antigens (antigens expressed by normal tissue) and non-self-antigens to eliminate malignancies that have evaded immunological recognition [62]. Initially, it was believed that this mechanism could explain the immunogenicity of tumors. However, the emergence of cancers that are not eliminated by the host immune system despite the generation of non-self-antigens cannot be explained by the proposed method, despite significant evidence to the contrary. Furthermore, it cannot account for the substantial production of foreign antigens by gut microorganisms that coexist harmoniously.

Theorizing that immunogenicity depends on the cancer cell presenting distinct antigen epitopes that are different from those to which the immune receptors are typically exposed, rather than whether an antigen is identified as non-self, Pradeu & Carosella sought to explain these discrepancies. They described how the epitope expression changed "abruptly." However, Pradeu & Carosella have convincingly shown that immunogenicity alone is insufficient to explain immunological tolerance caused by a change in epitope presentation. Depending on how the new antigen is presented in relation to the other frequently exposed antigens, it can lead to either immunity or tolerance. Immune tolerance occurs when the new antigen is introduced in small doses and with gradual alterations [63]. Therefore, immunogenicity requires tumor cells to express sufficient levels of antigens that are exclusively expressed by the tumor cells, as opposed to the normal cells from which the tumor formed. Additionally, the tumor cells must successfully convey antigens in a manner that results in immune activation, rather than immunological tolerance. Recent experiments and observations provide some evidence supporting this conclusion.

Antigenicity, or antigenic reactivity, refers to the ability of a virus to bind to specific antibody molecules. Enveloped viruses do not possess antigenicity through the antigenic sites or B-cell epitopes of the capsid proteins, which are recognized by the antibody binding sites. The continuity or discontinuity of a protein epitope is determined by the proximity of the amino acids that make up the polypeptide chain. The antigenic reactivity of most epitopes relies on the native conformation of the protein, as they consist of surface residues brought together by the folding of the peptide chain. Viral capsid quaternary structures, also known as neotype epitopes, are formed by viral capsids [64]. The development of neotype epitopes involves inter-subunit interactions that induce conformational changes in the capsid proteins, as well as the arrangement of residues from adjacent subunits, resulting in a specific epitope [65].

The host's response to biomaterials is influenced by the antigenicity and immunogenicity of the immune system, which serve different functions. The ability of a substance to attach to or interact with the byproducts of the body's ultimate cell-mediated reaction, such as B-cell or TCRs, is known as antigenicity. These antigens have structural components called antigenic determinants, or epitopes, which bind with B-cell receptors, also known as Abs or immunoglobulins. When a major histocompatibility complex (MHC) molecule combines with the linear amino acid sequences (epitopes) of a protein antigen, TCRs are able to identify them [66]. The body becomes more sensitive to foreign antigens when an "immunogen" initiates an immune response by activating the innate immune system before the adaptive (acquired) immune system. Biomaterials, particularly allogenic or xenogenic materials originating from nature, can function as immunogens. All antigenic materials are also immunogenic, while the opposite is not true. As a result, there are some molecules that are antigenic but not immunogenic by themselves. These molecules are known as "haptens," such as metal ions. They play a crucial role in postoperative metal hypersensitivity. Therefore, both immunogenicity and antigenicity have an impact on how the host immune system reacts to both natural and artificial biomaterials [67].

7.10 CURRENT DEVELOPMENTS

7.10.1 CELL-BASED CANCER VACCINES

There are currently ongoing clinical trials utilizing GVAX in combination with CRS-207, PD-1, or urelumab (NCT03190265, NCT02451982). A successful approach to cancer treatment involves the blockade of immune checkpoints along with cancer vaccinations. When PD-1/CTLA-4 was added to the GVAX vaccine, both the $CD8^+$/Treg ratios and the number of IFN-$^+$TNF-$^+$ $CD8^+$ tumor-infiltrating cells significantly increased. Additionally, a gene modification that has the potential to transform malignant cells into cells resembling cancer stem cells (CSCs) was incorporated into the comprehensive cancer cell vaccination. It has been demonstrated that AGI-101H, a gene-modified vaccine that mimics melanoma stem cells, has shown success in the treatment of melanoma patients. AGI-101H has shown promising results in two phase II studies conducted on melanoma patients, leading to increased survival rates [68].

7.10.2 VIRUS-BASED CANCER VACCINES

In preclinical and clinical trials, immune-stimulating adenovirus-based vaccinations or TAAs have been utilized to demonstrate significant activation of anti-tumor immunity. The VRP-HER2 vaccine, based on a viral vector expressing HER2, has shown promising results in preclinical and clinical testing. The $hHER2^+$ breast cancer mouse model exhibited improved tumor growth after receiving the VRP-HER2 vaccination. VRP-HER2 has also shown efficacy in breast cancer clinical trials. The patients' HER2-specific immune responses are closely associated with the length of their progression-free intervals. The combination of VRP-HER2 and anti-PD-1 therapy was explored to enhance overall efficacy, demonstrating the feasibility of the idea. Furthermore, the trial (NCT03632941) is still ongoing [69]. In the treatment of non-muscle invasive bladder cancer, nadofaragene firadenovec, a non-replicating adenovirus vector vaccine, has shown promising results. According to the outcomes of a clinical trial examining the drug's efficacy in treating BCG-resistant non-muscle invasive bladder cancer, 53.4% of patients achieved a complete response after the initial dose of Nadofaragene firadenovec [70].

Furthermore, it is important to consider BT-001, an oncolytic viral vaccine with the potential to express anti-CTLA4 Abs and GM-CSF. A previous clinical trial (NCT04725331) was conducted to evaluate its efficacy. In Phase II research, the oncolytic virus vaccination T-VEC was administered. Patients with stage IIIB-IVM1a disease exhibited objective/complete response rates of 32%/18%, compared to 28%/14% (112 participants) in the general population [71].

7.10.3 PEPTIDE-BASED CANCER VACCINES

The polypeptide IMU-131 and the diphtheria toxin are linked through B cell epitopes in the extracellular domain of HER2. IMU-131 has been shown to generate HER2-specific Abs and cellular responses in a published phase I clinical study. NeuVax, another peptide vaccine targeting HER2, is one of the CVs that has advanced to phase III clinical trials. Recent findings from a Phase III clinical study indicate that NeuVax alone did not have a noticeable impact on breast cancer. Furthermore, two phase II clinical trials (NCT02297898 and NCT01570038) are currently investigating the combination of NeuVax and Trastuzumab for the treatment of HER2-positive breast cancer [72].

Boston Biomedical has developed a novel cancer polypeptide vaccine called DSP-788. Peptides in DSP-7888 (WT1) activate helper T cells and CTLs with specificity for the Wilm tumor gene 1. Since WT1 is expressed in various hematological and solid malignancies, it is likely that DSP-7888 can target cancer cells expressing WT1. In individuals with recurrent or advanced malignancies, DSP-7888 has demonstrated good tolerability and no dose-limiting effects. Three phase I/II trials have assessed the safety and efficacy of DSP-7888 as a monotherapy, but no clinical outcomes have been reported yet (NCT02436252, NCT02750891, NCT02498665). Additionally, DSP-7888 is being evaluated in a phase II clinical study for individuals with acute leukemia (NCT04747002) and in a phase III clinical trial for individuals with recurrent glioblastoma in combination with Bevacizumab (NCT03149003) [73].

SurVaxM is another effective peptide vaccine for glioblastoma, significantly prolonging the survival time of patients with this condition. In phase II clinical trials, SurVaxM is being evaluated in combination with Pembrolizumab for the treatment of glioblastoma (NCT04013672) [74].

7.10.4 DNA VACCINE

Cervical cancer has been the focus of research on DNA vaccines. Currently, two Phase III clinical trials (NCT03185013, NCT03721978) are underway to evaluate the safety and effectiveness of VGX-3100, a DNA vaccine targeting HPV. The INOVIO Company has released positive findings from the first Phase III study of VGX-3100 for cervical precancerous lesions (NCT03185013). VGX-3100, along with the COVID-19 vaccine "ZycoV-D," has the potential to become the first DNA medication available for sale [75]. Another DNA vaccine for cervical cancer, GX-188E, combines multiple epitopes and specifically targets and activates DCs. In a Phase II trial of GX-188E for cervical cancer, 67% of the 52 patients showed a reduction in lesions after 36 weeks of treatment. Clinical trials of the combination treatment infused with GX-188E have shown a promising future [76].

7.10.5 MRNA VACCINES

TriMix mRNA has demonstrated favorable immunogenicity and tolerability in various clinical trials. Two recently published Phase II clinical trials have shown that TriMix and TAA mRNA vaccines induce robust $CD8^+$ T cell responses in patients with stage III or IV melanoma, resulting in significant tumor response rates. mRNA-252 serves as another example of an immuno-stimulant mRNA vaccination [77]. mRNA-2752, a lipid nanoparticle (LNP) developed by Moderna for lymphoma treatment, contains human OX40L, IL-23, and IL-36 mRNAs. It is currently being evaluated in a clinical trial (NCT03739931). Several TAA-encoding mRNA vaccine combinations have also been employed in various clinical trials for the treatment of metastatic melanoma. For instance, the mRNA vaccination BNT111 includes four TAAs (NY-ESO-1, MAGE-A3, tyrosinase, and TPTE). BNT111 demonstrates the broad application of non-mutated common tumor antigen targets and has shown success in treating melanoma patients with CPI. Neoantigen-encoding mRNA vaccines are the most widely used form of personalized vaccination. Customized mRNA CVs have proven to be safe and well-tolerated in clinical trials [78].

Clinical trials of personalized vaccines developed by BioNTech and Moderna have demonstrated promising anti-tumor benefits. For example, Moderna has developed mRNA-4157, a personalized vaccine that contains up to 34 neoantigens encoded in LNPs. In patients with unresectable solid tumors, mRNA-4157, either alone or in combination with Pembrolizumab, has shown strong anti-tumor efficacy. Clinical trials for mRNA-4157 are currently in phase II.

Using the mRNA-tailored platform, BioNTech has created the BNT122 neoantigen CV. BNT122 encodes up to 20 patient-specific neoantigens. In the prior Phase 1a/1b trial, BNT122 was administered to patients with solid tumor cells either alone or in combination with the anti-PD-L1 antibody Atezolizumab (NCT03289962). Experimental research has shown that BNT122 induces distinct clinical responses and T-cell responses that are specific to neoantigens. BNT122 is now undergoing a Phase II clinical trial for the treatment of colorectal cancer (CRC) (NCT04486378) [79].

7.11 CONCLUSION

The process of developing vaccines is difficult and complex. In comparison to conventional immunizations, protein and peptide vaccines offer a number of benefits. This method is safe and affordable compared to conventional vaccinations composed of inactivated toxins, dead or attenuated microorganisms, and recombinant components. Due to their simplicity and ease of synthesis, protein and peptide vaccines can be produced for comparatively little money. Furthermore, peptide vaccines do not include LPSs, lipids, and toxin components that are unnecessary but have high reactogenicity to the host. Additionally, peptide vaccines can combine epitopes from several antigens to create a single antigenic formulation that includes both T cells and B cells.Peptides provide a number of benefits, but when administered by themselves, they are often not very immunogenic. To address this issue, the next generation of adjuvants is needed. Future delivery techniques for peptide vaccines may focus on the combination of adjuvants in formulations, techniques to improve antigen uptake, innovative adjuvants to promote innate immunity, and adjuvants to generate specific adaptive immunological responses. However, there is still significant concern about the safety of adjuvants. The use of adjuvants and particle peptide vaccine delivery methods has safety concerns that can be resolved, potentially allowing peptide vaccine technology to advance into the next generation of subunit vaccinations.

As medicine progresses into a new era, illness management strategies are being designed at the level of genetic and protein data, which form the foundation of all biology. In this context, protein treatments are gaining increasing importance. The development of therapeutic CVs holds great promise, provided that careful selection of the best antigens and antigen delivery systems is made, along with the application of appropriate co-treatment. This selection should be informed by knowledge of the immuno-editing that has occurred and the immune-suppressive mechanisms that are specific to each cancer patient. Peptide vaccines, with their known structure, lower cost, ease of production and manipulation, and lack of batch-to-batch variance due to their synthetic nature, offer a favorable alternative.

KEYWORDS

- **bursa-derived cells**
- **cancer vaccines**
- **interferons**
- **monophosphoryl lipid**
- **protein-and peptide-based vaccines**
- **thymocyte cells**
- **tumor associated antigens**
- **vaccine delivery**

REFERENCES

1. Dunn, G. P., Old, L. J., & Schreiber, R. D., (2004). The immunobiology of cancer immunosurveillance and immunoediting. *Immunity, 21*(2), 137–148.
2. Hanahan, D., (2022). Hallmarks of cancer: New dimensions. *Cancer Discovery, 12*(1), 31–46.
3. Bukowski, K., Kciuk, M., & Kontek, R., (2020). Mechanisms of multidrug resistance in cancer chemotherapy. *International Journal of Molecular Sciences, 21*(9), 3233.
4. Wang, J. J., Lei, K. F., & Han, F. J., (2018). Tumor microenvironment: Recent advances in various cancer treatments. *Eur. Rev. Med. Pharmacol. Sci., 22*(12), 3855–3864.
5. Bullement, A., Cranmer, H. L., & Shields, G. E., (2019). A review of recent decision-analytic models used to evaluate the economic value of cancer treatments. *Applied Health Economics and Health Policy, 17*(6), 771–780.
6. Brisse, M., Vrba, S. M., Kirk, N., Liang, Y., & Ly, H., (2020). Emerging concepts and technologies in vaccine development. *Frontiers in Immunology, 11*, 583077.
7. Pollard, A. J., & Bijker, E. M., (2021). A guide to vaccinology: From basic principles to new developments. *Nature Reviews Immunology, 21*(2), 83–100.
8. Demaria, O., Cornen, S., Daëron, M., Morel, Y., Medzhitov, R., & Vivier, E., (2019). Harnessing innate immunity in cancer therapy. *Nature, 574*(7776), 45–56.
9. Carty, M., Guy, C., & Bowie, A. G., (2021). Detection of viral infections by innate immunity. *Biochemical Pharmacology, 183*, 114316.
10. Mantovani, A., & Netea, M. G., (2020). Trained innate immunity, epigenetics, and COVID-19. *New England Journal of Medicine., 383*(11), 1078–1080.
11. McGhee, J. R., Mestecky, J., Dertzbaugh, M. T., Eldridge, J. H., Hirasawa, M., & Kiyono, H., (1992). The mucosal immune system: From fundamental concepts to vaccine development. *Vaccine, 10*(2), 75–88.
12. Bonilla, F. A., & Oettgen, H. C., (2010). Adaptive immunity. *Journal of Allergy and Clinical Immunology, 125*(2), S33–40.
13. Moser, M., & Leo, O., (2010). Key concepts in immunology. *Vaccine, 28*, C2–13.
14. Smith, B. T., (2008). *Concepts in Immunology and Immunotherapeutics*. ASHP.
15. Lopes, A., Vandermeulen, G., & Préat, V., (2019). Cancer DNA vaccines: Current preclinical and clinical developments and future perspectives. *Journal of Experimental & Clinical Cancer Research., 38*(1), 1–24.
16. Muyldermans, S., (2021). A guide to: Generation and design of nanobodies. *The FEBS Journal, 288*(7), 2084–2102.
17. Domínguez-Andrés, J., Joosten, L. A., & Netea, M. G., (2019). Induction of innate immune memory: The role of cellular metabolism. *Current Opinion in Immunology, 56*, 10–16.
18. Ahrends, T., & Borst, J., (2018). The opposing roles of CD 4+ T cells in anti-tumor immunity. *Immunology, 154*(4), 582–592.
19. Patel, C. H., Leone, R. D., Horton, M. R., & Powell, J. D., (2019). Targeting metabolism to regulate immune responses in autoimmunity and cancer. *Nature Reviews Drug Discovery, 18*(9), 669–688.
20. Trabolsi, A., Arumov, A., & Schatz, J. H., (2019). T cell–activating bispecific antibodies in cancer therapy. *The Journal of Immunology, 203*(3), 585–592.

21. Volz, A., & Sutter, G., (2017). Modified vaccinia virus Ankara: History, value in basic research, and current perspectives for vaccine development. *Advances in Virus Research, 97*, 187–243.
22. Liu, M. A., (2011). DNA vaccines: An historical perspective and view to the future. *Immunological Reviews, 239*(1), 62–84.
23. Arnon, R., & Van, R. M. H., (1992). Structural basis of antigenic specificity and design of new vaccines. *The FASEB Journal, 6*(14), 3265–3274.
24. Kumai, T., Kobayashi, H., Harabuchi, Y., & Celis, E., (2017). Peptide vaccines in cancer—Old concept revisited. *Current Opinion in Immunology, 45*, 1–7.
25. D'Amico, C., Fontana, F., Cheng, R., & Santos, H. A., (2021). Development of vaccine formulations: Past, present, and future. *Drug Delivery and Translational Research, 11*(2), 353–372.
26. Negahdaripour, M., Nezafat, N., Eslami, M., Ghoshoon, M. B., Shoolian, E., Najafipour, S., Morowvat, M. H., et al., (2018). Structural vaccinology considerations for in silico designing of a multi-epitope vaccine. *Infection, Genetics and Evolution, 58*, 96–109.
27. Thundimadathil, J., (2012). Cancer treatment using peptides: Current therapies and future prospects. *Journal of Amino Acids, 2012.*
28. Cerezo, D., Pena, M. J., Mijares, M., Martínez, G., Blanca, I., De Sanctis, J. B., (2015). Peptide vaccines for cancer therapy. *Recent Patents on Inflammation & Allergy Drug Discovery, 9*(1), 38–45.
29. Nagorsen, D., Scheibenbogen, C., Marincola, F. M., Letsch, A., & Keilholz, U., (2003). Natural T cell immunity against cancer. *Clinical Cancer Research, 9*(12), 4296–4303.
30. Davis, M. M., & Bjorkman, P. J., (1988). T-cell antigen receptor genes and T-cell recognition. *Nature, 334*(6181), 395–402.
31. Ma, M., Liu, J., Jin, S., & Wang, L., (2020). Development of tumor peptide vaccines: From universalization to personalization. *Scandinavian Journal of Immunology, 91*(6), e12875.
32. Ribas, A., Butterfield, L. H., Glaspy, J. A., & Economou, J. S., (2003). Current developments in cancer vaccines and cellular immunotherapy. *Journal of Clinical Oncology, 21*(12), 2415–2432.
33. Gkountidi, A. O., Garnier, L., Dubrot, J., Angelillo, J., Harlé, G., Brighouse, D., Wrobel, L. J., et al., (2021). MHC class II antigen presentation by lymphatic endothelial cells in tumors promotes intratumoral regulatory T cell–suppressive functions. *Cancer Immunology Research, 9*(7), 748–764.
34. Zhao, X., Zhao, R., & Nie, G., (2022). Nanocarriers based on bacterial membrane materials for cancer vaccine delivery. *Nature Protocols, 17*(10), 2240–2274.
35. Pearson, A. D., Rossig, C., Mackall, C., Shah, N. N., Baruchel, A., Reaman, G., Ricafort, R., et al., (2022). Paediatric strategy forum for medicinal product development of chimeric antigen receptor T-cells in children and adolescents with cancer: ACCELERATE in collaboration with the European Medicines Agency with participation of the Food and Drug Administration. *European Journal of Cancer, 160*, 112–133.
36. Hiam-Galvez, K. J., Allen, B. M., & Spitzer, M. H., (2021). Systemic immunity in cancer. *Nature Reviews Cancer, 21*(6), 345–359.
37. Runge, S., & Rosshart, S. P., (2021). The mammalian metaorganism: A holistic view on how microbes of all kingdoms and niches shape local and systemic immunity. *Frontiers in Immunology, 12*, 2628.

38. Ichim, C. V., (2005). Revisiting immunosurveillance and immunostimulation: Implications for cancer immunotherapy. *Journal of Translational Medicine, 3*(1), 1–3.
39. Sears, A. K., Perez, S. A., Clifton, G. T., Benavides, L. C., Gates, J. D., Clive, K. S., Holmes, J. P., et al., (2011). AE37: A novel T-cell-eliciting vaccine for breast cancer. *Expert Opinion on Biological Therapy, 11*(11), 1543–1550.
40. Palucka, K., Ueno, H., Fay, J., & Banchereau, J., (2011). Dendritic cells and immunity against cancer. *Journal of Internal Medicine, 269*(1), 64–73.
41. Cohen-Armon, M., Visochek, L., Katzoff, A., Levitan, D., Susswein, A. J., Klein, R., Valbrun, M., & Schwartz, J. H., (2004). Long-term memory requires polyADP-ribosylation. *Science, 304*(5678), 1820–1822.
42. Dorshkind, K., Montecino-Rodriguez, E., & Signer, R. A., (2009). The ageing immune system: Is it ever too old to become young again? *Nature Reviews Immunology, 9*(1), 57–62.
43. Davis, H. L., Weeranta, R., Waldschmidt, T. J., Tygrett, L., Schorr, J., & Krieg, A. M., (1998). CpG DNA is a potent enhancer of specific immunity in mice immunized with recombinant hepatitis B surface antigen. *The Journal of Immunology, 160*(2), 870–876.
44. Effros, R. B., (1999). Ageing and the immune system. In: *Novartis Foundation Symposium* (pp. 130–145). Chichester; New York; John Wiley.
45. Valiathan, R., Ashman, M., & Asthana, D., (2016). Effects of ageing on the immune system: Infants to elderly. *Scandinavian Journal of Immunology, 83*(4), 255–266.
46. Cervantes-Villagrana, R. D., Albores-García, D., Cervantes-Villagrana, A. R., & García-Acevez, S. J., (2020). Tumor-induced neurogenesis and immune evasion as targets of innovative anti-cancer therapies. *Signal Transduction and Targeted Therapy, 5*(1), 1–23.
47. Holmström, M. O., Cordua, S., Skov, V., Kjær, L., Pallisgaard, N., Ellervik, C., Hasselbalch, H. C., & Andersen, M. H., (2020). Evidence of immune elimination, immuno-editing, and immune escape in patients with hematological cancer. *Cancer Immunology, Immunotherapy, 69*(2), 315–324.
48. Li Santi, A., Napolitano, F., Montuori, N., & Ragno, P., (2021). The urokinase receptor: A multifunctional receptor in cancer cell biology. therapeutic implications. *International Journal of Molecular Sciences, 22*(8), 4111.
49. Hao, Q., Wu, Y., Wu, Y., Wang, P., & Vadgama, J. V., (2022). Tumor-derived exosomes in tumor-induced immune suppression. *International Journal of Molecular Sciences, 23*(3), 1461.
50. Howard, M. A., (2020). Reversing tumor induced immune suppression: Role of the secretory autophagy pathway. *Cancer Research, 80*(16), 996.
51. De Brito, R. C., Cardoso, J. M., Reis, L. E., Vieira, J. F., Mathias, F. A., Roatt, B. M., Aguiar-Soares, R. D., et al., (2018). Peptide vaccines for leishmaniasis. *Frontiers in Immunology, 9*, 1043.
52. Disis, M. L., Bernhard, H., Shiota, F. M., et al. (1996). Granulocyte-macrophage colony-stimulating factor: an effective adjuvant for protein and peptide-based vaccines. *Blood. 88*(1), 202–210.
53. Alving, C. R., Peachman, K. K., Rao, M., & Reed, S. G., (2012). Adjuvants for human vaccines. *Current Opinion in Immunology, 24*(3), 310–315.
54. Dubensky, Jr. T. W., & Reed, S. G., (2010). Adjuvants for cancer vaccines. In: *Seminars in Immunology* (Vol. 22, No. 3, pp. 155–161). Academic Press.
55. Torchilin, V. P., & Lukyanov, A. N., (2003). Peptide and protein drug delivery to and into tumors: Challenges and solutions. *Drug Discovery Today, 8*(6), 259–266.

56. Mishra, B., & Bonde, G. V., (2020). Transdermal drug delivery. In: *Controlled Drug Delivery Systems* (pp. 239–275). CRC Press.
57. Farzad, N., Barati, N., Momtazi-Borojeni, A. A., Yazdani, M., Arab, A., Razazan, A., Shariat, S., et al., (2019). P435 HER2/neu-derived peptide conjugated to liposomes containing DOPE as an effective prophylactic vaccine formulation for breast cancer. *Artificial Cells, Nanomedicine, and Biotechnology, 47*(1), 664–672.
58. Tornesello, A. L., Tagliamonte, M., Tornesello, M. L., Buonaguro, F. M., & Buonaguro, L., (2020). Nanoparticles to improve the efficacy of peptide-based cancer vaccines. *Cancers, 12*(4), 1049.
59. Kersten, G. F., & Crommelin, D. J., (2003). Liposomes and ISCOMS. *Vaccine, 21*(9, 10), 915–920.
60. World Health Organization, (1999). *Guidelines for the Production and Quality Control of Synthetic Peptide Vaccines* (p. 889). WHO Technical Report Series.
61. Blankenstein, T., Coulie, P. G., Gilboa, E., & Jaffee, E. M., (2012). The determinants of tumor immunogenicity. *Nature Reviews Cancer, 12*(4), 307–313.
62. Jhunjhunwala, S., Hammer, C., & Delamarre, L., (2021). Antigen presentation in cancer: Insights into tumor immunogenicity and immune evasion. *Nature Reviews Cancer, 21*(5), 298–312.
63. Escors, D., (2014). Tumor immunogenicity, antigen presentation, and immunological barriers in cancer immunotherapy. *New Journal of Science, 2014*.
64. Meraviglia-Crivelli, D., Zheleva, A., Barainka, M., Moreno, B., Villanueva, H., & Pastor, F., (2022). Therapeutic strategies to enhance tumor antigenicity: Making the tumor detectable by the immune system. *Biomedicines, 10*(8), 1842.
65. Zhou, G., Lu, Z., McCadden, J. D., Levitsky, H. I., & Marson, A. L., (2004). Reciprocal changes in tumor antigenicity and antigen-specific T cell function during tumor progression. *The Journal of Experimental Medicine, 200*(12), 1581–1592.
66. Wang, S., He, Z., Wang, X., Li, H., & Liu, X. S., (2019). Antigen presentation and tumor immunogenicity in cancer immunotherapy response prediction. *Elife, 8*, e49020.
67. Bai, Y., Jiang, H., Zhang, Y., Dou, L., Liu, M., Yu, W., Wen, K., et al., (2021). Hydrophobic moiety of capsaicinoids haptens enhancing antibody performance in immunoassay: Evidence from computational chemistry and molecular recognition. *Journal of Agricultural and Food Chemistry, 69*(34), 9957–9967.
68. Santos, P. M., & Butterfield, L. H., (2018). Dendritic cell–based cancer vaccines. *The Journal of Immunology, 200*(2), 443–449.
69. Crosby, E. J., Gwin, W., Blackwell, K., Marcom, P. K., Chang, S., Maecker, H. T., Broadwater, G., et al., (2019). Vaccine-induced memory CD8+ T cells provide clinical benefit in HER2 expressing breast cancer: A mouse to human translational study. *Clinical Cancer Research, 25*(9), 2725–2736.
70. Packiam, V. T., Richards, J., Schmautz, M., Heidenreich, A., & Boorjian, S. A., (2021). The current landscape of salvage therapies for patients with bacillus Calmette-Guérin unresponsive nonmuscle invasive bladder cancer. *Current Opinion in Urology, 31*(3), 178–187.
71. Long, G., Dummer, R., Johnson, D., Michielin, O., Martin-Algarra, S., Treichel, S., Chan, E., Diede, S., & Ribas, A. (2020). Long-term analysis of MASTERKEY-265 phase 1b trial of talimogene laher parepvec (T-VEC) plus pembrolizumab in patients with unresectable stage IIIB-IVM1c melanoma. *Journal for Immuno Therapy of Cancer, 8*:doi: 10.1136/jitc-2020-SITC2020.0429.

72. Clifton, G. T., Kemp, B. P. M., Hale, D. F., Myers, J. W., Brown, T. A., Holmes, J. P., Vreeland, T. J., et al., (2019). Abstract P2-09-01: Subgroups analysis of a multicenter, prospective, randomized, blinded phase 2b trial of trastuzumab+ nelipeptimut-S (NeuVax) vs trastuzumab for prevention of recurrence in breast cancer patients. *Cancer Research, 79*(4_Supplement), P2–09.
73. Sugiyama, H. (2022). WT1 cancer vaccine for the treatment and prevention. *Medical Research Archives, [online], 10*(4).
74. Ahluwalia, M. S., Reardon, D. A., Abad, A. P., Curry, W. T., Wong, E. T., Belal, A., Qiu, J., et al., (2019). SurVaxM with standard therapy in newly diagnosed glioblastoma: Phase II trial update. *J. Clin. Oncol., 37*(Suppl 15), 2016.
75. Bhuyan, P. K., Dallas, M., Kraynyak, K., Herring, T., Morrow, M., Boyer, J., Duff, S., et al., (2021). Durability of response to VGX-3100 treatment of HPV16/18 positive cervical HSIL. *Human Vaccines & Immunotherapeutics, 17*(5), 1288–1293.
76. Hollenberg, R. K., Krieger, D. R., Samuels, R., Kraynyak, K., Sylvester, A., Morrow, M., Boyer, J., et al., (2020). Safety and immunogenicity of VGX-3100 formulations in a healthy young adult population. *Human Vaccines & Immunotherapeutics, 16*(6), 1404–1412.
77. De Keersmaecker, B., Claerhout, S., Carrasco, J., Bar, I., Corthals, J., Wilgenhof, S., Neyns, B., & Thielemans, K., (2020). TriMix and tumor antigen mRNA electroporated dendritic cell vaccination plus ipilimumab: Link between T-cell activation and clinical responses in advanced melanoma. *Journal for Immunotherapy of Cancer, 8*(1).
78. Sahin, U., Oehm, P., Derhovanessian, E., Jabulowsky, R. A., Vormehr, M., Gold, M., Maurus, D., et al., (2020). An RNA vaccine drives immunity in checkpoint-inhibitor-treated melanoma. *Nature, 585*(7823), 107–112.
79. Dolgin, E., (2019). Unlocking the potential of vaccines built on messenger RNA. *Nature, 574*(7778), S10.

CHAPTER 8

Stem Cell-Based Cancer Vaccines as Promising Immuno-Therapy: Current Progress and Challenges

MEHUL R. CHORAWALA,[1] HUMZAH POSTWALA,[1] YESHA SHAH,[1] PALAK A. SHAH,[2] BHUPENDRA PRAJAPATI,[3] and VISHVAS N. PATEL[1]

[1]*Department of Pharmacology, L. M. College of Pharmacy, Opp. Gujarat University, Navrangpura, Ahmedabad, Gujarat, India*

[2]*Department of Pharmacology, K. B. Institute of Pharmaceutical Education and Research, Gandhinagar, Gujarat, India*

[3]*Shree S.K. Patel College of Pharmaceutical Education and Research, Ganpat University, Mehsana, Gujarat, India*

ABSTRACT

The main characteristic by which tumor cells differ from normal cells is their ability for indefinite proliferation and spread to different organs via invasion and metastasis. Various similarities have been identified between embryonic stem cells (ESCs) and cancerous cells. Stem cell-based vaccines can thus aid in developing a robust prophylactic immunity against cancer cells by presenting various tumor antigens to the immune system. Induced pluripotent stem cells (iPSCs), in particular ESCs, may resemble tumor cells in their freedom from various growth restrictions. Immunotherapy is currently recognized as a paradigm-shifting therapeutic strategy. Cancer vaccines (CVs) fall under the category of immunotherapy, with a special focus on stem cell-based vaccines targeting various solid tumors with a promising approach. These vaccines work by inducing anti-tumor immunity in the tumor microenvironment (TME), and they could be clinically valuable when

Cancer Vaccination and Challenges, Volume 1: Strategies for Therapeutic Cancer Vaccine Development.
Rishabha Malviya, Bhupendra Prajapati, and Sonali Sundram (Eds.)

combined with other targeted therapies to achieve better clinical outcomes. Despite significant efforts in developing rational cancer regimens, the issue of resistance due to cancer stem cells (CSCs) remains a challenge. Stem cell-based CVs have the potential to elicit marked anti-tumor responses, potentially inhibiting the metastasis of primary tumors and inducing humoral as well as cytotoxic immune responses against CSCs. These stem cell-based vaccines hold significant therapeutic and prophylactic potential in cancer management, addressing the problem of drug resistance. While they have shown promising results in preclinical studies, further exploration is needed for their use in humans. Therefore, a novel approach targeting cancer with stem cell-based vaccines might prove useful in overcoming this obstacle.

8.1 INTRODUCTION

Cancer is a non-communicable disease that originates from the Greek word 'Karkinos,' meaning a mass of malignant cells or the crab. It results in abnormal cellular growth when normal cells bypass the body's regulatory processes, leading to the malignant or uncontrolled growth of these cells [1]. It is considered a serious and life-threatening condition, posing a significant public health concern [2]. Approximately 70% of all cancer deaths occur in low- or middle-income countries, primarily due to inadequate medical services resulting in delayed diagnosis and poorer prognosis [3]. With increased life expectancy and the prevalence of modern unhealthy lifestyles, there has been an epidemiological rise in cancer cases over the years [4]. In India, about 80% of cancer patients are diagnosed at a late stage of the disease, leading to worsened clinical outcomes [5].

Currently, management strategies for cancer include surgery, chemotherapy, radiation therapy, as well as other targeted therapies that focus on either the immune system or hormonal system. Immune targeted therapies are centered around the concept of immune surveillance, which allows the recognition of abnormal tumor cells by the host immune cells with minimal impact on normal dividing cells. Antibodies (Abs) that act on various targets, such as cytotoxic T-lymphocyte-associated protein 4 (CTLA-4), programmed death-ligand 1 (PD-L1), as well as vaccines targeting oncogenic viruses (virus-associated cancer) like hepatitis B virus (HBV) and human papillomavirus (HPV), are included in immune targeted therapies [6]. Cancer vaccines (CVs) belong to the subgroup of active immunomodulatory therapy as they assist the body's native immune cells in recognizing and destroying tumors [7]. These vaccines are considered a promising approach to immunotherapy as they can generate long-term anticancer effects by influencing antigen-specific T cell immune responses [8]. They typically function by stimulating or modifying immune responses through nonspecific immunomodulation [6]. CVs can be either preventive or therapeutic [9]. Preventive vaccines are designed to provide prophylaxis against viruses that are significant risk factors in the development of cancer. The United States Food and Drug Administration (USFDA) has approved vaccines against two viruses, namely HPV and HBV, which are leading causes of cervical and hepatic carcinoma, respectively, responsible for nearly 25% of cases in lower-income countries [10, 11]. Another type of vaccine, known as therapeutic or curative vaccines, can be used to treat cancers with non-infectious causes by directly stimulating immune-mediated attacks on cancerous cells. There are FDA-approved therapeutic vaccines available for cervical, prostate, lung, breast, head and neck cancers, pancreatic cancers, and glioblastoma [12]. Conventional therapies often fail to target certain tumor cells that possess the ability to survive in various environments within the human body through rapid proliferation and self-renewal [13].

Apart from vaccination, other immunotherapy strategies, despite showing noteworthy effects, face various issues such as high cost and other immune-related side effects [14]. Thus, the concept of cancer vaccination was eventually developed. In recent years, various strategies for cancer vaccination have been developed, falling into four basic categories: cell-based, protein-based, viral/bacterial-based, and gene-based vaccines [15, 16].

Cell-based vaccines include vaccines derived from immune cells such as dendritic cells (DCs) or natural killer cells (NK-cell). Despite their advantages of high immunogenicity and controlled antigen presentation, these vaccines are very costly and difficult to prepare, with a high risk of leukapheresis-related complications such as vascular injury and electrolyte imbalance [17]. On the other hand, protein or peptide-based vaccines are easier to prepare, have minimal toxicity and low immunogenicity, mainly due to restricted recognition of specific types of human leukocyte antigen (HLA) subtypes [15].

Viral or bacterial-based vaccines, however, carry a higher risk of toxicities and increased chances of opportunistic infections due to their high immunogenicity [18]. Gene-based vaccines, including deoxyribonucleic acid/ribonucleic acid (DNA/RNA)-based vaccines, induce HLA nonspecific cellular and humoral immunity. However, they require very specific and challenging storage conditions and show poor immunogenicity in humans [19].

The use of DNA plasmid-based vaccines has proven to provide significant safety and efficacy through broad and specific immune activation. However, the rebound immunosuppression developed by the tumor results in a lessened therapeutic effect of this type of vaccine [20]. The main reason for the drawback of DNA-based vaccines includes the resistance that tumors acquire during development. Several important factors contribute to the development of resistance, including: (i) changes in epitopes that are assumed to be recognized by the host immune cells; (ii) exhaustion of T-cells; (iii) overexpression of immunosuppressive cells such as regulatory T-cells (Tregs), myeloid-derived suppressor cells (MDSCs), and tumor-associated macrophages (TAMs); and (iv) release of various cytokines including transforming growth factor β (TGF-β) and interleukin-10 (IL-10) [21]. In addition to these theoretical drawbacks, preclinical and clinical trials of current vaccines have not shown significant promising impact on disease outcomes as a standalone therapy [20]. The merits and demerits of different CVs are summarized in Table 8.1. Therefore, a newer approach is required to develop a CV that overcomes these limitations and provides better outcomes, with a focus on CSCs, a major cell type responsible for cancer relapse and the development of resistance to chemo-radiation [22]. CSCs contribute to tumor formation through differentiation into various cells by expressing different oncogenic surface markers [23]. The interactions between CSCs and the TME lead to various alterations in the tumor milieu, such as epithelial mesenchymal transition (EMT), which has been well-documented and strongly associated with the phenotype of cancer cells [24]. Oncofetal antigens, including carcinoembryonic antigen (CEA), alpha fetoprotein (AFP), and human chorionic antigen (HCG), are typically synthesized during the fetal stage but are expressed in cancer [25]. These oncofetal antigen-based vaccines have potential therapeutic applications, but they are associated with limitations such as ethical concerns and allergenicity. Consequently, various stem cells, such as embryonic stem cells (ESCs), have been developed, sharing similarities and common factors with CSCs. These ESCs can be explored as targets in various cancers due to their ability to deliver multiple antigens and induce an immunogenic response against tumor cells. Induced pluripotent stem cells (iPSCs), a subtype of stem cells resembling tumor cells like ESCs, can be reprogrammed from patients' stem cells and serve as valuable targets [26, 27]. This reprogramming triggers an immune response against CSCs through tumor antigens [28]. Preliminary studies have investigated the effectiveness of iPSC-derived vaccines; however, these vaccines cannot be employed as standalone therapeutic approaches. Therefore, they would be utilized in combination with conventional therapy to improve patient outcomes and overcome the problem of anticancer drug resistance [29].

Based on this hypothesis or concept, various targets can be studied, such as oncofetal peptide vaccines, ESCs-based vaccines, adult stem cells (ASCs)-based vaccines, iPSCs-based vaccines, and more. This chapter summarizes the immunogenicity of iPSCs, ESCs, and ASCs, along with current research on stem cell-based vaccines in different types of cancers. This chapter also provides a summary of various preclinical and clinical trials that have been conducted and are currently ongoing. This information may assist our scientific community in making a decision regarding the inclusion of stem cell-based vaccines in the regimen of anticancer drugs.

8.2 IMMUNOGENICITY OF STEM CELLS

Stem cells can be classified based on their potency, which includes the following categories: (i) totipotent stem cells, which have the ability to differentiate into all possible types of stem cells, such as the morula stage of the embryo; (ii) pluripotent stem cells, which can differentiate into most cell types, including the blastocyst mass of cell types; (iii) multipotent stem cells, which transform into closely related hematopoietic stem cells (HSCs); (iv) oligopotent stem cells, which give rise to acute lymphoid and myeloid cells; and (v) unipotent stem cells, which can renew themselves into muscle stem cells. Stem cells are also categorized based on their origin, such as neural stem cells (NSCs), HSCs, mesenchymal stem cells (MSCs), and ESCs (*see* Figure 8.1).

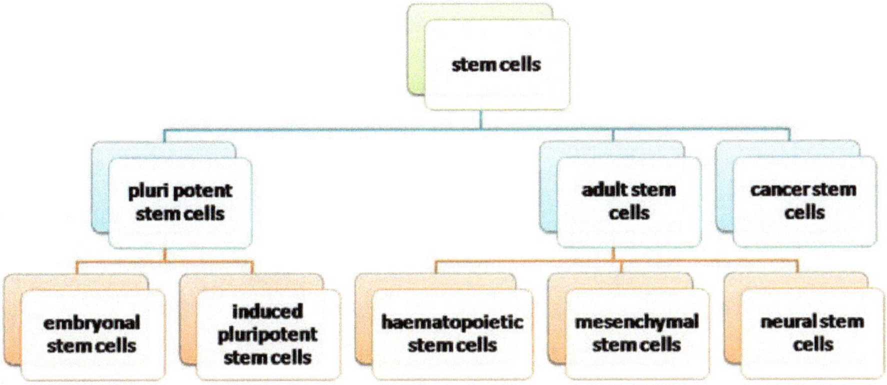

FIGURE 8.1 Classification of the stem cells.

Note: Stem cells can be classified based on their ability to differentiate into specific cell types. Embryonic stem cells and iPSCs, as their names suggest, are referred to as pluripotent stem cells. On the other hand, adult stem cells are known as multipotent stem cells. Cancer stem cells can exhibit pluripotent or multipotent characteristics depending on the type, origin, and nature of the tumor.

Cancer stem cells (CSCs) are cells whose activity can be blocked at the initial stages of development. CSCs are mainly classified as tumor-initiating cancer (TIC) cells and tumor-migrating cancer (TMC) cells. TICs are of epithelial origin and have the ability to grow independently, while TMCs are of mesenchymal origin and have the ability to migrate and invade other tissues. The initiating CSCs, also known as para-embryonic stem cells (P-ESCs), are responsible for tumor initiation, which requires somatic reprogramming via mesenchymal-epithelial transition (MET). CSCs are categorized into three subgroups based on their similarity to embryonal or embryonic cells, including primary, secondary, and tertiary CSCs. Primary CSCs have structural similarities with the epiblast cells of the human embryo. Due to the overactivity of several transcription factors, such as NANOG, Octamer-binding transcription factor 4 (OCT4), Inhibitor of DNA binding-1 (ID-1), and SRY-Box Transcription Factor 2 (SOX2) gene expression, primary CSCs are stuck in a pluripotent stage and exhibit characteristics like self-rejuvenation, similar to epiblast cells [30, 31]. Furthermore, secondary CSCs, which resemble the hypoblast cells of the human embryo, are organized for tumor growth and progression by promoting angiogenesis, invasion, and tissue proliferation resulting from various mutations, overexpression, and autocrine/paracrine signaling [32, 33]. When the TME becomes limiting for further growth and proliferation of the cancer due to factors such as hypoxia and other autocrine/paracrine signaling stressors, a new phenotypic cell characterized as tertiary CSCs emerges. These cells are responsible for cancer migration and invasion to other sites, resulting in micro- and macro-metastasis [31, 34].

Various oncofetal proteins, including CEA, alpha-fetoprotein (AFP), HCG, glypican 3, cancer testis antigen, immature laminin receptor, placental alkaline phosphatase, and IGF-2 mRNA binding protein (IMP) family, are generally present in CSCs. Targeting these antigens could lead to the development of oncofetal protein-based anti-CVs [26]. These antigens play a role in initial tumor development and in turning tumor cells into chemo-radiation resistant cells, making them ideal targets for antitumor immunotherapy [35].

8.2.1 EMBRYONAL STEM CELLS (ESCS)

Embryonal stem cells (ESCs) are essentially derived from the inner cell mass (ICM) of embryonal origin, possessing the remarkable ability of extensive growth, proliferation, and adaptability to various environments within the

TABLE 8.1 Merits and Demerits of Different Approaches for Cancer Vaccines

Type of Vaccine	Advantages	Disadvantages
	Cell-based Vaccine	
• Dendritic cell-based vaccine. • Antigen presenting cell-based vaccine.	• Controlled antigen presentation. • High immunogenicity. • Targeting multiple tumor antigens. • Useful for tumors with multiple or unknown target antigens.	• Risk of leukapheresis. • Costly and difficult to prepare. • Downregulation of MHC Class I • Limited clinical efficacy.
	Protein/Peptide-based Vaccine	
• Protein subunit vaccine. • Peptide based vaccine.	• Better induction of humoral responses. • Comparatively safe and cost effective. • Easy production and storage.	• Restricted to a particular HLA subtype. • Low and variable immune responses thus need adjuvants. • Less or no Th and CTL epitope cells. • Inefficient delivery.
	Viral/Bacterial-based Vaccine	
• Whole inactivated vaccine. • Live attenuated vaccine. • Viral vector vaccine	• High immunogenicity. • Availability of variety of vectors.	• Higher toxicities and chances of opportunistic infection. • Chances of preexisting or neutralizing antibodies.
	Gene-based Vaccine	
• DNA-based vaccine. • mRNA-based vaccine.	• Broad HLA humoral and cellular immunity allowing multiple immunizations. • Self-adjuvant. • Low toxicity profile. • Capable of lymph node migration. • Personalized vaccine possible via specific Th and CTL epitopes. • Cost effective.	• Low immunogenicity. • Difficult storage conditions as it is susceptible to degradation. • Difficult administration. • Immune memory driven rebound immunosuppression.
	Stem Cell-based Vaccine	
• Embryonal stem cells.	• Low cost and has been establish and characterized since a long time. • MHC downregulation possible. • Multiple antigen delivery possible.	• High mutation rate and risk of carcinogenesis (tetratoma). • Low immunogenicity. • Ethical issues and religious concerns in use.

TABLE 8.1 *(Continued)*

Type of Vaccine	Advantages	Disadvantages
• IPSCs	• Less toxic. • Easy to use in any blood group and less HLA compatibility issues. • Donor cells easily obtained.	• Graft versus host disease. • Very costly. • No proper standardization. • Lower reprogramming efficacy.
• Adult stem cells	• Easy isolation and *in vitro* culture. • High stability.	• Uncertain effectiveness. • High toxicity due to retention in lungs, occlusion of macrovesicles. • Alloantibody production.

Abbreviations: MHC: Major histocompatibility complex; HLA: human leukocyte antigen; Th: helper T-cells; and CTL: cytotoxic T-lymphocytes.

human body [36]. A parallel can be drawn between the development of human embryos and cancer. Embryonal stem cells and CSCs share several characteristics and common factors, such as increased telomerase activity that confers immortality to CSCs by maintaining telomere length [26, 37]. While human ESCs originate from epithelial cells, they undergo a transition from epithelial to mesenchymal cells, also known as an EMT. The main distinction between ESCs and CSCs lies in genomic homeostasis, which is observed in the former but not the latter [38] (Figure 8.2).

ESCs also undergo de-differentiation and re-differentiation, along with various oncogenic mutations taking place, leading to malignancy [39]. When transplanted, ESCs trigger an allogenic immune response in host cells, mainly due to the presence of major histocompatibility complex (MHC) and minor histocompatibility (miH) antigens. Despite the expression of two subclasses, class I and class II MHC, being very low and negligible, respectively [40, 41].

ESCs-based vaccines are preferred over oncofetal peptide-specific vaccines due to their ability to deliver multiple antigens. A study conducted in 2012 by Kavitha et al. [42] using murine-derived ESCs as a vaccine in mice showed that the immune response produced was not due to the allogeneic response, but rather due to cluster of differentiation (CD) 8^+ effector T cells. The tumor protective effect requires a certain level of $CD8^+$ T cells. Depletion of these cells leads to the complete loss of tumor protective activity [42].

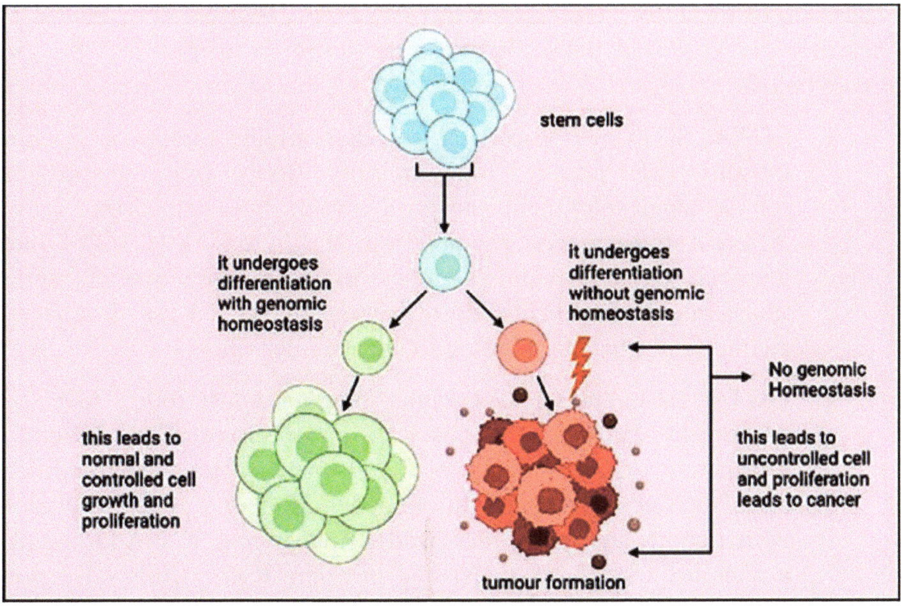

FIGURE 8.2 Genomic instability in cancer stem cells (CSCs).

Note: Under normal disease-free conditions, stem cells undergo differentiation to form various cells that undergo normal growth and proliferation. This normal growth is maintained by genomic homeostasis during differentiation. However, in the presence of disease or risk factors that predispose to cancer, genomic homeostasis is disrupted. As a result, stem cells experience genomic instability during differentiation, leading to uncontrolled proliferation and tumor formation. This is why stem cells are referred to as cancer-initiating cells and represent a promising target for cancer treatment.

8.2.2 INDUCED PLURIPOTENT STEM CELLS (IPSCS)

IPSCs were first discovered in 2006, bringing a novel perspective to the field of stem cell utilization in medicine [43]. Initially, much focus was placed on the regenerative potential of IPSCs, with concerns regarding their immunogenicity. However, this same property can be explored for its antitumor activity as immunotherapy [44]. IPSCs can be categorized as autologous and xenogeneic IPSCs. Among these, autologous IPSCs have shown a wider range of tumor antigens and can be better explored as CVs. They are formed by exposing human somatic cells to four oncogenic Yamanaka transcription factors, namely Sox2, Oct3/4, c-Myc, and Kruppel-like factor (KLF4), in order to change their transcriptional state to resemble tumor cells [45]. At least one of these four factors is known to be overexpressed in almost all cancer types [46]. The details of these transcription factors for CSCs are given below:

1. **SOX4:** SOX, also known as sex determining region on Y chromosome (SRY)-related high mobility group box, is responsible for cancer development and proliferation. It is categorized into 8 different subcategories, ranging from SOX A to SOX H. SOX4 plays a crucial role in providing stemness to CSCs. It promotes transcription associated with histone deacetylase (HDAC)-1 and enhances stemness in CSCs by binding to its promoter site [1].
2. **OCT4:** Also known as octamer binding transcription factor, is universally expressed in most CSCs and serves as a marker for pluripotency. The regulation of OCT4 involves epigenetic modifications such as DNA methylation, acetylation, and phosphorylation, which contribute to the increased stemness of CSCs. Additionally, a complex framework consisting of micro-RNAs (miRNAs), non-coding RNAs (ncRNAs), and small interfering RNAs (siRNAs) provides multiple targets for CSCs [2].
3. **c-Myc:** It is a regulatory gene that controls the essential transcription factor responsible for cell proliferation, angiogenesis, and cell cycle progression under normal circumstances. Any irregularities in its transcription lead to genomic instability and uncontrolled growth, proliferation, and cell cycle progression in tumor cells. C-Myc also contributes to chemo-resistance by increasing chromosomal aberration, genomic instability, and activating various downstream signaling genes [3]. In various studies, some lesser oncogenic genes were replaced by the c-Myc proto-oncogene [47, 48].
4. **KLF4:** Also known as Krupple-like factor 4, is a zinc finger transcription factor involved in proliferation, differentiation, and migration. KLF4 is one of the four factors responsible for the stemness of CSCs [4]. Decreased expression of KLF4 results in lung and colon malignancy, explaining its putative role. Conversely, in head and neck cancer, its upregulation promotes tumor survival and progression. The role of KLF4 in CSCs is still to be explored, and further categorization is needed [4, 5].

IPSCs can be reprogrammed to differentiate into pluripotent stem cells through external inductions. IPSCs share epigenetic and transcriptomic similarities with pluripotent tumor cells [49]. One major advantage of using IPSCs instead of ESCs is the ability to bypass MHC mismatches and allogenic immunity [50]. While both ESCs and IPSCs resemble tumor cells in many ways, IPSCs exhibit greater pluripotency, making them a better option [51]. Additionally, IPSCs have a lower incidence of toxicities associated with cytokine release syndrome, a common issue in immunotherapies such as chimeric antigen receptor-T cell (CAR-T cell) therapy and PD-L1 inhibitors [52, 53].

A study by Kooreman and colleagues in 2018 demonstrated a significant genetic overlap between IPSCs, ESCs, and cancer tissues [54]. Vaccines generated from IPSCs have the ability to modulate the host immune response against malignant cells by stimulating the synthesis of various inflammatory mediators and suppressing other immune cells with tumor-promoting activities [55]. IPSCs exhibit antitumor activity by inducing interferon (INF)-γ and IL-4 producing splenocytes [56]. Several studies have shown significant anti-tumor effects after immunizing tumor cells with fetal tissue [57]. Stemness is a quantification of a cell's ability to differentiate [58]. The indices of stemness, when calculated, were found to be highest in the tumor cells that had metastasized or invaded further from the primary site, while being significantly lower at the primary tumor site [59]. A study was conducted to calculate the stemness indices, which revealed a significant positive correlation between tumor transcriptomic signatures (associated with stemness) and iPSCs [60]. In addition to developing other immunotherapy targeted approaches, iPSCs can also aid in generating human hematopoietic cells for targeting various cancers [61].

Undifferentiated iPSCs exert their immune response through granzyme-B, interferon-γ (IFN-γ), and perforin. On the other hand, autologous iPSCs induce their immunogenic effect in a different manner. A study conducted in 2014 by Almeida et al. [49] using mouse models described the immune response of iPSC-derived endothelial cells as being mediated by a tolerogenic response, characterized by prominent expression of IL-10. This mechanism of immune response by autologous iPSCs significantly differs from the aforementioned immune response of undifferentiated iPSCs [49]. This distinction can be attributed to the tumorigenicity and immunogenicity of iPSCs, which are relevant to their potential as a stem cell-based vaccine for cancer [62]. Therefore, besides their application as a stem cell-based vaccine, iPSCs can also contribute to the elucidation and understanding of the molecular mechanisms underlying tumor initiation and progression [63, 64].

Instead of using IPSC-derived vaccines alone, a study combined them with toll-like receptor (TLR)-9 agonists, CpG Oligodeoxynucleotide (CpG ODN), which helped enhance the antitumor activity of the vaccine by accelerating the maturation of antigen-presenting cells (APCs) [54]. This IPSC plus CpG vaccine led to the rejection of tumor tissue in mice, indicating tumor-specific immune stimulation mediated by T-cells [65, 66]. In 2007, Gąbka-Buszek et al. [67] reported that hyper IL-6 (HIL6) melanoma tumor cells were exposed to a combination of mixed vaccines, where a melanoma vaccine derived from whole cells was combined with stem cells. This resulted in enhanced anti-tumor activity by inhibiting Treg activation, promoting dendritic activation, and antigen presentation [67]. HIL6-modified tumor cells showed significant improvement in terms of disease-free survival (DFS) as well as long-term overall survival (OS) [68]. Various studies have concluded that, as an alternative to directly using IPSCs, the regeneration of other immune cells such as DCs, T-cells, or natural killer-cells (NK-cells) from IPSCs is also capable of exhibiting significant tumor cytotoxicity [69, 70]. In 2018, Kooreman et al. [54] conducted a study using irradiated IPSCs as prophylactic vaccines in various mouse-based cancer models. This resulted in reduced recurrence and metastasis by lowering the levels of T helper (Th)-17 cells and increasing the levels of $Gr1^{+}CD11b^{+}$ myeloid cells [54].

As mentioned above, the IPSCs are derived from somatic cells, such as skin fibroblasts or peripheral blood mononucleated cells (PBMCs), extracted from whole blood. These cells are reprogrammed into IPSCs using the four Yamanaka factors [26]. After irradiation, they are combined with CpG ODN and other molecules, such as immunostimulants including monophosphoryl lipid A (MPL), a bacterial-derived non-toxic molecule responsible for the activation of macrophages. Additionally, various saponins and cytokines can be used as immunostimulatory adjuvants to modify and redirect immune responses [71]. When these IPSCs are injected into an organism as a vaccine, they lead to a reduction in tumor size and volume mainly through two pathways: IPSC-mediated direct presentation and IPSC-mediated cross-presentation [72].

The use of IPSCs as a whole cell-based vaccine leads to the activation of naïve or memory $CD4^{+}$ cells directly, without involving APCs. This activation is carried out by tumor antigen-mediated activation of T-cell receptors (TCRs) on $CD4^{+}$ cells, which are directly responsible for tumor regression. On the other hand, in cross-presentation, the activation of $CD8^{+}$ as well as $CD4^{+}$ naive and memory cells is mediated by APC presentation of the tumor antigen to the $CD8^{+}$ and $CD4^{+}$ cells, subsequently leading to their activation and immunogenic response in tumor regression [72]. Thus, IPSCs, after entering into the tumor, lead to upregulated APC presentation and demonstrate their antitumor action by activating tumor-specific Abs as well as cytotoxic T cells (Figure 8.3).

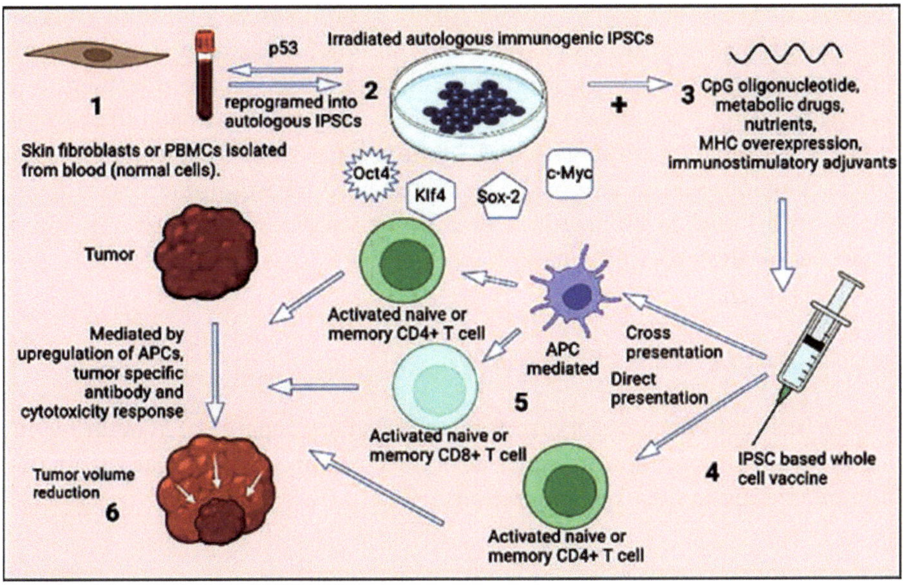

FIGURE 8.3 Derivation of IPSCs from human fibroblasts or peripheral blood mononucleated cells and its mechanism of actions in tumor size reduction.

Note: (1) Human skin fibroblasts or PBMCs are irradiated and treated with the four Yamanaka transcription factors, namely OCT4, KLF4, Sox-2, and c-Myc. (2) Immunogenic autologous iPSCs are formed. (3) These iPSCs are then combined with CpG oligonucleotide, as well as other nutrients, metabolic drugs, and immunostimulatory molecules. (4) This mixture is formulated into a vaccine known as iPSC-based whole cell vaccine. (5) When injected into the host, this iPSC vaccine exerts its anticancer activity through direct presentation to activated or memory $CD4^+$ cells, or cross-presentation to activated or memory $CD8^+$ or $CD4^+$ cells mediated by antigen-presenting cells. (6) This leads to tumor volume reduction mediated by tumor-specific antibody release. [PBMC: Peripheral blood mononucleated cells; iPSC: induced pluripotent stem cells; OCT4: octamer binding transcription factor; KLF4: Kruppel-like factor; Sox-2: sex determining region Y-box 2; MHC: major histocompatibility complexes; APC: antigen-presenting cells; and CD: cluster of differentiation].

Apart from autologous IPSCs, the use of allogenic IPSCs as cancer immunotherapy is still under research. Autologous IPSCs have the drawback of a lengthy and complicated manufacturing process, which can be bypassed by using allogenic IPSCs. Several studies have shown promising results indicating the potential use of allogenic IPSCs in vaccination strategies. In 2016, Kindy et al. conducted a study [74] using semi-allogenic IPSCs irradiated with tumor cells as an antiCV against glioblastoma in murine models. The results demonstrated improved OS and a 5-to-10-fold reduction in tumor volume in vaccinated mice [74]. Despite the impressive theoretical background, IPSC-based vaccines alone do not exhibit strong antitumor effects in patients. However, they can synergize with other anticancer strategies or be used as an adjuvant alongside other treatments. Based on the aforementioned studies, it can be concluded that IPSCs play a significant role in the exploration of vaccination strategies and immunotherapy in antitumor therapies. In conclusion, whole cell-based vaccines derived from autologous IPSCs have been found to be useful as preventive vaccines and as an adjuvant to therapeutic strategies for cancer treatment [54].

8.2.3 ADULT STEM CELLS (ASCS)

ASCs belong to the category of stem cells that are capable of generating new cells for specific groups of tissues or organs. Their uses in cancer vaccination are mentioned in detail in subsections.

8.2.3.1 MESENCHYMAL STEM CELLS (MSCS)

MSCs, a subtype of ASCs, are multipotent in nature with a mesodermal origin and possess a specific stimulatory effect on the host immune system. MSCs are primarily acknowledged for their anti-inflammatory activity. However, certain evidence from the past suggests the involvement of MSCs in the stimulation of adaptive immunity. This makes them a promising therapeutic option as prophylactic and therapeutic vaccines for cancers of viral or non-viral origin, which is considered one of the most important therapeutic discoveries of the current era. They have the potential to activate tumor-specific immunity [75].

To enhance the anticancer effects of MSCs, certain modifications are made, such as transfection with plasmids, which subsequently leads to more long-term and stable expression of the proteins [76]. The modified MSCs express viral antigens such as TLR ligands, which later stimulate the production of antigen-specific Abs [77]. MSCs essentially promote the expression of IFN-α or IFN-γ, leading to apoptosis in tumor cells [78]. In a study conducted by Wei et al. in 2011 [79], a vaccine based on MSCs expressing the combined E6/E7 protein was investigated in a murine model. The mice were injected with metastatic fibrosarcoma cells. The results demonstrated that mice exposed to the E7 protein-modified MSCs vaccine exhibited reduced tumor progression compared to mice receiving the MSCs vaccine alone [79].

Furthermore, MSCs serve as foreign antigens to immuno-inflammatory cells, aiding in the promotion of IFN-γ expression. This, in turn, leads to the protection of tumor or cancerous cells through the upregulation of MHC-I and -II, which are recognized by cytotoxic $CD8^+$ and $CD4^+$ T cells, respectively. Consequently, this modulates effector cells [80].

In another study, significant protection against lung and colon malignancies in mice was demonstrated using a vaccine derived from primary adipose tissue stem cells and adapted E7 antigen. These vaccines act by reducing the expression of vascular endothelial growth factor (VEGF) and CD31, thereby inhibiting angiogenesis and inducing apoptosis in malignant cells [81].

8.2.3.2 HEMATOPOIETIC STEM CELLS (HSCS)

HSCs are multipotent primitive ASCs derived from cord and placenta and could transform into all types of mature blood cells. These type of stem cells have not much been explored as vaccines for different cancers. Allogenic HSCs, however, have shown a considerable impact on adaptive immunity [82]. Autologous peripheral blood stem cells were given as an infusion post-chemotherapy, which did not show promising results due to a high relapse rate [83]. A study showed a group of patients receiving IL-2 incubated HSCs, which demonstrated an antigen-specific cell-based immune response [84]. Autogenic as well as allogenic HSCs are also used to improve the survival of several cancers, such as multiple myeloma, various lymphomas, leukemias, neuroblastoma, ovarian cancers, and many more subtypes [85, 86].

8.2.3.3 NEURAL STEM CELLS (NSCS)

NSCs, a subtype of ASCs of the nervous system that are develop into regeneration of neurons and glial cells. It has been reported that the combination of DC lysate and glioma stem cells vaccine induces the formation of cytotoxic T-cells and subsequent development of antitumor immunity in a murine model of malignant glioma [87]. In another study, NSCs were found to modulate HER2 positive breast cancer cells that had metastasized to the brain. Intracranial injection of NSCs promoted the production of anti-HER2

Abs that specifically bind to tumor cells over-expressing HER2, leading to the inhibition of phosphoinositide 3-kinases/Akt strain transforming (PI3K-Akt) signaling and improving OS in murine models [88].

8.3 PRECLINICAL AND CLINICAL DATA OF STEM CELLS-BASED VACCINE IN SPECIFIC CANCERS

8.3.1 LUNG CANCER (LC)

LC is considered the leading cause of cancer morbidity and mortality worldwide. Despite being the gold standard for treatment, surgical resection of the tumor is associated with local and distant recurrence and poorer responses [89]. Various clinical and preclinical trials are being conducted to develop newer immunotherapy strategies, such as stem cell-based cancer vaccination, for the successful treatment of LC.

An *in vivo* study in mice with induced lung adenocarcinoma demonstrated more significant tumor suppression when pre-immunized with human-iPSCs in combination with CpG, resulting in more robust antigen presentation and cytotoxic T-cell responses. Therefore, it was concluded that IPSCs have tumor-protective properties compared to unvaccinated mice [62].

Another study revealed the antitumor susceptibility of tumor-initiating cells (TICs) in LC, showing that these cells bypass immune surveillance by modifying their MHC-I expression. Therefore, improving MHC-I expression on malignant cells can enhance the sensitivity of the tumor to T-cell cytotoxicity [90].

A CSC-DC vaccine was tested in LC squamous cell tissues given in murine models as an adjuvant after surgical resection of the tumor, and therapy was augmented with anti-PD-L1 inhibitors. The results were promising, showing a reduction in tumor metastasis and an improvement in survival rate [91].

Allogenic murine-derived ESCs were used as a prophylactic vaccine for preventing lung tumors in mouse models, in combination with immunostimulatory therapy. The tumor-protective effect of the vaccine was partly mediated by adaptive and humoral immune responses, specifically showing CD8-mediated cytotoxic responses [42].

A phase I and II clinical trial (NCT02115958) was conducted on 40 patients with nasopharyngeal cancer to assess the safety and efficacy of CSC-loaded DC vaccines for clinical use. The primary objective of the study was to determine the safety of administering the stem cell-based vaccine in human subjects, with a secondary objective of measuring immune responses to the vaccine [92]. A similar clinical trial (NCT02084823) for lung neoplasm was also conducted to evaluate the applicability of stem cell-based vaccines in clinical practice. The results showed only grade I local and systemic adverse events (AEs), with injection site reaction being the most common, all of which were symptomatically relieved. To assess immune responses, lymphocyte counts were measured in control and comparator groups, revealing a significant increase in lymphocyte count in the group of patients receiving the vaccine, along with an elevated level of Th-1 type cytokines. In addition to immune markers, the binding of CSC into the plasma of vaccinated patients was also significantly higher than in the unvaccinated group [93].

8.3.2 OVARIAN CANCER

Ovarian cancer is one of the most aggressive gynecological cancers, with an OS rate of 43.6% in the past three decades. This is due to advancements in the screening of other gynecological cancers using biomarkers [94]. Significant efforts have been made to develop effective screening methods and models for ovarian cancer, utilizing various biomarkers such as the epitope of mucin 16 (MUC16) or cancer antigen 125 (CA125) [95]. The tumor of ovarian cancer has been found to contain a stem cell niche that interacts with the TME and the immune system in a dynamic manner, playing a crucial role in tumor growth, proliferation, and metastasis [96].

One of the most effective approaches to inhibit the growth of tumor cells is through vaccination against tumor-associated antigens (TAAs) [97]. DC-based vaccines have been used against CSCs. While this vaccination has shown benefits in mouse models of squamous cell cancer, its efficacy in ovarian cancer is yet to be proven [98]. Oncolytic virotherapy (OVT) targets cancer-initiating stem cells by blocking chemokine (CXC motif) ligand 2 (Cxcl2) [99]. Only a few trials have been conducted targeting the stem cell niche in ovarian cancer. One study demonstrated the efficacy of metformin in preventing stem cell-induced chemo-resistance in ovarian cancer, possibly through DNA methylation [100]. Another study generated Abs against the stem cell niche, resulting in anti-tumor activity in ovarian cancer [101]. A study conducted by Zhang et al. in 2012 [102] evaluated the potential of ESCs as a CV to generate an antitumor effect. They demonstrated cross-reactivity between ovarian tumor cells and ESCs through western blot analysis and found that ESCs presented a significant number of markers, antigens, and genes responsible for tumor growth. However, there was no difference in cytokine levels. Overall, ESCs showed great therapeutic potential for developing an ESCs-based vaccine for ovarian cancer [103]. DCs carrying specific peptides, such as NANOG peptide, instead of entire tumor lysate, were able to elicit a cytotoxic T cell response, which can be stored as immune memory. In the future, this approach could provide action against destroying CSCs generated via plasticity [104].

A Phase 1/2 study (NCT02178670) conducted by Ning et al. in 2012 [101] aimed to demonstrate the anti-tumor activity of DC-based CSCs vaccine. It was found to induce immunogenicity towards CSCs and specific lysis using B cells stimulated by the CSCs-containing DC vaccine [101].

8.3.3 PANCREATIC CANCER

Pancreatic cancer is a lethal and highly challenging disease to treat, characterized by the formation of malignant (cancer) cells in the tissues of the pancreas. It is the leading cause of death among different cancer subtypes, with a very poor prognosis of less than 5% for a five-year OS rate [2]. Due to the inherent resistance of this type of tumor to gemcitabine, various novel approaches, including stem cell-based vaccines as immunotherapy, are being developed [105]. Pancreatic cancer is known for its high heterogeneity, with CSCs acting as tumor-initiating cells (TICs) and expressing high levels of CD9, which contribute to tumor growth [106]. Below, we provide a summary of the various preclinical and clinical trials conducted for stem cell-based vaccines.

As mentioned earlier, there are significant molecular similarities between human iPSCs and TIC stem cells. A study was conducted using a murine model of pancreatic ductal adenocarcinoma, where a combination vaccine of autologous IPSCs and CpG was introduced. The results demonstrated a significant protection against tumor growth in mice, primarily through the induction of memory T- and B-cell responses and the inhibition of Treg-mediated immunosuppressive responses by $CD4^+$ T-cells. The combination of IPSCs and CpG was reported to work synergistically by releasing TAAs via IPSCs and activating antigen-presenting cells (APCs) through CpG. Approximately 75% of the mice showed significant protection against tumor formation [55]. Additionally, genetically engineered IPSCs have been developed as a vaccine by introducing virus-infected, reprogrammed somatic cells-derived tumor cells into mice. This approach resulted in delayed tumor development in mice when injected with pancreatic tumor cells, providing a foundation for the development of prophylactic IPSC-based pancreatic CVs for individuals with high-risk factors [107].

To investigate the safety of stem cell-based vaccines in humans, a clinical trial (NCT02074046) was conducted. This phase I/II trial involved a total of 40 recruited patients with pancreatic cancer who were administered a vaccine derived from CSCs with loaded DCs. The primary objective of the study was to evaluate the safety of the vaccine in humans. Secondary outcomes, such as immune responses induced by the vaccine, were also measured [108]. Another ongoing phase II clinical trial (NCT01088789) involves a total of 72 patients and utilizes an allogenic pancreatic tumor cell vaccine transfected with granulocyte-macrophage colony stimulating factor (GM-CSF). This trial has shown promising results and aims to assess the safety and feasibility of vaccination for pancreatic cancers in humans. The final outcome will be determined based on disease-free OS as well as any specific AEs associated with the vaccination.

8.3.4 HEPATOCELLULAR CANCER

Hepatocellular carcinoma (HCC) is the sixth most prevalent cancer worldwide and is one of the most challenging to treat. Conventional management strategies, including transarterial chemoembolization, have not yielded satisfactory results [109]. Additionally, newer immunotherapy modalities have shown effectiveness in only 10%–20% of patients [110]. The FDA has approved a prophylactic antitumor vaccine for liver cancer targeting HBV [111].

The main mechanism of tumor rejection involves the immune response generated by embryonal products, suggesting the potential use of stem cells as vaccines for different cancers. A whole-cell-based vaccine using iPSCs has been patented for various cancers, including HCC [26]. A study using irradiated HCC cells as a whole-cell vaccine, combined with lymphocyte-induced artificial Th9 cells, demonstrated a protective effect against HCC in a mouse model. The beneficial effect was mainly attributed to increase the differentiation and transcription factor levels, resulting in reduced tumor size and growth [112]. Another study investigating the preventive efficacy of CSCs based vaccines for liver metastasis showed resistance to tumor development in approximately 50% of vaccinated mice. The vaccinated mice exhibited a reduction in the size of the primary tumor and prevention of metastasis [113].

A Phases 1 and 2 clinical trial (NCT02089919) has been conducted, enrolling 40 patients with HCC. These patients were given CSCs-loaded DC vaccines at varying doses in different cohorts, with the aim of evaluating the feasibility of using stem cell-based vaccines in humans. The primary objective was to assess the safety of these vaccines, while the secondary objective was to measure the immunogenicity of the vaccine.

8.3.5 COLORECTAL CANCER (CRC)

CRC is classified as the third most prevalent cancer, with 0.9 million deaths worldwide in 2020 [114]. The OS rate for CRC is approximately 65%, which decreases to 15% in cases of distant metastatic cancer sites [115]. In CRC, the stem cell niche is composed of Paneth cells and vascular endothelial cells [116]. These cells provide important growth factors and also regulate pathways such as wingless related integration site (Wnt), notch, and hedgehog (Hh), which are responsible for growth, proliferation, and metastasis [117]. Clinical trials have shown progress in the development of a CRC stem cell-based vaccine, demonstrating safety and slowed tumor development.

Some studies have been conducted using attenuated colorectal CSCs, but they did not yield promising results. However, the regulatory role of the p53 and p21 genes has been highlighted and targeting them simultaneously may further increase the efficacy of the vaccine [118, 119]. The DC-associated vaccine has shown potential in inducing an antigenic response through innate and adaptive immunity. The CSCs stroma has been implicated in growth, proliferation, and tumorigenesis, and targeting the TME of colorectal CSCs could be an important approach for the development of a CV [120]. The CV, targeting various biomarkers such as CD24, CD26, and CD29, could play a decisive role in the development of CVs [121]. Another study examined the effect of a vaccine derived from IPSCs induced by transfection of *in vitro* translational RNA, which successfully induced a cytotoxic-T lymphocyte immune response against neoantigens in an *in-vitro* tumor spheroid model for CRC [122]. A vaccine obtained from colon CSCs $CD44^+$ has been formulated into defective ribosomal products-containing autophagosome-rich blebs (DRibbles) vaccine. Interestingly, when tested in an animal model of CRC, such a vaccine upregulated $CD8^+$ in splenic lymphocytes and activated MHC-I and -II complexes, thus producing strong cytotoxic effects [123]. A study combined MUC1, a cell surface antigen associated with colon tumors, with colon CSCs and integrated them into a vaccine given therapeutically in mice infected with colon cancer (CC) cells. The study concluded that MUC1 might play a critical role in expressing the antitumor immunogenicity of colon CSCs [124].

A Phase 1/2 clinical trial (NCT02176746) has been conducted to demonstrate the efficacy of the DCs-associated CSCs vaccine in inducing immunity against CSCs and B-cells associated binding and lysis generated from PBMCs [125]. The primary outcome measure of the trial was the safety of using stem cell vaccines in humans. Different doses of the vaccines were administered in different cohorts to determine a safe dose. The secondary outcome was measured based on the immune response of the vaccine on the cancer.

8.3.6 BREAST CANCER

Breast cancer is one of the most common malignancies, causing cancer-associated morbidity and mortality in women [126]. Despite early diagnosis and targeted therapies leading to a significant improvement in prognosis, many mutated tumors are resistant to these conventional therapies. Vaccinations for breast cancer can be cell-based, peptide-based, or gene-targeted vaccines. These vaccines work by targeting the stem cells as TICs [127]. A study demonstrated the use of IPSCs-based vaccine combined with non-genetically engineered CpG as a prophylactic in a murine model of orthotropic breast cancer. This approach resulted in an effective immunogenic response, with the highest IgG binding and upregulation of antigen presentation in Th1 and cytotoxic cells [54]. IPSCs have proven to elicit tumor-protective effects as prophylactic vaccines. Triple-negative breast cancer (TNBC) is known to have stem cell-like properties due to the over-expression of hypoxia-inducible factor-1α (HIF-1α). A study conducted a HIF-1α-based vaccine to target breast CSCs *in vivo* in murine models, which demonstrated properties that downregulate tumor growth through active immunization of tumor cells [128]. Another study derived a vaccine from breast progenitor cells and used it as a therapeutic vaccine in a breast cancer model. These vaccines are generally known to be downregulated by regulatory T (Treg) cells. However, when previously sensitized with lymph nodal T-cells, they showed significant tolerance to the Treg inhibitory action. Thus, a vaccine sensitized with T-cells demonstrated immunogenic tumor destruction effects with much more significance [129]. A clinical trial (NCT02063893) was conducted, recruiting 40 breast cancer patients to evaluate the safety, immunogenicity, and feasibility of different doses of CSCs-based vaccine.

8.3.7 MULTIPLE MYELOMA

Multiple myeloma, also known as plasma cell myeloma, is a rare cancer of plasma cells, a type of white blood cell that normally produces Abs. It accounts for around 10% of all liquid cancers. Despite recent advances in treatment, the rate of relapses is very high, with a survival rate of 8–9 months [130]. CAR-T cell-based therapy has recently gained attention in the management of multiple myeloma. This therapy involves extracting T-cells from the patient's blood and engineering them to target specific immune system markers, such as CD-19, -38, and -138, which are prevalent on the surface of liquid tumors. These engineered T-cells are then administered back to the patients, resulting in reduced tumor growth. CAR-T cell therapy has been approved for various B cell-based tumors [131].

A study demonstrated the efficacy of an HSC-derived DC vaccine, which showed multiple benefits such as enhanced antigenic uptake by antigen-presenting cells (APCs) and increased T-cell stimulation and migration mediated by cytotoxic T-lymphocyte-associated antigen-4 (CTLA-4) blockage on the T-cells. Therefore, it can be suggested that the combination of CTLA-4 blockers with this HSC-based vaccine may have a synergistic effect in multiple myeloma [132]. Another study highlights the importance of MSCs in multiple myeloma. Administering an MSCs-based vaccine from a non-diseased host could be significantly beneficial in improving disease outcomes [133].

In a Phase II clinical trial (NCT00019097), autologous stem cells (where the donor and recipient are the same) were administered to patients with multiple myeloma. A total of 60 patients have been given an autologous tumor cell vaccine in combination with sargramostim, a well-known recombinant granulocyte-macrophage-colony-stimulating factor (GM-CSF).

8.3.8 GLIOMAS

Glioma, also known as cancer or malignancy of the glial cells, is the most common type of brain tumor. Low-grade gliomas, classified as Grade I and II, are less aggressive and typically found in younger patients. On the other hand, high-grade gliomas, Grade III and IV, are more commonly seen in older adults. Glioblastoma, an aggressive form of glioma, is categorized as a stage IV glioma [134]. It has the lowest survival rates among all primary brain tumors, with an average of only 1.5 years [135].

To address the limitations of conventional cell-based glioma vaccines, a study developed a vaccine called STDENVANT. This vaccine combines glioma stem cells, CpG oligonucleotide, and DCs. In a murine model of gliomas, this therapeutic vaccine demonstrated significant efficacy. It upregulated PD-1/PD-L1, leading to an increase in Treg cells in the CNS and lymph nodes. As a result, it improved the outcomes of tumor treatment [136]. MSCs primed with TLR-4 have been demonstrated to exhibit greater stability compared to regular MSCs. Additionally, these MSCs can be formulated into micro-vesicles to enhance the immunogenic response against glioma tumor cells [137]. Another study reported the effectiveness of a DC-based vaccine transfected with RNA material from CSCs in a glioma model. Rats vaccinated with this modified vaccine showed an improved OS of 10–15 days compared to those vaccinated with a conventional DC-based vaccine. Therefore, immunotherapy using a modified vaccine that combines DCs and CSCs could be more effective in glioma treatment [138]. Sox2 is known to be an important factor in maintaining normal NSCs. Thus, vaccinating mice with CSCs after Sox2 conditional deletion mutation leads to a delay in tumor development and improved survival [139]. A DC vaccine made using the cell lysate of glioma stem cells, targeting glioma-CSCs, resulted in significant antitumor immunity mediated by cytotoxic T-lymphocytes through the surface activation of CD8, CD11, and MHC-II complexes. Glioma stem cells pulsed DC vaccine has also been shown to induce IFN-γ levels *in vitro* [87].

A Phase 1/2 clinical trial (NCT00846456) has tested a DC-based vaccine transfected with CSCs-mRNA that was injected intradermally in 20 recruited patients after receiving six-week chemoradiation post-surgery. The primary objective of the conventional study was to establish the safety of the CSCs-based vaccine in humans. The recruited patients showed no autoimmune-related or any other specific adverse effects due to the vaccine. In parallel, the secondary objective of the study was to determine the effectiveness of the vaccine, mainly based on disease progression and survival rate. Finally, they demonstrated that the progression-free survival (PFS) in vaccinated patients was 2.9 times longer compared to matched controls [140].

8.4 SUMMARY OF CLINICAL TRIALS CONDUCTED USING STEM CELL-BASED VACCINES IN VARIOUS CANCERS

To date, no stem cell-derived vaccines have been approved for the management of cancer. However, due to promising *in-vitro* results, numerous preclinical and clinical studies are being conducted worldwide to explore the use of stem cell-based vaccines in the management of various types of cancer. These vaccines are currently in different phases of preclinical/ clinical development, and the details of the ongoing clinical trials are listed in Table 8.2.

8.5 SHORTCOMINGS OF STEM CELL-BASED CANCER VACCINES

There are various risk factors associated with stem cell-based therapy, depending on the site of administration, types of cells utilized, and the method of administration. Stem cell-based CVs are supposed to be assessed based on theoretical and potential risks. AEs and clinical outcomes need to be assessed rigorously. There may be various side effects attributed to stem cell therapy, as stated in subsections.

8.5.1 TUMORIGENESIS

The IPSCs and ESCs, which are utilized in the development of whole or entire cell-based CVs, bear significant resemblance to the CSCs [143]. During the embryonal stage, these ESCs are enveloped by surrounding cells that serve as checkpoints, thereby preventing their uncontrolled growth and proliferation. However, when utilized as CVs and cultured in an external environment without these surrounding cells, ESCs may undergo uncontrolled proliferation, leading to the development of malignancies such as teratoma and teratocarcinoma [144–146].

8.5.2 GRAFT VERSUS HOST DISEASE

The allogeneic stem cell-based vaccine, in certain cases, may also trigger graft-versus-host disease through the development of B-cell derived Abs and T-cell mediators against the foreign antigen [147]. It can also lead to abnormal immune responses, which can cause damage to end organs, secondary cancers, and infections [148]. Allograft rejection may result in the loss of vaccine efficacy and elicit adverse reactions. However, ESCs and MSCs are considered to be immunologically privileged due to their very low immunogenic potential. Although, upon administration to an unusual site and differentiation, their immunogenicity might increase [149, 150].

TABLE 8.2 List of Clinical Studies/Trials for Stem Cell-based Vaccines in Management of Different Cancers

Trial Name or No.	Phase	Status	No. of Patients	Disease Condition	Intervention	Primary Endpoints	Secondary Endpoints
NCT02115958 [92]	I/II	Completed	40	Nasopharyngeal cancer	Cancer stem cell vaccine (CSC loaded DC vaccine)	No. of patients with AEs	Immune responses of the vaccine.
NCT02074046 [108]	I/II	Completed	40	Pancreatic cancer	Cancer stem cell vaccine (CSC loaded DC vaccine).	No. of patients with AEs.	Immune responses of the vaccine.
NCT02176746	I/II	Completed	40	Colorectal cancer	Cancer stem cell vaccine (CSC loaded DC vaccine).	No. of patients with AEs.	Immune responses of the vaccine.
NCT02084823 [93]	I/II	Completed	40	Lung cancer	Cancer stem cell vaccine (CSC loaded DC vaccine).	No. of patients with AEs.	Immune responses of the vaccine.
NCT02178670	I/II	Completed	40	Ovarian cancer	CSCs-DC vaccine	No. of patients with AEs.	Immune responses of the vaccine.
NCT02089919	I/II	Completed	40	Hepatocellular cancer	Cancer stem cell vaccine (CSC loaded DC vaccine).	No. of patients with AEs.	Immune responses of the vaccine.
NCT02063893	I/II	Completed	40	Breast cancer	Cancer stem cell vaccine (CSC loaded DC vaccine).	No. of patients with AEs.	Immune responses of the vaccine.
NCT00846456	I/II	Completed	20	Glioblastoma, brain malignancy.	Dendritic cell vaccine with mRNA derived from tumor stem cells.	No. of patients with AEs.	Time to disease progression, survival time.
NCT00019097	II	Completed	60	Multiple myeloma	Autologous tumor cell vaccine.	–	–
NCT00490529 [141]	II	Completed	59	Mantle cell lymphoma	CpG-MCL vaccine	Freedom from molecular residual disease.	Time to progression, overall survival, detection of tumor specific $CD8^+$ and $CD4^+$ T-cells.
NCT03334305 [142]	I	Recruiting	10	Malignant glioma	Autologous HSCs.	No. of patients with AEs.	Immune responses of the vaccine, progression free survival, overall survival.

Abbreviations: CSC: Cancer stem cells; DC: dendritic cells; AE: adverse events; mRNA: mitochondrial ribonucleic acid; and MCL: mantle cell lymphoma.

8.5.3 AUTOIMMUNITY

The whole-cell-based vaccines contain both cancer-specific as well as normal cellular antigens [151]. Hence, they can elicit an immune reaction towards the normal tissue of the body. However, a study conducted by Nigel et al. in 2018 [54] on IPSCs did not show any autoimmune reaction toward normal cells [54]. Further research should be conducted in the future to explore the safety of whole-cell-based vaccines.

8.5.4 VIRAL INTEGRATION AND TRANSGENE REACTIVATION

The xenogeneic or allogenic-based iPSCs are produced using retroviruses and lentiviruses. Consequently, this raises ethical concerns similar to the gene therapy employed for x-linked immunodeficiency disorders [152]. These viruses, along with their vectors, can integrate with oncogenes and potentially lead to the formation of new tumors (neotumorigenesis). In essence, this can reactivate transgenes or their factors, such as c-Myc, which can contribute to the development of malignancies [153]. Additionally, iPSCs can induce epigenetic changes in chromatin, further increasing the tumorigenic potential of the cells and reactivating tumor mediators like c-Myc [154]. However, more research is needed to fully understand their role in malignancy formation. Newer approaches have been developed to address this limitation, such as using small molecules to generate iPSCs and utilizing non-viral vectors for integration. Nevertheless, this method carries a risk of malignancy development depending on the mutation site [155].

8.6 WAY FORWARD

Based on the current availability of literature, a major pathogenic process of cancer can be explained by CSCs, also known as TIC cells [156]. The CSCs are not only responsible for the initiation and proliferation of tumor cells, but they are also the main cause for the development of resistance towards gross chemotherapy and other targeted therapies [157]. The resistance mechanism of CSCs is mainly mediated by various pathways, including Hh signaling, TGF-β, Notch pathway, and Wnt/β-catenin. The activation of these pathways is mediated via antigenic surface markers present on CSCs, such as CD133 [158]. The activation of Hh pathway signaling is mainly carried out via the overexpression of Hh signaling genes using CSCs, leading to chemotherapy resistance [159]. By concluding that the main reason for the uncontrolled growth of the tumor is the presence of CSCs, they can act as therapeutic targets through the production of Abs against CSCs [160]. Firstly, specific protein or genomic markers need to be identified and understood that would differentiate between non-stem cancerous cells, normal stem cells, and CSCs. Furthermore, if the surface antigens of the CSCs are targeted via vaccine-induced antibody response, it might not result in an instant reduction in tumor volume, but the subsequent absence of CSCs would prevent further tumor proliferation [161]. The surface antigens on these CSCs include CD133, which acts as an antigenic marker in CSCs of various tumors [158]. Apart from the surface markers, various protein molecules such as aldehyde dehydrogenase 1 (ALDH-1) are expressed intracellularly in CSCs of different tumors [162]. A cluster of side population (SP) cells secretes a non-surface marker dye molecule, which can help differentiate them from normal stem cells [163]. IPSCs, as a vaccine, are capable of overcoming many obstacles of oncofetal or ESC-based vaccines [73]. These stem cell-based vaccines cannot elicit an antitumor response and show a clinically significant effect when used alone [164]. Therefore, they can be used as adjuvants to other immunostimulatory agents to achieve optimal responses. Additionally, they can also be used in combination with conventional chemotherapy to achieve tumor size reduction. This combination is not limited by the issue of tumor recurrence, as the tumor will be eradicated in the absence of CSCs. Currently, IPSCs or any other stem cell-derived vaccines are derived from the patient's own tissues, re-engineered, and administered back as a vaccine. Another approach to this can be the development of a universal vaccine that is capable of inducing an immune response against a broad category of tumors [73]. As mentioned in the above text, research is currently underway to explore the potential of repurposing stem cells as CVs through animal and human trials. The aim is to demonstrate the feasibility of using these vaccines in clinical practice. However, there are still many unexplored mechanisms and safety concerns that need to be addressed before incorporating them into the treatment regimen.

8.7 CONCLUSION

Cancer is a complex condition that affects a significant portion of the global population. The rise in cancer cases and the challenges in achieving positive clinical outcomes have not been accompanied by the development of new management strategies or improvements in the effectiveness of existing therapies. Current therapies have various limitations, such as intolerable side effects and the development of resistance. Therefore, there is a need for a new approach to achieve optimal clinical outcomes.

Several therapies are being discovered, including gene-based therapies, hormonal therapies, targeted therapies, and vaccination. Among these, stem cell-based vaccination, which is discussed in detail in this text, holds particular importance as it addresses a common pathology shared by all tumors. CSCs represent a small but influential population within tumors. Given the molecular and genomic similarities between fetal stem cells and TICs, they can be considered ideal targets for the development of an anticancer vaccine. This can be achieved using different types of stem cells, including ESCs, iPSCs, and ASCs.

In conclusion, directly targeting tumor-initiating stem cells could be a significant advancement in cancer therapy.

KEYWORDS

- **cancer stem cells**
- **colorectal cancer**
- **embryonic stem cells**
- **immunotherapy**
- **induced pluripotent stem cells**
- **transgene reactivation**
- **tumorigenesis**
- **vaccines**

REFERENCES

1. Weinberg, A. D., Jackson, P. M., DeCourtney, C. A., Cravatt, K., Ogo, J., Sanchez, M. M., et al., (2010). *Progress in Addressing Disparities Through Comprehensive Cancer Control, 21*(12), 2015–2021.
2. Siegel, R. L., Miller, K. D., & Jemal, A., (2019). Cancer statistics. *CA Cancer J. Clin., 69*(1), 7–34.

3. Colditz, G. A., & Wei, E. K., (2012). Preventability of cancer: The relative contributions of biologic and social and physical environmental determinants of cancer mortality. *J. Annu. Rev. Public Health, 33*, 137.
4. Fitzmaurice, C., Akinyemiju, T. F., Al Lami, F. H., Alam, T., Alizadeh-Navaei, R., Allen, C., et al., (2018). *Global, Regional, and National Cancer Incidence, Mortality, Years of Life Lost, Years Lived with Disability, and Disability-Adjusted Life-Years for 29 Cancer Groups, 1990 to 2016: A Systematic Analysis for the Global Burden of Disease Study, 4*(11), 1553–1568.
5. Takiar, R., Nadayil, D., & Nandakumar, A., (2010). Projections of number of cancer cases in India (2010–2020) by cancer groups. *Asian Pac. J. Cancer Prev., 11*(4), 1045–1049.
6. Naran, K., Nundalall, T., Chetty, S., & Barth, S., (2018). Principles of immunotherapy: Implications for treatment strategies in cancer and infectious diseases. *J. Front. Microbiol., 9*, 3158.
7. Qin, L., Zhang, H., Zhou, Y., Umeshappa, C. S., & Gao, H., (2021). Nanovaccine-based strategies to overcome challenges in the whole vaccination cascade for tumor immunotherapy. *Small, 17*(28), 2006000.
8. Shae, D., Baljon, J. J., Wehbe, M., Becker, K. W., Sheehy, T. L., & Wilson, J. T., (2020). At the bench: Engineering the next generation of cancer vaccines. *J. Leukoc. Biol., 108*(4), 1435–1453.
9. Palucka, K., & Banchereau, J., (2013). Dendritic-cell-based therapeutic cancer vaccines. *J. Immunol., 39*(1), 38–48.
10. Ogale, P., Nikam, V. S., Gautam, M., Gairola, S., & Jadhav, S., (2021). Vaccines as immunotherapy. *Immunotherapy–A Novel Facet of Modern Therapeutics* (pp. 31–61). Springer.
11. Agius, M., Klodowska-Duda, G., Maciejowski, M., Potemkowski, A., Sweeny, S., Li, J., et al., (2015). *Safety and Tolerability of MEDI-551 in Patients with Relapsing Forms of Multiple Sclerosis: Results from a Phase 1 Randomized, Placebo-Controlled, Escalating Intravenous and Subcutaneous Dose Study, 21*(S11), 235–236.
12. Shafabakhsh, R., Pourhanifeh, M. H., Mirzaei, H. R., Sahebkar, A., Asemi, Z., & Mirzaei, H., (2019). Targeting regulatory T cells by curcumin: A potential for cancer immunotherapy. *J. Pharmacol. Res., 147*, 104353.
13. Chism, S. E., Burton, R. C., & Warner, N. L., (1978). Immunogenicity of oncofetal antigens: A review. *J. Clinical Immunology and Immunopathology, 11*(3), 346–373.
14. Cebon, J., (2018). Perspective: Cancer vaccines in the era of immune checkpoint blockade. *J. Mamm. Genome, 29*(11), 703–713.
15. Obara, W., Kanehira, M., Katagiri, T., Kato, R., Kato, Y., & Takata, R., (2018). Present status and future perspective of peptide-based vaccine therapy for urological cancer. *J. Cancer Sci., 109*(3), 550–559.
16. Russell, S. J., & Barber, G. N., (2018). Oncolytic viruses as antigen-agnostic cancer vaccines. *Cancer Cell, 33*(4), 599–605.
17. Shang, N., Figini, M., Shangguan, J., Wang, B., Sun, C., Pan, L., et al., (2017). *Dendritic Cells Based Immunotherapy, 7*(10), 2091.
18. Chauchet, X., Wang, Y., Polack, B., & Gouëllec A., (2013). Live-attenuated bacteria as a cancer vaccine vector AU-Toussaint, Bertrand. *Expert Rev. Vaccines, 12*(10), 1139–1154.

19. Gatti-Mays, M. E., Redman, J. M., Collins, J. M., & Bilusic M., (2017). Cancer vaccines: Enhanced immunogenic modulation through therapeutic combinations. *J. Hum. Vaccin. Immunother., 13*(11), 2561–2574.
20. Lopes, A., Vandermeulen, G., & Préat V., (2019). Cancer DNA vaccines: Current preclinical and clinical developments and future perspectives. *J. Exp. Clin. Cancer Res., 38*(1), 1–24.
21. Aldous, A. R., & Dong, J. Z., (2018). Personalized neoantigen vaccines: A new approach to cancer immunotherapy. *Bioorg. Med. Chem., 26*(10), 2842–2849.
22. Chen, W., Dong, J., Haiech, J., Kilhoffer, M. C., & Zeniou M., (2016). Cancer stem cell quiescence and plasticity as major challenges in cancer therapy. *Stem Cells Int., 2016.*
23. Visvader, J. E., & Lindeman, G. J., (2012). Cancer stem cells: Current status and evolving complexities. *J. Cell Stem. Cell, 10*(6), 717–728.
24. Mani, S. A., Guo, W., Liao, M. J., Eaton, E. N., Ayyanan, A., Zhou, A. Y., et al., (2008). *The Epithelial-Mesenchymal Transition Generates Cells with Properties of Stem Cells, 133*(4), 704–715.
25. Tran, L., Judor, J. P., Gauttier, V., Geist, M., Hoffman, C., Rooke, R., et al., (2012). *The Immunogenicity of the Tumor-Associated Antigen α-Fetoprotein is Enhanced by a Fusion with a Transmembrane Domain, 2012.*
26. Ouyang, X., Telli, M. L., & Wu, J. C., (2019). Induced pluripotent stem cell-based cancer vaccines. *J. Front Immunol., 10*, 1510.
27. Chowell, D., Morris, L. G., Grigg, C. M., Weber, J. K., Samstein, R. M., Makarov, V., et al., (2018). *Patient HLA Class I Genotype Influences Cancer Response to Checkpoint Blockade Immunotherapy, 359*(6375), 582–587.
28. Wang, L., Pegram, M. D., & Wu, J. C., (2019). Induced pluripotent stem cells as a novel cancer vaccine. *Expert Opin. Biol. Ther., 19*(11), 1191–1197.
29. Zheng, Q., Zheng, Y., Chen, J., You, J., Zhu, Y., Liu, Y., et al., (2017). *A Hepatic Stem Cell Vaccine is Superior to An Embryonic Stem Cell Vaccine in the Prophylaxis and Treatment of Murine Hepatocarcinoma, 37*(3), 1716–1724.
30. Samanta, T., & Kar, S., (2019). Dynamical reorganization of transcriptional events governs robust Nanog heterogeneity. *J. Phys. Chem. B., 123*(25), 5246–5255.
31. Manzo, G., (2019). Similarities between embryo development and cancer process suggest new strategies for research and therapy of tumors: A new point of view. *Front Cell Dev. Biol., 7*, 20.
32. Pires, I., Ózsvàri, B., & Ahmad, A., (2021). Giovanni Manzo. Wicha MSJAiCSCB.
33. Springer, C., Zakhartchenko, V., Wolf, E., & Simmet, K., (2021). Hypoblast formation in bovine embryos does not depend on NANOG. *Journal Cell, 10*(9), 2232.
34. Ferrer, A. I., Trinidad, J. R., Sandiford, O., Etchegaray, J. P., & Rameshwar, P., (2020). Epigenetic dynamics in cancer stem cell dormancy. *J. Cancer Metastasis Rev., 39*(3), 721–738.
35. Mahmoodi, S., Nezafat, N., Negahdaripour, M., & Ghasemi, Y., (2019). A new approach for cancer immunotherapy based on the cancer stem cell antigens properties. *J. Curr. Mol. Med., 19*(1), 2–11.
36. Hassani, S. N., Moradi, S., Taleahmad, S., Braun, T., & Baharvand, H., (2019). Transition of inner cell mass to embryonic stem cells: Mechanisms, facts, and hypotheses. *Cellular and Molecular Life Sciences: CMLS., 76*(5), 873–892.
37. Penev, A., Markiewicz-Potoczny, M., Sfeir, A., & Denchi, E. L., (2022). Stem cells at odds with telomere maintenance and protection. *J. Trends Cell Biol., 2022.*

38. Manzo, G., (2020). Defined mathematical relationships among cancer cells suggest modular growth in tumor progression and highlight developmental features consistent with a para-embryonic nature of cancer. *Front Cell Dev. Biol., 2020*, 804.
39. Walcher, L., Kistenmacher, A. K., Suo, H., Kitte, R., Dluczek, S., Strauß, A., et al., (2020). Cancer stem cells—Origins and biomarkers: Perspectives for targeted personalized therapies. *Front. Immunol., 11*, 1280.
40. Müller, L., Kschischo, M., Vokuhl, C., Stahl, D., & Gütgemann, I., (2022). Stemness correlates inversely with MHC class I expression in pediatric small round blue cell tumors. *J. Cancers (Basel), 14*(15), 3584.
41. Grinnemo, K. H., Kumagai-Braesch, M., Mânsson-Broberg, A., Skottman, H., Hao, X., Siddiqui, A., et al., (2006). *Human Embryonic Stem Cells are Immunogenic in Allogeneic and Xenogeneic Settings, 13*(5), 712–724.
42. Yaddanapudi, K., Mitchell, R. A., Putty, K., Willer, S., Sharma, R. K., Yan, J., et al., (2012). *Vaccination with Embryonic Stem Cells Protects Against Lung Cancer: Is a Broad-Spectrum Prophylactic Vaccine Against Cancer Possible? 2012.*
43. Takahashi, K., Tanabe, K., Ohnuki, M., Narita, M., Ichisaka, T., Tomoda, K., et al., (2007). *Induction of Pluripotent Stem Cells from Adult Human Fibroblasts by Defined Factors, 131*(5), 861–872.
44. Qiao, Y., Agboola, O. S., Hu, X., Wu, Y., & Lei, L., (2020). Tumorigenic and immunogenic properties of induced pluripotent stem cells: A promising cancer vaccine. *J. Stem Cell Reviews and Reports, 16*(6), 1049–1061.
45. Takahashi, K., & Yamanaka, S., (2006). Induction of pluripotent stem cells from mouse embryonic and adult fibroblast cultures by defined factors. *Cell, 126*(4), 663–676.
46. Tian, Y., Luo, A., Cai, Y., Su, Q., Ding, F., Chen, H., et al., (2010). *MicroRNA-10b Promotes Migration and Invasion Through KLF4 in Human Esophageal Cancer Cell Lines, 285*(11), 7986–7994.
47. Okita, K., Matsumura, Y., Sato, Y., Okada, A., Morizane, A., Okamoto, S., et al., (2011). *A More Efficient Method to Generate Integration-Free Human iPS Cells, 8*(5), 409–412.
48. Maekawa, M., & Yamanaka, S., (2011). Glis1, a unique pro-reprogramming factor, may facilitate clinical applications of iPSC technology. *Cell Cycle*, 3613–3614. Taylor & Francis.
49. De Almeida, P., Meyer, E. H., Kooreman, N. G., Diecke, S., Dey, D., Sanchez-Freire, V., et al., (2014). *Transplanted Terminally Differentiated Induced Pluripotent Stem Cells are Accepted by Immune Mechanisms Similar to Self-tolerance, 5*(1), 1–12.
50. Nishikawa, S. I., Goldstein, R. A., & Nierras, C. R., (2008). The promise of human induced pluripotent stem cells for research and therapy. *Nat. Rev. Mol. Cell Biol., 9*(9), 725–729.
51. Narsinh, K. H., Sun, N., Sanchez-Freire, V., Lee, A. S., Almeida, P., Hu, S., et al., (2011). *Single Cell Transcriptional Profiling Reveals Heterogeneity of Human Induced Pluripotent Stem Cells, 121*(3), 1217–1221.
52. Chalan, P., Di Dalmazi, G., Pani, F., De Remigis, A., Corsello, A., & Caturegli, P., (2018). Thyroid dysfunctions secondary to cancer immunotherapy. *J. Endocrinol. Invest., 41*(6), 625–638.
53. Bonifant, C. L., Jackson, H. J., Brentjens, R. J., & Curran, K. J., (2016). Toxicity and management in CAR T-cell therapy. *Mol. Ther. Oncolytics, 3*, 16011.

54. Kooreman, N. G., Kim, Y., De Almeida, P. E., Termglinchan, V., Diecke, S., Shao, N. Y., et al., (2018). *Autologous iPSC-Based Vaccines Elicit Anti-tumor Responses in Vivo, 22*(4), 501–513. e7.
55. Ouyang, X., Liu, Y., Zhou, Y., Guo, J., Wei, T. T., Liu, C., et al., (2021). *Antitumor Effects of iPSC-based Cancer Vaccine in Pancreatic Cancer, 16*(6), 1468–1477.
56. Li, Y., Zeng, H., Xu, R. H., Liu, B., & Li, Z., (2009). Vaccination with human pluripotent stem cells generates a broad spectrum of immunological and clinical responses against colon cancer. *Stem Cells, 27*(12), 3103–3111.
57. Brewer, B. G., Mitchell, R. A., Harandi, A., Eaton, J. W., (2009). Embryonic vaccines against cancer: An early history. *Exp. Mol. Pathol., 86*(3), 192–197.
58. Ge, Y., Gomez, N. C., Adam, R. C., Nikolova, M., Yang, H., Verma, A., et al., (2017). *Stem Cell Lineage Infidelity Drives Wound Repair and Cancer, 169*(4), 636–650. e14.
59. Clarke, M. F., (2019). Clinical and therapeutic implications of cancer stem cells. *N. Engl. J. Med., 380*(23), 2237–2245.
60. Malta, T. M., Sokolov, A., Gentles, A. J., Burzykowski, T., Poisson, L., Weinstein, J. N., et al., (2018). *Machine Learning Identifies Stemness Features Associated with Oncogenic Dedifferentiation, 173*(2), 338–354. e15.
61. Sachamitr, P., Hackett, S., & Fairchild, P. J., (2014). Induced pluripotent stem cells: Challenges and opportunities for cancer immunotherapy. *Front. Immunol., 5*, 176.
62. Wang, J., Shao, L., Wu, L., Ma, W., Zheng, Y., Hu, C., et al., (2020). *Expression Levels of a Gene Signature in hiPSC Associated with Lung Adenocarcinoma Stem Cells and its Capability in Eliciting Specific Antitumor Immune-Response in a Humanized Mice Model, 11*(6), 1603–1612.
63. Araki, R., Uda, M., Hoki, Y., Sunayama, M., Nakamura, M., Ando, S., et al., (2013). *Negligible Immunogenicity of Terminally Differentiated Cells Derived from Induced Pluripotent or Embryonic Stem Cells, 494*(7435), 100–104.
64. Kim, J. J., (2015). Applications of iPSCs in cancer research: Supplementary issue: Stem cell biology. *BMI, 10*, S20065.
65. Inui, S., Minami, K., Ito, E., Imaizumi, H., Mori, S., Koizumi, M., et al., (2017). *Irradiation Strongly Reduces Tumorigenesis of Human Induced Pluripotent Stem Cells, 58*(4), 430–438.
66. Katsukawa, M., Nakajima, Y., Fukumoto, A., Doi, D., Takahashi, J., (2016). Fail-safe therapy by gamma-ray irradiation against tumor formation by human-induced pluripotent stem cell-derived neural progenitors. *Stem Cells Dev., 25*(11), 815–825.
67. Rose-John, S., Waetzig, G. H., Scheller, J., Grötzinger, J., & Seegert, D., (2007). The IL-6/sIL-6R complex as a novel target for therapeutic approaches. *Expert Opin. Ther. Targets, 11*(5), 613–624.
68. Mackiewicz, J., Burzykowski, T., Iżycki, D., & Mackiewicz, A., (2018). Re-induction using whole cell melanoma vaccine genetically modified to melanoma stem cells-like beyond recurrence extends long term survival of high risk resected patients-updated results. *J. Immunother. Cancer, 6*(1), 1–8.
69. Kitadani, J., Ojima, T., Iwamoto, H., Tabata, H., Nakamori, M., Nakamura, M., et al., (2018). *Cancer Vaccine Therapy Using Carcinoembryonic Antigen-Expressing Dendritic Cells Generated from Induced Pluripotent Stem Cells, 8*(1), 1–11.
70. Kitayama, S., Zhang, R., Liu, T. Y., Ueda, N., Iriguchi, S., Yasui, Y., et al., (2016). *Cellular Adjuvant Properties, Direct Cytotoxicity of Re-differentiated Vα24 Invariant NKT-Like Cells from Human Induced Pluripotent Stem Cells, 6*(2), 213–227.

71. Crunkhorn, S., (2018). iPSC-based cancer vaccine. *Nature Reviews Drug Discovery, 17*(4), 242.
72. Bernardes De, J. B., Neves, B., Ferreira, M., & Nobrega-Pereira, S., (2020). Strategies for cancer immunotherapy using induced pluripotency stem cells-based vaccines. *Cancers, 12*, 3581.
73. Manoutcharian, K., Guzman, V. J., & Gevorkian, G., (2021). Neoantigen cancer vaccines: Real opportunity or another illusion? *Arch. Immunol. Ther. Exp. (Warsz), 69*(1), 1–5.
74. Kindy, M. S., Yu, J., Zhu, H., Smith, M. T., & Gattoni-Celli, S., (2016). A therapeutic cancer vaccine against GL261 murine glioma. *J. Transl. Med., 14*(1), 1–9.
75. Shammaa, R., El-Kadiry, A. E. H., Abusarah, J., & Rafei, M., (2020). Mesenchymal stem cells beyond regenerative medicine. *Front. Cell Dev. Biol., 8*, 72.
76. Choi, J., Yoo, S., Park, S., Kang, Y., Kim, W., Oh, I., et al., (2008). *Mesenchymal Stem Cells Overexpressing Interleukin-10 Attenuate Collagen-Induced Arthritis in Mice, 153*(2), 269–276.
77. Tomchuck, S. L., Norton, E. B., Garry, R. F., Bunnell, B. A., Morris, C. A., Freytag, L. C., et al., (2012). *Mesenchymal Stem Cells as a Novel Vaccine Platform, 2*, 140.
78. Ren, C., Kumar, S., Chanda, D., Chen, J., Mountz, J. D., & Ponnazhagan, S., (2008). Therapeutic potential of mesenchymal stem cells producing interferon-α in a mouse melanoma lung metastasis model. *Stem Cells, 26*(9), 2332–2338.
79. Wei, H. J., Wu, A. T., Hsu, C. H., Lin, Y. P., Cheng, W. F., Su, C. H., et al., (2011). *The Development of a Novel Cancer Immunotherapeutic Platform Using Tumor-Targeting Mesenchymal Stem Cells and a Protein Vaccine, 19*(12), 2249–2257.
80. Van, M. K. M., Van't, W. E. J. T., Lages, M. J., Dekker, B., Nikolic, T., & Roep, B. O., (2019). Activated mesenchymal stromal cells process and present antigens regulating adaptive immunity. *Front Immunol., 10*, 694.
81. Lu, J. H., Peng, B. Y., Chang, C. C., Dubey, N. K., Lo, W. C., Cheng, H. C., et al., (2018). *Tumor-Targeted Immunotherapy by Using Primary Adipose-Derived Stem Cells and an Antigen-Specific Protein Vaccine, 10*(11), 446.
82. Welniak, L. A., Blazar, B. R., Murphy, W. J., (2007). Immunobiology of allogeneic hematopoietic stem cell transplantation. *Annu. Rev. Immunol., 25*, 139–170.
83. Gutierrez-Delgado, F., Holmberg, L., Hooper, H., Appelbaum, F., Livingston, R., Maziarz, R., et al., (2000). *High-Dose Busulfan, Melphalan and Thiotepa as Consolidation for Non-inflammatory High-Risk Breast Cancer, 26*(1), 51–59.
84. Holmberg, L. A., Oparin, D. V., Gooley, T., & Sandmaier, B. M., (2003). The role of cancer vaccines following autologous stem cell rescue in breast and ovarian cancer patients: Experience with the STn-KLH vaccine (Theratope®). *Clin. Breast Cancer, 3*, S144–S51.
85. Edward A. Copelan, (2006). *Hematopoietic Stem-Cell Transplantation, 354*(17), 1813–1826.
86. Khaddour, K., Hana, C. K., & Mewawalla, P., (2021). Hematopoietic stem cell transplantation. *StatPearls [internet]*. StatPearls Publishing.
87. Ji, B., Chen, Q., Liu, B., Wu, L., Tian, D., Guo, Z., et al., (2013). *Glioma Stem Cell-Targeted Dendritic Cells as a Tumor Vaccine Against Malignant Glioma, 54*(1), 92–100.
88. Kanojia, D., Balyasnikova, I. V., Morshed, R. A., Frank, R. T., Yu, D., Zhang, L., et al., (2015). *Neural Stem Cells Secreting Anti-HER2 Antibody Improve Survival in a*

Preclinical Model of HER2 Overexpressing Breast Cancer Brain Metastases, 33(10), 2985–2994.

89. Dubinett, S., & Sharma, S., (2009). Towards effective immunotherapy for lung cancer: Simultaneous targeting of tumor-initiating cells and immune pathways in the tumor microenvironment. *Journal Immunotherapy, 1*(5), 721–725.
90. Morrison, B. J., Steel, J. C., & Morris, J. C., (2018). Reduction of MHC-I expression limits T-lymphocyte-mediated killing of cancer-initiating cells. *BMC Cancer, 18*(1), 1–10.
91. Hu, Y., Lu, L., Xia, Y., Chen, X., Chang, A. E., Hollingsworth, R. E., et al., (2016). *Therapeutic Efficacy of Cancer Stem Cell Vaccines in the Adjuvant Setting Therapeutic Efficacy of CSC Vaccine in the Adjuvant Setting, 76*(16), 4661–4672.
92. Lin, M., Yuan, Y., Xu, J., Cai, X., Liu, S., Niu, L., et al., (2015). *Safety and Efficacy Study of Nasopharyngeal Cancer Stem Cell Vaccine, 165*(1), 26–31.
93. Lin, M., Li, S. Y., Xu, K. C., Liu, Z. P., Mu, F., Yuan, Y. Y., et al., (2015). *Safety and Efficacy Study of Lung Cancer Stem Cell Vaccine, 62*(1), 16–22.
94. Society ac. Ovarian Cancer. [Ovarian cancer is one of the most aggressive gynecological cancers, with an OS rate of 43.6% in the past three decades].
95. Skates, S. J., Greene, M. H., Buys, S. S., Mai, P. L., Brown, P., Piedmonte, M., et al., (2017). *Early Detection of Ovarian Cancer Using the Risk of Ovarian Cancer Algorithm with Frequent CA125 Testing in Women at Increased Familial Risk–Combined Results from Two Screening Trials Ovarian Cancer Early Detection: ROCA in Increased Risk Women, 23*(14), 3628–3637.
96. Alvero, A. B., Chen, R., Fu, H. H., Montagna, M., Schwartz, P. E., Rutherford, T., et al., (2009). *Molecular Phenotyping of Human Ovarian Cancer Stem Cells Unravels the Mechanisms for Repair and Chemoresistance, 8*(1), 158–166.
97. Jain, S., Annett, S. L., Morgan, M. P., & Robson, T., (2021). *The Cancer Stem Cell Niche in Ovarian Cancer and Its Impact on Immune Surveillance, 22*(8), 4091.
98. Dashti, A., Ebrahimi, M., Hadjati, J., Memarnejadian, A., & Moazzeni, S. M., (2016). Dendritic cell based immunotherapy using tumor stem cells mediates potent antitumor immune responses. *Cancer Lett., 374*(1), 175–185.
99. Gil, M., Komorowski, M. P., Seshadri, M., Rokita, H., McGray, A. R., Opyrchal, M., et al., (2014). *CXCL12/CXCR4 Blockade by Oncolytic Virotherapy Inhibits Ovarian Cancer Growth by Decreasing Immunosuppression and Targeting Cancer-Initiating Cells, 193*(10), 5327–5337.
100. Brown, J. R., Chan, D. K., Shank, J. J., Griffith, K. A., Fan, H., Szulawski, R., et al., (2020). *Phase II Clinical Trial of Metformin as a Cancer Stem Cell–Targeting Agent in Ovarian Cancer, 5*(11).
101. Ning, N., Pan, Q., Zheng, F., Teitz-Tennenbaum, S., Egenti, M., Yet, J., et al., (2012). *Cancer Stem Cell Vaccination Confers Significant Antitumor Immunity, 72*(7), 1853–1864.
102. Zhang, Z. J., Chen, X. H., Chang, X. H., Ye, X., Li, Y., & Cui, H., (2012). Human embryonic stem cells–a potential vaccine for ovarian cancer. *Asian Pac. J. Cancer Prev., 13*(9), 4295–4300.
103. Zhang, Z. J., Chen, X. H., Chang, X. H., Ye, X., Li, Y., & Cui, H., (2012). Human embryonic stem cells—A potential vaccine for ovarian cancer. *Asian Pacific Journal of Cancer Prevention: APJCP, 13*(9), 4295–4300.

104. Wefers, C., Schreibelt, G., Massuger, L. F., De Vries, I. J. M., & Torensma R., (2018). Immune curbing of cancer stem cells by CTLs directed to NANOG. *Front. Immunol., 9*, 1412.
105. Seicean, A., Petrusel, L., & Seicean R., (2015). New targeted therapies in pancreatic cancer. *World J. Gastroenterol., 21*(20), 6127.
106. Wang V. M.Y., Ferreira, R. M., Almagro, J., Evan, T., Legrave, N., Zaw, T. M., et al., (2019). *CD9 Identifies Pancreatic Cancer Stem Cells and Modulates Glutamine Metabolism to Fuel Tumor Growth, 21*(11), 1425–1435.
107. Lu, S., Zhang, Z., Du, P., Chard, L. S., Yan, W., El Khouri, M., et al., (2020). *A Virus-Infected, Reprogrammed Somatic Cell–Derived Tumor Cell (VIReST) Vaccination Regime Can Prevent Initiation and Progression of Pancreatic Cancer A Novel Prophylactic Vaccination for Pancreatic Cancer, 26*(2), 465–476.
108. Lin, M., Yuan, Y. Y., Liu, S. P., Shi, J. J., Long, X. A., Niu, L. Z., et al., (2015). *Prospective Study of the Safety and Efficacy of a Pancreatic Cancer Stem Cell Vaccine, 141*(10), 1827–1833.
109. Bray, F., Ferlay, J., Soerjomataram, I., Siegel, R. L., Torre, L. A., & Jemal, A., (2018). Global cancer statistics 2018: GLOBOCAN estimates of incidence and mortality worldwide for 36 cancers in 185 countries. *CA: A Cancer Journal for Clinicians, 68*(6), 394–424.
110. Waidmann, O., (2018). Recent developments with immunotherapy for hepatocellular carcinoma. *Expert Opin. Biol. Ther., 18*(8), 905–910.
111. Qin, Y., & Liao, P., (2018). Hepatitis B virus vaccine breakthrough infection: Surveillance of S gene mutants of HBV. *Acta Virol., 62*(2), 115–121.
112. Chen, J., Ding, Y., Huang, F., Lan, R., Wang, Z., Huang, W., et al., (2021). *Irradiated Whole-Cell Vaccine Suppresses Hepatocellular Carcinoma Growth in Mice Via Th9 Cells, 21*(5), 1–9.
113. Duarte, S., Momier, D., Baqué, P., Casanova, V., Loubat, A., Samson, M., et al., (2013). *Preventive Cancer Stem Cell-Based Vaccination Reduces Liver Metastasis Development in a Odel, 31*(3), 423–432.
114. Xi, Y., & Xu, P., (2021). Global *Rat Colon Carcinoma Syngeneic* Mcolorectal cancer burden in 2020 and projections to 2040. *Transl. Oncol., 14*(10), 101174.
115. Mattiuzzi, C., Sanchis-Gomar, F., & Lippi, G., (2019). Concise update on colorectal cancer epidemiology. *Annals of Translational Medicine, 7*(21), 609.
116. Pflügler, S., Svinka, J., Scharf, I., Crncec, I., Filipits, M., Charoentong, P., et al., (2020). *IDO1+ Paneth Cells Promote Immune Escape of Colorectal Cancer, 3*(1), 1–14.
117. Lupo, B., Sassi, F., Pinnelli, M., Galimi, F., Zanella, E. R., Vurchio, V., et al., (2020). *Colorectal Cancer Residual Disease at Maximal Response to EGFR Blockade Displays a Druggable Paneth Cell–Like Phenotype, 12*(555), eaax8313.
118. Güllülü, Ö., Hehlgans, S., Rödel, C., Fokas, E., & Rödel, F., (2021). Tumor suppressor protein p53 and inhibitor of apoptosis proteins in colorectal cancer—A promising signaling network for therapeutic interventions. *Journal Cancers, 13*(4), 624.
119. Liebl, M. C., & Hofmann, T. G., (2021). The role of p53 signaling in colorectal cancer. *Cancers (Basel), 13*(9), 2125.
120. Szaryńska, M., Olejniczak, A., Kobiela, J., Łaski, D., Śledziński, Z., & Kmieć, Z., (2018). Cancer stem cells as targets for DC-based immunotherapy of colorectal cancer. *Scientific Reports, 8*(1), 12042.

121. Kopenhaver, J., Crutcher, M., Waldman, S. A., & Snook, A. E., (2021). The shifting paradigm of colorectal cancer treatment: A look into emerging cancer stem cell-directed therapeutics to lead the charge toward complete remission. *Expert Opin. Biol. Ther., 21*(10), 1335–1345.
122. Maruoka, S., Ojima, T., Iwamoto, H., Kitadani, J., Tabata, H., Tominaga, S., et al., (2022). *Tumor RNA Transfected DCs Derived from iPS Cells Elicit Cytotoxicity Against Cancer Cells Induced from Colorectal Cancer Patients in Vitro, 12*(1), 1–14.
123. Fu, C., Tian, G., Duan, J., Liu, K., Zhang, C., Yan, W., et al., (2021). *Therapeutic Antitumor Efficacy of Cancer Stem Cell-Derived DRibble Vaccine on Colorectal Carcinoma, 18*(14), 3249.
124. Guo, M., You, C., Dong, W., Luo, B., Wu, Y., Chen, Y., et al., (2020). *The Surface Dominant Antigen MUC1 is Required for Colorectal Cancer Stem Cell Vaccine to Exert Anti-tumor Efficacy, 132*, 110804.
125. Roy, S., Sethi, T. K., Taylor, D., Kim, Y. J., & Johnson, D. B., (2020). Breakthrough concepts in immune-oncology: Cancer vaccines at the bedside. *Journal of Leukocyte Biology, 108*(4), 1455–1489.
126. Abbaspour, M., & Akbari, V., (2022). Cancer vaccines as a targeted immunotherapy approach for breast cancer: An update of clinical evidence. *Expert Rev. Vaccines, 21*(3), 337–353.
127. Pallerla, S., Abdul, A. Ur. M., Comeau, J., & Jois, S., (2021). Cancer vaccines, treatment of the future: With emphasis on her2-positive breast cancer. *Int. J. Mol. Sci., 22*(2), 779.
128. Cecil, D. L., Slota, M., O'Meara, M. M., Curtis, B. C., Gad, E., Dang, Y., et al., (2017). *Immunization Against HIF-1α Inhibits the Growth of Basal Mammary Tumors and Targets Mammary Stem Cells in Vivo HIF-1α Vaccine Targets Breast Cancer Stem Cells, 23*(13), 3396–3404.
129. Wang, L. X., & Plautz, G. E., (2012). T cells sensitized with breast tumor progenitor cell vaccine have therapeutic activity against spontaneous HER2/neu tumors. *Breast Cancer Res. Treat., 134*(1), 61–70.
130. Munshi, N. C., Anderson Jr, L. D., Shah, N., Madduri, D., Berdeja, J., Lonial, S., et al., (2021). *Idecabtagene Vicleucel in Relapsed and Refractory Multiple Myeloma, 384*(8), 705–716.
131. Sidana, S., & Shah, N., (2019). CAR T-cell therapy: Is it prime time in myeloma? *Hematology Am. Soc. Hematol. Educ. Program., 3*(21), 3473–3480.
132. Shinde, P., Melinkeri, S., Santra, M. K., Kale, V., & Limaye, L., (2019). Autologous hematopoietic stem cells are a preferred source to generate dendritic cells for immunotherapy in multiple myeloma patients. *Front Immunol., 10*, 1079.
133. Reagan, M. R., & Ghobrial, I. M., (2012). Multiple myeloma mesenchymal stem cells: Characterization, origin, and tumor-promoting effects mesenchymal stem cells in multiple myeloma. *Clin. Cancer Res., 18*(2), 342–349.
134. Weng, W., Chen, X., Gong, S., Guo, L., & Zhang, X., (2018). Preoperative neutrophil–lymphocyte ratio correlated with glioma grading and glioblastoma survival. *Neurol. Res., 40*(11), 917–922.
135. Xu, L. W., Chow, K. K., Lim, M., & Li, G., (2014). Current vaccine trials in glioblastoma: A review. *J. Immunol. Res., 2014*.
136. Zhu, S., Lv, X., Zhang, X., Li, T., Zang, G., Yang, N., et al., (2019). *An Effective Dendritic Cell-Based Vaccine Containing Glioma Stem-Like Cell Lysate and CpG Adjuvant for an Orthotopic Mouse Model of Glioma, 144*(11), 2867–2879.

137. Kukia, N. R., Alipanah-Moghadam, R., Delirezh, N., & Mazani, M., (2018). Mesenchymal stromal stem cell-derived microvesicles enhance tumor lysate pulsed dendritic cell stimulated autologous T lymphocyte cytotoxicity. *Asian Pac. J. Cancer Prev., 19*(7), 1895.
138. Bagheri, V., Abbaszadegan, M. R., Memar, B., Motie, M. R., Asadi, M., Mahmoudian, R. A., et al., (2019). *Induction of T Cell-Mediated Immune Response by Dendritic Cells Pulsed with mRNA of Sphere-Forming Cells Isolated from Patients with Gastric Cancer, 219*, 136–143.
139. Barone, C., Pagin, M., Serra, L., Rigoldi, L., Giubbolini, S., Badiola-Sanga, A., et al., (2018). *Sox2 Functions in Neural Cancer Stem Cells: The Importance of the Context, 2*, 18–26.
140. Vik-Mo, E. O., Nyakas, M., Mikkelsen, B. V., Moe, M. C., Due-Tønnesen, P., Suso, E. M. I., et al., (2013). *Therapeutic Vaccination Against Autologous Cancer Stem Cells with mRNA-Transfected Dendritic Cells in Patients with Glioblastoma, 62*(9), 1499–1509.
141. Faham, M., Brody, J., Kohrt, H. E., Czerwinski, D. K., & Levy, R., (2013). High-throughput sequencing of T cell receptors detects signatures of T cell repertoire that predict MRD status in patients with mantle cell lymphoma undergoing immunotransplantation. *Journal Blood, 122*(21), 4251.
142. Wildes, T. J., Flores, C. T., Dean, B. D., Grippin, A., Dyson, K., Mitchell, D. A., (2019). Adoptive cellular therapy overcomes tumor-induced dysregulation of myelopoiesis. *Cancer Research, 79*(13_Supplement), 1194.
143. Zhu, P., & Fan, Z., (2018). Cancer stem cells and tumorigenesis. *Biophys. Rep., 4*(4), 178–188.
144. La Noce, M., Paino, F., Mele, L., Papaccio, G., Regad, T., Lombardi, A., et al., (2018). *HDAC2 Depletion Promotes Osteosarcoma's Stemness Both in Vitro and in Vivo: A Study on a Putative New Target for CSCs Directed Therapy, 37*(1), 1–15.
145. Lezmi, E., & Benvenisty, N., (2022). The tumorigenic potential of human pluripotent stem cells. *Stem Cells Translational Medicine, 2022.*
146. Andrews, P. W., (2021). Human pluripotent stem cells: Tools for regenerative medicine. *Biomater. Transl., 2*(4), 294.
147. Ozdemir, Z. N., & Bozdağ, S. C., (2018). Graft failure after allogeneic hematopoietic stem cell transplantation. *Transfusion and Apheresis Science, 57*(2), 163–167.
148. Chu, D. T., Nguyen, T. T., Tien, N. L. B., Tran, D. K., Jeong, J. H., Anh, P. G., et al., (2020). *Recent Progress of Stem Cell Therapy in Cancer Treatment: Molecular Mechanisms and Potential Applications, 9*(3), 563.
149. Wang, L. T., Liu, K. J., Sytwu, H. K., Yen, M. L., & Yen, B. L., (2021). Advances in mesenchymal stem cell therapy for immune and inflammatory diseases: Use of cell-free products and human pluripotent stem cell-derived mesenchymal stem cells. *Stem Cells Transl. Med., 10*(9), 1288–1303.
150. Han, Y., Li, X., Zhang, Y., Han, Y., Chang, F., & Ding, J., (2019). Mesenchymal stem cells for regenerative medicine. *Journal Cells, 8*(8), 886.
151. Pers, Y. M., Jorgensen, C., & Khoury, M., (2021). The role of metabolism in MSC-mediated immunomodulation. *Front. Immunol., 12*, 751865.
152. Nayerossadat, N., Maedeh, T., & Ali, P., (2012). *Viral and Nonviral Delivery Systems for Gene Delivery, 1*(1), 27.

153. Dhenni, R., & Phan, T. G., (2020). The geography of memory B cell reactivation in vaccine-induced immunity and in autoimmune disease relapses. *Immunol. Rev., 296*(1), 62–86.
154. Barati, M., Akhondi, M., Mousavi, N. S., Haghparast, N., Ghodsi, A., Baharvand, H., et al., (2021). *Pluripotent Stem Cells: Cancer Study, Therapy, and Vaccination, 17*(6), 1975–1992.
155. Herberts, C. A., Kwa, M. S. G., & Hermsen, H. P. H., (2011). Risk factors in the development of stem cell therapy. *Journal of Translational Medicine, 9*(1), 29.
156. Biserova, K., Jakovlevs, A., Uljanovs, R., & Strumfa, I., (2021). Cancer stem cells: Significance in origin, pathogenesis, and treatment of glioblastoma. *Journal Cells, 10*(3), 621.
157. Maccalli, C., Rasul, K. I., Elawad, M., & Ferrone, S., (2018). The role of cancer stem cells in the modulation of anti-tumor immune responses. *Seminars in Cancer Biology, 2018*. Elsevier.
158. Aghajani, M., Mansoori, B., Mohammadi, A., Asadzadeh, Z., & Baradaran, B., (2019). New emerging roles of CD133 in cancer stem cell: Signaling pathway and miRNA regulation. *J. Cell Physiol., 234*(12), 21642–21661.
159. Zhang, H., Hu, L., Cheng, M., Wang, Q., Hu, X., & Chen, Q., (2020). The hedgehog signaling pathway promotes chemotherapy resistance via multidrug resistance protein 1 in ovarian cancer. *Oncol. Rep., 44*(6), 2610–2620.
160. Razi, E., Radak, M., Mahjoubin-Tehran, M., Talebi, S., Shafiee, A., Hajighadimi, S., et al., (2020). *Cancer Stem Cells as Therapeutic Targets of Pancreatic Cancer, 34*(2), 202–212.
161. Shrestha, S., Banstola, A., Jeong, J. H., Seo, J. H., & Yook, S., (2022). Targeting cancer stem cells: Therapeutic and diagnostic strategies by the virtue of nanoparticles. *J. Control Release, 348*, 518–536.
162. Han, J., Won, M., Kim, J. H., Jung, E., Min, K., Jangili, P., et al., (2020). *Cancer Stem Cell-Targeted Bio-imaging and Chemotherapeutic Perspective, 49*(22), 7856–7878.
163. Ahmed, N., Kadife, E., Raza, A., Short, M., Jubinsky, P. T., & Kannourakis, G., (2020). Ovarian cancer, cancer stem cells and current treatment strategies: A potential role of magmas in the current treatment methods. *Journal Cells, 9*(3), 719.
164. Masae Heront-Kishi, Afag Asgarova, Christophe Desterke, Diana Chaker, Marie-Ghislaine de Goër de Herve, Ali G Turhan, Annelise Bennaceur-Griscelli, & Frank Griscelli (2020). *Pharmacologically Modified Pluripotent Stem Cell-Based Cancer Vaccines with Anti-metastatic Potential, 2020.*

CHAPTER 9

Preventive Cancer Vaccine

SACHIN P. BHATT,[1] AVANI N. JOSHI,[1] and BHUPENDRA G. PRAJAPATI[2]

[1]*Gyanmanjari Pharmacy College, Bhavnagar, Gujarat, India*

[2]*Shree S.K. Patel College of Pharmaceutical Education and Research, Ganpat University, Mehsana, Gujarat, India*

ABSTRACT

Despite recent advances in diagnosis, prevention, and treatment, cancer continues to be the world's most common cause of death and affects millions of people globally. The development of cancer vaccines (CVs) is advancing significantly as a result of the US FDA's clearance of several preventative CVs, including Gardasil (Merck), Cervarix (Glaxosmithkline), and the therapeutic vaccine Sipulencel-T (Provenge). The idea of treating cancer via cellular immunotherapy has gained support thanks to the approval of these vaccinations. Recent results indicate that immunotherapy may be an effective tool in the fight against cancer. Immunotherapy more precisely targets cancer cells than conventional surgery, chemotherapy, or radiotherapy, providing patients the chance to have higher response rates, a better quality of life, or possibly a cure for their condition. CVs may be created to target tumor-specific antigens (TSAs), also known as neoantigens, tumor-associated antigens (TAAs), cancer germline antigens, virus-associated antigens, or TAAs. With various types of adjuvants, the CVs could be cell-based (such as the dendritic cell (DC) vaccine provenge (sipuleucel-T) targeting prostatic acid phosphatase for metastatic prostate cancer (PC)), peptide/protein-based, or gene (DNA/RNA)-based. Neoantigens are tumor-specific compounds that can be presented by MHC molecules and recognized by T cells, making them ideal immunological targets for enhancing treatment specificity and reducing the risk of nonspecific autoimmunity. The CV has the potential to

Cancer Vaccination and Challenges, Volume 1: Strategies for Therapeutic Cancer Vaccine Development.
Rishabha Malviya, Bhupendra Prajapati, and Sonali Sundram (Eds.)

treat the condition by targeting specific private epitopes and shared antigens. By reducing the frequency and duration of vaccinations, the feasibility of CVs can be improved without compromising effectiveness. However, the extensive length of *in vivo* research and the multitude of potential variations in vaccination procedures have hindered their optimization. Current CV trials are limited by the failure of immunologic assays to track clinically significant indicators of response. Recent advancements in the measurement and phenotyping of circulating tumor cells (CTCs) in breast cancer patients suggest that CTCs may play a role in assessing the efficacy of vaccine-based immunotherapy. Certain types of head and neck cancers, as well as cervical, vulvar, vaginal, and anal cancers, have all been associated with the human papillomavirus (HPV). Therefore, effective HPV vaccination offers an opportunity to reduce the morbidity and mortality associated with HPV.

9.1 INTRODUCTION

Cancer is a condition in which abnormal cells multiply uncontrollably and can infiltrate surrounding tissues. Almost any location in the body, which contains trillions of cells, can become the site of cancer. Human cells typically divide in order to expand, proliferate, and create new cells as needed by the body. New cells replace old ones when they die as a result of aging or damage. The three primary types of tumors are as follows:

1. **Benign (Noncancerous):** These tumors are benign. They do not spread to other areas of the body or infiltrate neighboring tissue.
2. **Pre-Malignant:** Although the cells in these tumors are not currently cancerous, they have the potential to develop into cancer.
3. **Malignant:** This type of cancer is present in tumors [1]. Breast, lung, colon, rectum, and PC are the most prevalent cancers. About one-third of cancer deaths are caused by smoking, having a high body mass index, consuming alcohol, eating few fruits and vegetables, and not exercising [2]. Cancer cells differ from healthy cells in several ways. For instance, cancer cells grow without signals instructing them to do so. Normal cells only expand when they receive such signals. Cancer cells also disregard signals that would ordinarily instruct cells to cease dividing or to undergo programmed cell death, known as apoptosis. They invade surrounding regions and spread to other body parts, while most normal cells do not travel across the body and cease growing when they come into contact with other

cells. Blood vessels should expand near tumors. These blood vessels transport waste from tumors and provide oxygen and nutrients to the tumors. Cancer cells also try to evade the immune system, which frequently eliminates damaged or abnormal cells [3].

9.1.1 DIFFERENT TYPES OF CANCER/CARCINOMA

Various types of cancer are summarized in Figure 9.1.

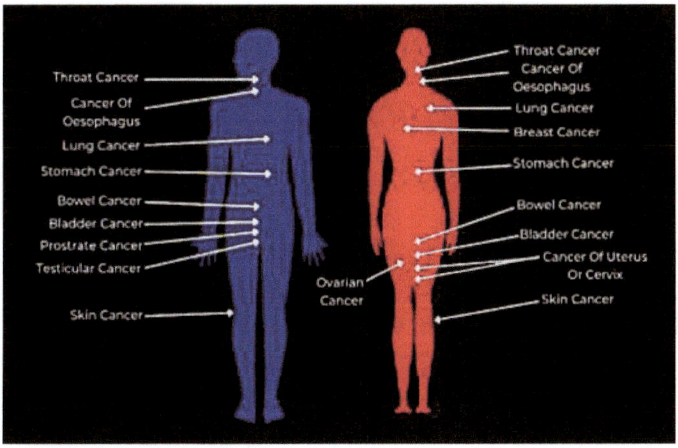

FIGURE 9.1 Metastatic organotropism.

Source: Adopted from: Fares et al. [4].

9.1.2 ETIOLOGY OF CANCER

Numerous cancers, particularly in adults, have been linked to frequent exposures or risk factors, as previously mentioned. A risk factor is something that could increase the likelihood of developing a disease. Although a risk factor doesn't always cause the disease, it can weaken the body's defenses. The following mechanisms and risk factors have been proposed as potential cancer triggers:

1. **Lifestyle Factors:** Smoking, consuming a high-fat diet, and working with hazardous chemicals are among the lifestyle choices that may increase the risk of developing certain adult cancers [5].
2. **Family History:** Genetics, inheritance, and family history may play significant roles in some childhood malignancies. It is possible for multiple cases of cancer to occur within a family.

3. **Exposures to Certain Viruses:** Certain childhood malignancies, such as Hodgkin and non-Hodgkin lymphomas, have been associated with an increased chance of developing the Epstein-Barr virus and HIV, the virus that causes AIDS. The virus can induce changes in a cell, leading to the formation of an altered cell, which eventually develops into a cancer cell that continues to divide and multiply [6].
4. **Environmental Exposures:** Research has been conducted on the direct connection between childhood malignancies and exposures to pesticides, fertilizers, and electrical lines [7].

9.1.3 CANCER EPIDEMIOLOGY

Based on the most recent population-based data for cancer incidence and mortality, as well as the anticipated numbers of cases and fatalities in 2018 from the World Health Organization (WHO) and Global Cancer Observatory (GLOBOCAN), a comprehensive overview of cancer incidence will be provided. Additionally, an estimation of the total number of cancer deaths that have been prevented up until 2019 will be presented. This is attributed to the continuous decline in cancer death rates since the early 1990s [8] (Figure 9.2).

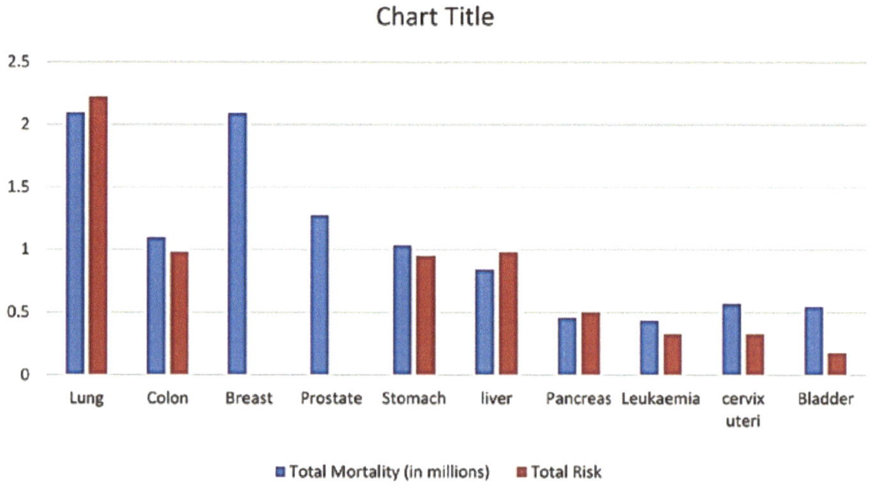

FIGURE 9.2 Estimated epidemiological data on cancer, overall death rate, and mortality rate in millions.

Source: Modified from: WHO [8].

9.1.4 *CANCER AND ITS MOLECULAR MECHANISM*

Cancer and its molecular mechanisms are summarized in Figure 9.3.

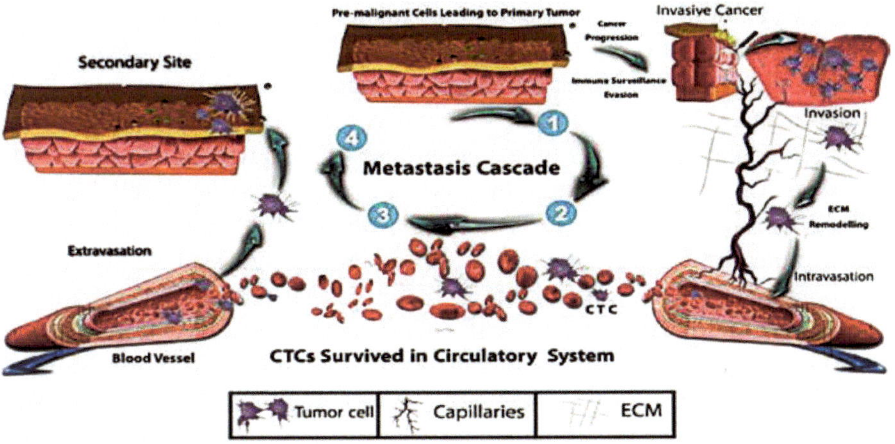

FIGURE 9.3 CTCs mechanism.

Source: Modified from: Chen et al. [9].

9.1.5 *SIGNS AND SYMPTOMS OF CANCERS*

Signs and symptoms of cancer are shown in Figure 9.4.

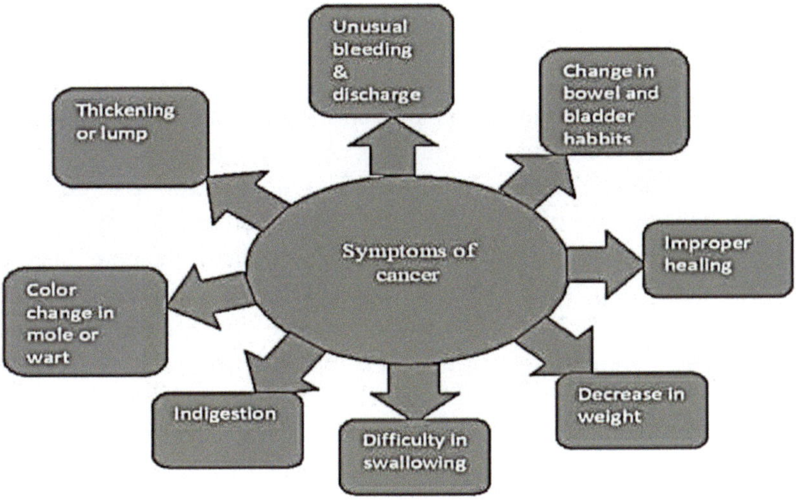

FIGURE 9.4 Symptoms of cancer.

Source: Modified from: Sawicki et al. [10].

9.1.6 CANCER VACCINE (CV)

New vaccination technologies have the ability to stop many common ailments, and vaccinations have been employed to stop and treat many disorders for more than 200 years. Prophylactic or preventative cancer vaccinations may reduce the incidence of the disease by inducing a particular immunological reaction that will attack cancer before it may manifest. There are several techniques to create CVs for delivery, counting by fusing with antigen-presenting cells (APCs), encapsulating in liposomes, or incorporating stabilizing agents like polyethylene glycol (PEG) and pan-DR helper T-lymphocyte epitope. CVs can contain a wide range of different ingredients, such as peptides and carbohydrates (PADRE). Currently, five CVs that protect against either the human papillomavirus (HPV) or the hepatitis B virus (HBV) are being used in clinics to prevent a variety of cancers, including oral and cervical cancer. Prophylactic CVs can induce humoral (B cell, or antibody-mediated), cellular (T cell), or a combination of the two types of adaptive responses. These cancer prevention vaccines are currently being studied and possess the capacity to both stop or retard the growth of tumors as well as decline the cases of several malignancies. When it comes to eliciting an adaptive immune response, each vaccine has benefits and drawbacks [11].

9.2 PREVENTIVE CANCER VACCINE

Nearly 10 million people died from cancer in 2018, making it the second leading cause of death worldwide after heart disease. Furthermore, the inadequate state of current cancer prevention medications compromises the quality of life for individuals at high risk. As a result, there is an urgent need for innovative cancer therapy options [12]. One alternative to conventional medications is the use of CVs. The term "immunotherapies" typically refers to the management of chronic illnesses, while the term "vaccine" specifically describes preventive immunological treatments. To differentiate between these two modalities, we will use the phrases "therapeutic vaccines" and "prophylactic vaccinations" in addition to the term "vaccines" to encompass all immune interventions. The development of both existing and novel vaccines can benefit from the implementation of modeling approaches. The ultimate objective of a vaccination is to activate immune system cells, such as T lymphocytes that directly target tumor cells or B lymphocytes that produce antibodies (mAbs). In order for this activation to occur, antigens on the surface of the cell must be recognized by Abs in their original protein state [13]. The ultimate outcome of a vaccination is the activation of immune system cells, such as T lymphocytes that eliminate tumor cells or B lymphocytes that generate Abs. On the surface of the cell, antigens must be recognized by Abs in their original protein state. Once attached, Abs can kill tumor cells by either complement-mediated cytotoxicity or antibody-dependent cellular cytotoxicity. T lymphocytes identify proteins on the surface of identified cells as major histocompatibility complex (MHC) complexes with peptides of various sizes. Recent studies have shown that the vaccine strategy may potentially be helpful for both cancer prevention and treatment. Thus, the use of tumor vaccines to elicit an immune response against tumor forms with weak immunogenicity is possible. Abs, dendritic cells (DCs), and the deployment of cytotoxic T lymphocytes (CTLs) are all well-known methods of cancer immunotherapy [13]. Cancer immune therapy often includes both apathetic and vital treatments, and it can be targeted or non-targeted. Instead of primarily using the patient's immune system to fight cancer cells, a passive immunological response uses immune cells or other substances that are first created outside the body and then administered to the patient (*ex vivo*). By supplying an instant and powerful immune cell force, this procedure is thought to support and increase the patient's own immune system's ability to combat tumor growth. When a patient's immune system is severely compromised, passive immune therapy can be beneficial since it avoids the need to activate endogenous immunity. Natural killer (NK) cells are needed in passive immunotherapy in the form of Abs in order to induce antibody-dependent cell-mediated cytotoxicity. The goal of a CV is to either initiate a de novo immune response against tumor antigens or to reawaken an earlier, less potent anti-tumor immunity. In contrast to the severe side effects associated with chemotherapy, relatively modest side effects (local erythema, flu-like symptoms) may develop as a result of cancer-specific activation of the host immune system. Cancer vaccination has primarily been studied in a therapeutic environment, that is, in patients who have already established a malignancy, as opposed to preventative immunization, which aims to guard against potential future exposure to an infectious agent. Since many infectious diseases have been significantly reduced or eradicated during the past century, vaccination for disease prevention has had a significant impact on the burden of infectious diseases globally. However, the high expectations for therapeutic cancer vaccination have not yet been reached. This is mostly because the immune-suppressive milieu was little recognized until a few years ago [15] (Table 9.1).

TABLE 9.1 List of the Primary Cancer Immunotherapies

Sl. No.	Cancer Immune-Therapy	Mode of Action	References
1.	Monoclonal antibodies	Immune checkpoints are regulatory molecules that maintain immune system balance in normal conditions. They play a dual role in protecting healthy tissues from excessive immune responses and activating lymphocytes to effectively eliminate infections. This is achieved by providing T cells with a series of co-stimulatory or co-inhibitory signals through specific receptors.	[16]
2.	Adoptive cell transfer	Effective adoptive T-cell therapy (ACT) involves the therapeutic use of transplanted T cells to eliminate cancer cells. The use of chimeric antigen receptor T cells is one of the key ACT strategies. By allowing T cells to bind target cell surface antigens via a single-chain variable fragment (scFv) recognition domain, CAR T cells facilitate MHC-unrestricted tumor cell killing.	[17]
3.	Oncolytic virotherapy	Oncolytic viruses (OVs) are gaining popularity as potential clinical therapeutic agents. Most recently, the United State Food and Drug Administration approved an OV based on the Herpes simplex virus type 1 (HSV-1) for the treatment of melanoma. Oncolytic virotherapy (OVT), which primarily operates through a dual mode of action, is being enhanced through various combination techniques, leveraging the significant potential of OVs in combating cancer.	[18]
4.	Immunosuppressive agent	Based on current immunopathophysiological information and extrapolations of treatment options for primary autoimmune illnesses, immunosuppressive drugs are particularly recommended for severe or steroid-refractory cases.	[19]
5.	Intratumoral PRR-agonists	The intravenous route of administration is frequently used to administer a variety of anticancer immunotherapeutic drugs, such as monoclonal immune checkpoint blocking antibodies, toll-like receptor (TLR) agonists, cytokines, and immunostimulatory mRNA.	[2]
6.	Cancer vaccination	cancer vaccines work to maintain the proper level of interaction between the host immune system and tumor cells. The discovery of tumor antigens with the potential to strengthen anti-tumor immunity has been made possible by recent improvements in our understanding of the nature of tumor-mediated tolerogenicity and antigen presentation.	[21]

9.3 STRATEGIES FOR PREVENTIVE CANCER VACCINATION

CVs can be divided into two primary categories. There are two types of vaccines: therapeutic vaccines used to treat cancer that has already spread by enhancing the immune system's antitumor capability, and preventative vaccines designed to prevent normal cells in healthy individuals from developing cancer. The key challenges in cancer vaccination include finding the right antigen, adjuvant, and immune reactions. It has been demonstrated that peptide, protein, whole tumor cells, DC, DNA, and viral vaccines can all induce tumor-antigen-specific immunity, each with different strengths and limitations [22].

9.3.1 PEPTIDE-BASED VACCINES

A peptide CV is a type of subunit vaccine that utilizes a specific peptide from the complex tumor antigen for vaccination. Tumor-specific peptides are identified based on their ability to bind to specific HLA-I or HLA-II molecules. Targeting HLA class I molecules offers the advantage that a significant portion of the population (up to 35–50%) may possess an individual copy of HLA-I [23]. The peptides that bind to HLA class I contain unique sequences, which increases the number of sequences required for the vaccine to ensure appropriate population coverage. Recent research has shown that a single HLA-DR peptide can bind to multiple HLA-DR subtypes. Consequently, developing a variety of vaccines may require fewer HLA class II peptides compared to HLA class I peptides. Peptide-based CVs have several advantages over protein vaccines, including lower costs, simplicity in synthesis, formulation, and delivery. However, the emergence of antigen-loss variations in tumor cells can pose a challenge when using specific peptide CVs. The use of polyvalent tumor antigens, such as allogeneic whole tumor cell vaccines, may help mitigate the risk of antigen loss variations and improve population coverage [24].

9.3.2 WHOLE-CELL VACCINATIONS FOR TUMORS

Whole tumor cell vaccination is a successful method of administering tumor cells that have been killed or attenuated, often in combination with co-stimulatory substances such as cytokines. This process falls into three major categories of vaccines: autologous, allogeneic, and gene-modified. Unlike specialized peptide vaccinations, whole tumor cell vaccination allows the immune system to naturally recognize the majority of immunogenic tumor-specific antigens (TSAs), eliminating the need for MHC restriction-specific epitope discovery [25]. However, the presence of co-stimulatory adjuvants could potentially lead to autoimmune reactions if self-tolerance to normal molecules breaks down. Studies have shown that genetically modifying dormant melanoma cells to secrete immunostimulatory molecules, such as GM-CSF, enhances tumor antigen presentation by attracting DCs and macrophages, resulting in the production of tumor-specific CD4 and CD8 T cells, NKT cells, and Abs for effective tumor rejection. Currently, research is underway to investigate the use of autologous whole tumor cell methods as treatment options for metastatic non-small cell lung cancer (NSCLC) and acute myeloid leukemia [26].

9.3.3 HEAT-SHOCK PROTEIN VACCINES

From a patient's tumor cells, heat-shock proteins (HSPs) express a variety of unknown immunogenic tumor antigens that can be isolated and used as a polyvalent autologous CV formulation. The role of HSPs in tumor antigen presentation and cross-presentation is still largely unknown. According to certain studies, HSPs can bind to and deliver tumor antigens to APCs via HLA-I and II molecules, activating CD8 and CD4 T lymphocytes specific to the tumor. The capacity of HSPs to act as adjuvants and activate both innate and adaptive immune responses underlies the efficacy of HSP vaccinations [27]. When surface HSP receptors such as c-type lectin receptors (CLRs) and scavenger receptors are activated, DCs undergo a maturation process that enables them to act as powerful APCs. Late phase II clinical trials using the Hsp7-HPV 16 fusion protein vaccine in women with advanced cervical cancer demonstrated robust immune responses and significant lesion reduction [28].

9.3.4 VACCINATIONS BASED ON ANTI-IDIOTYPIC ANTIBODIES

Upon injection of tumor-specific monoclonal antibodies (mAbs), autologous Abs against the initial vaccination may develop. These latter Abs, also referred to as anti-idiotypic Abs, are specific to the original mAb's idiotopes [29]. The vaccines under investigation are made of immunoglobulins (Ig) generated from tumors that contain tumor-specific idiotypes. Anti-idiotypic vaccines are significant because they can effectively immunize against non-protein antigens, including lipid or carbohydrate antigens specific to tumors, in order to elicit an effective T cell memory [30].

9.3.5 DENDRITIC CELL (DC)-BASED VACCINES

Even in patients with advanced illness who have tried every available treatment option, DC therapy is a current, healthy, and promising type of cancer treatment and preventative medication. For tumor biologists, using DCs in cancer medications to trigger the most efficient tumor antigen-specific response makes perfect sense because they are the most skilled activators of naive T cells [31]. DCs transduced with viral vectors such as lentiviruses, retroviruses, fowlpox, adenoviruses, and alphaviruses, whole-killed and protein extract of tumor cells, and DCs fused with tumor cells using hybrid technology are only a few of the DC-based CVs that have been created to date. Different functional T CD4 subsets can be activated or suppressed differently by each functional subpopulation of DC [32].

9.3.6 DNA AND VIRAL VECTOR-BASED VACCINES

By injecting genetically modified DNA encoding the appropriate antigens, DNA vaccination offers several distinctive advantages over other vaccination processes. These advantages include the capacity to elicit a wide range of immunological responses, ease of production, low cost, and the lack of a need for knowledge regarding the HLA-I and II genotypes. Similar to protein-based vaccinations, DNA vaccines rely on antigen-presenting cells (APCs) to digest and display the antigen. The goal of DNA vaccines is to introduce a specific tumor antigen gene into the body using a bacterial vector [33]. The first observation of a plasmid-induced immune response occurred when mice were inoculated with a plasmid encoding human growth hormone, resulting in the production of Abs. In reality, tumor antigen-expressing host cells invade the lymph node, secrete the specific tumor antigen, and then, after being picked up by DCs and presented to T cells or identified by B cells, initiate a variety of immunological reactions. Unlike peptide-based vaccinations that rely on a single arm of immunity, DNA vaccines activate a wide array of effector arms of the immune system. By incorporating and encoding DNA into viral vectors, DNA vaccination platforms can be designed to be more immunogenic. For years, live recombinant viral vaccines have been investigated as potential cancer vaccinations. Viral vectors offer a viable method for delivering tumor antigens since they simulate a real infection and provide the danger signals needed to fully activate DC. A number of viral vectors have been developed based on poxviruses, including avian poxviruses, fowlpox, modified vaccinia virus Ankara (MVA), canarypox, recombinant adenoviruses, and herpes virus. The first trial with viral vectors was conducted using vaccinia over two decades ago [34].

9.3.7 PARTICLE-BASED VACCINE

Microparticles, emulsions, immune-stimulating complexes (ISCOM), liposomes, virosomes, and virus-like particles (VLPs) are increasingly being used in vaccination formulations as carriers that deliver the cancer antigen to DC in a more immunogenic form. One advantage of particle platforms is their ability to carry multiple copies of the tumor antigen on the surface, which can be as potent as those internalized by DCs, macrophages, and B cells [35]. Another benefit of particle platforms is their similarity in size to pathogens. Furthermore, to enhance DC activation and elicit the most effective adaptive immune response, danger signals and co-stimulators such as heat shock proteins (HSP) and unmethylated CpG DNA should be included with the tumor antigens during formulation [36].

9.3.8 COMBINATIONAL THERAPY

Combining CVs with other immunotherapeutic and non-immunotherapeutic techniques, such as mAbs, chemotherapy, radiation therapy, and surgery, is likely to result in the most successful cancer treatment. Following a cancer vaccination (referred to as the "prime") with other immunotherapeutics and/or non-immunotherapeutics (referred to as the "boost") may result in the best-in-class cancer therapy regimen [37]. There are numerous preclinical and clinical contexts for combinational therapy. Pretreatment of tumor cells with Trastuzumab increases the cytotoxicity of HER-2 specific T lymphocytes. It is hypothesized that the antibody causes HER-2 to be internalized and degraded, increasing the amount of HER-2-MHC presented and ultimately enhancing the activation and growth of HER-2-specific T-cells. Clinical trials are currently underway to examine the possibility of increased efficacy with this combination. Immune-based therapies that target CD25 and utilize substances and procedures to deplete or inhibit circulating Treg, such as cyclophosphamide, anti-CTLA4, and *Denileukin diftitox* [38] (Figure 9.5).

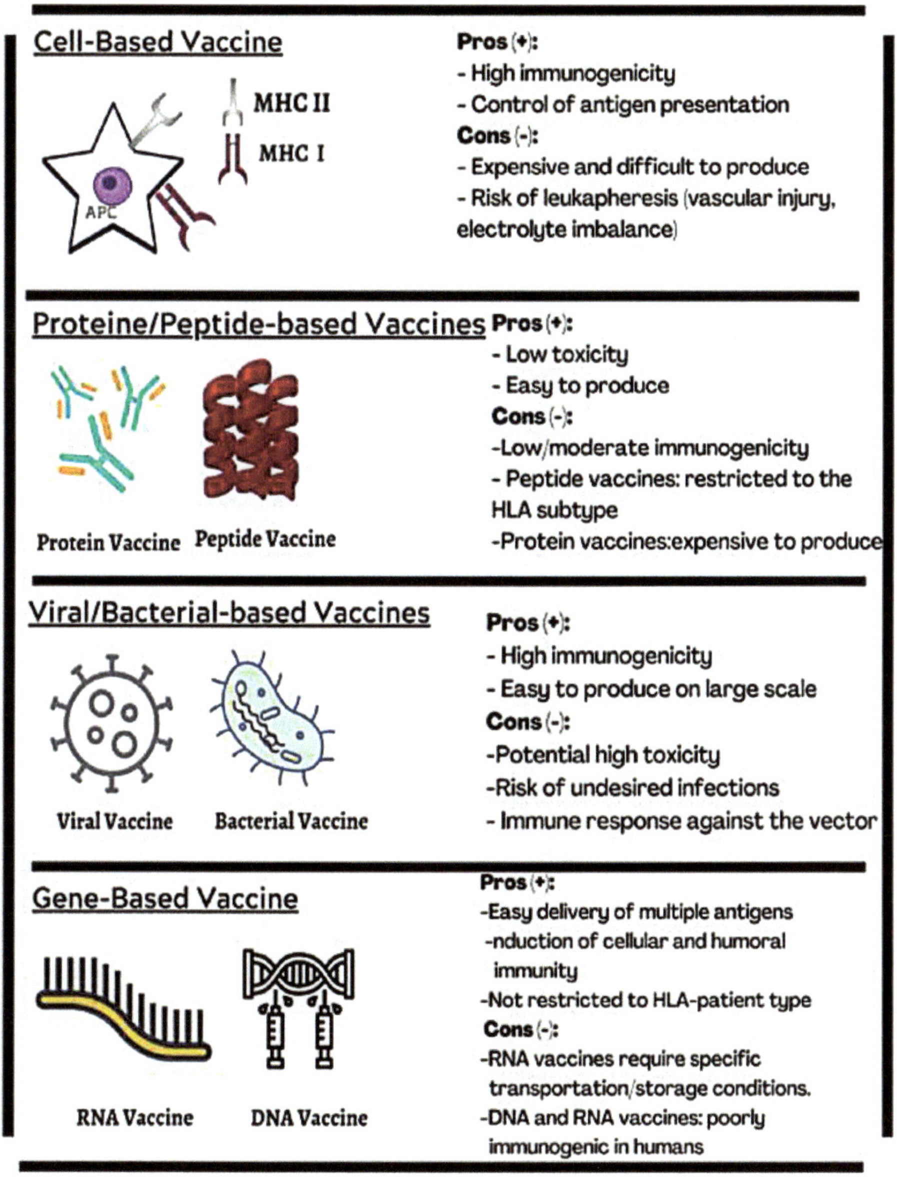

FIGURE 9.5 Strategies to develop cancer vaccine.

Source: Modified from: Singh et al. [39].

9.4 CASE STUDY OF PREVENTIVE CANCER VACCINE

9.4.1 HPV VACCINE FOR CERVICAL CANCER

It is widely recognized that cervical cancer screening can reduce the occurrence and mortality of the disease [40]. Various nations employ different screening methods for cervical cancer. Organized cervical screening programs have the potential to be more effective than opportunistic screening due to their ability to achieve high participation rates at scheduled intervals, ensuring equal access and excellent diagnostic criteria [41, 42]. High-income countries such as the UK, Australia, Canada, Finland, and Singapore are examples of nations that have implemented coordinated programs for cervical cancer screenings. On the other hand, Eastern European countries experience higher cervical cancer incidence and mortality rates due to opportunistic screening, inadequate screening coverage, and insufficient vaccination coverage [43]. Opportunistic cervical cancer screening, characterized by Romanowsky staining and often limited or undiscovered coverage, is prevalent in most Central Asian countries, the Caucasus region, the Russian Federation, and the Western parts of the former Soviet Union [44]. Developed countries have significantly reduced cervical cancer mortality and prevalence over the past 40 years, primarily due to the introduction of cytological smear screening [45]. For instance, in the US, the implementation of coordinated cervical cancer screening initiatives and an 83% screening rate have led to a decrease in cervical cancer deaths and incidence since the introduction of the Pap smear cytology test in the 1950s and 1960s [46]. Between 6% and 8% of the population is covered by screening programs in underdeveloped nations [47]. Recently, international screening recommendations have been published, and HPV testing is now included whenever possible [48]. The effectiveness of cervical cancer screening programs varies based on resource availability, execution strategies, laboratory and pathology testing quality, and community awareness, despite significant advancements in disease understanding and efficient screening [49]. Screening and HPV genotyping require expensive materials and specialized personnel, making it challenging for many impoverished countries [50]. Even in wealthy nations with robust healthcare systems and established cervical cancer screening methods, population coverage is not ideal [51]. The frequency of screening exams also varies between nations and age groups [52]. For women between the ages of 21 and 29 in countries like England and the USA, screening is recommended every three years, while for those between the ages of 30 and 65, it is suggested every five years [53, 54]. The objective of early detection screening is well known, according to the findings of a population-based survey of adults in England between the ages of 50 and 70. However, screening has not been able to prevent cancer across all demographic groups [55]. There are, however, a few exceptions. For example, the Central Asian nation of Kyrgyzstan does not have a cervical cancer screening program. In South Africa, a national cervical screening strategy was established in 2000, which allows for three free cervical smear tests to be conducted every 10 years, starting at the age of 30 [56]. Despite organized screening programs in some countries, underserved, uninsured, and underrepresented populations still do not fully benefit from screening due to issues such as cost, access, anxiety, discomfort with the screening process, fear of cancer, or low health literacy. These factors are all associated with poor outcomes for cervical cancer [57].

9.5 LIST OF FDA APPROVED PREVENTIVE CANCER VACCINE

Vaccines are drugs that support the body's immune system. They can teach the immune system to identify and eliminate dangerous microorganisms and cells. You receive a variety of vaccinations throughout your life to protect against common illnesses. There are also cancer vaccinations available. Both cancer prevention and cancer treatment vaccines are available. HPV, a relatively prevalent sexually transmitted infection, is responsible for certain malignancies. The HPV vaccine offers defense against the HPV strains that frequently result in these malignancies. Although it can be administered as early as age 9, HPV vaccination is advised for preteens between the ages of 11 and 12. Everyone up to the age of 26 is advised to get the HPV vaccine if they haven't previously. The HPV vaccine is not advised for anyone over the age of 26. However, some adults between the ages of 27 and 45 who have not yet received the HPV vaccine may opt to do so after discussing their risk for developing new HPV infections and the potential advantages of vaccination with their doctor [58] (Table 9.2).

- **Side Effects:** Vaccinations are typically safe, necessary, and likely to reduce morbidity and mortality. However, they do carry a potential risk of triggering autoimmune disorders. Like any medication, vaccines can have adverse effects. Mild side effects such as arm pain, redness, or swelling at the injection site, fever, feeling weak or dizzy, vomiting, fatigue, headache, joint or muscle aches, and autonomic dysfunction can occur after HPV vaccinations. These symptoms are usually temporary and subside within a day or two. It is worth noting that fainting, although more common among adolescents, can occur after any vaccine, including the HPV vaccine [64].

TABLE 9.2 List of FDA Approved Cancer Vaccine

Sl. No.	Name of Vaccine	Use	Company	References
1.	**HPV vaccine**	If administered before girls or women are exposed to the virus, this vaccination can prevent the majority of cases of cervical cancer.	CERVAVAC, Brentford, UK	[59]
2.	Hepatitis B vaccine	Preventing persistent hepatitis B infections prevents liver cancer caused by the hepatitis B virus.	Serum Institute of India – GeneVac-B – Hepatitis B Vaccine (rDNA) I.P.	[60]
3.	Sipuleucel-T	A prostate cancer vaccination is used to treat the disease after it has spread to other regions of the body.	Dendreon Corporation	[61]
4.	Gardasil 9	If the vaccination is administered before girls or women are exposed to the virus, it can prevent the majority of occurrences of cervical cancer.	Merck & Co, Inc., Kenilworth, NJ, USA,	[62]
5.	T-VEC	T-VEC is created using a herpesvirus variant that has been altered in the lab to attack and destroy cancer cells while sparing healthy ones.	Talimogene laher-parepvec (T-VEC; IMLYGIC®, Amgen Inc.)	[63]

Source: Adopted from: Margolis et al. [63].

9.5.1 CERVARIX VACCINE

Cervarix is a vaccine against specific strains of the HPV, which can cause cancer. Cervical cancer, a cancer that develops from the cervix and has the potential to infiltrate or spread to other parts of the body, is the second most prevalent cancer in women worldwide. It is estimated that there are 4,93,000 new cases and 2,74,000 fatalities from cervical cancer each year. The majority of cases, approximately 83%, occur in underdeveloped nations, with Nigeria accounting for 15% of all female cancer cases [38].

Following HPV infections, precursor lesions can develop, which may progress to cervical intraepithelial neoplasia and cervical cancer. Cervical cancer is more prevalent in Africa compared to previous years. In 2012, the survival rates for cervical cancer were 70% in the US, 66% in Western Europe, and only 21% in sub-Saharan Africa. Researchers examined the variables related to the use of Cervarix among students at the Schools of Nursing and Midwifery in Anambra State, Nigeria, due to the higher occurrence and death rates of cervical cancer. The study's findings could assist the government and other private organizations in designing initiatives to increase the availability and use of the Cervarix vaccination. This, in turn, would support the development of Nigeria's nursing and midwifery curriculum and emphasize the importance of the cervical CV [21].

9.5.2 HPV VACCINE

It is challenging to elicit a T-cell response that is specific for tumor antigens and does not attack normal cells. In many types of cancer, tumor cells activate genes that are only expressed during embryogenesis, resulting in the development of an antigenic target. However, many tumors are caused by mutations in genes that, under normal physiological circumstances, are essential for maintaining cellular homeostasis. Since mutations develop between the sites of original tumors, metastases, or recurrences, these malignancies are more challenging to treat. Neoantigens may be produced as a result of these alterations. Targeting "neoantigens" has the benefit because T-cells specialized for them are, by definition, not anergic or tolerant. Neoantigens can be produced by intrinsic mutation or by traditional cancer treatment modalities like chemotherapy or radiation [67]. Without first identifying neoantigens, immune-based therapy for malignancies brought on by infectious infections like HPV can be directed against the pathogen itself. Delivery methods that have been tested so far include naked DNA, HPV peptides, HPV proteins, live attenuated bacterium vectors, live attenuated viral vectors, and inactivated viral vectors [68]. HPV types 6, 11, 16, 18, 31, 33, 45, 52, and 58 infection can be managed and prevented with the 9-valent human papillomavirus vaccination (9vHPV). It belongs to the group of drugs called vaccines. The most common sexually transmitted viruses (STVs) to cause precancerous and cancerous lesions are HPVs. The US Food and Drug Administration (FDA) has licensed the 9-valent human papillomavirus (HPV) vaccine (9vHPV). Recombinant 9vHPV is produced from the L1 VLP carcinogenic protein subunit component of the types of HPV 6, 11, 16, 18, 31, 33, 45, 52, and 58. Inactive HPV L1 VLPs in the vaccine, as indicated by a 2016 immunogenicity study, elicit the production of neutralizing Abs against various HPV strains and trigger a strong humoral immune response that provides protection against HPV-related diseases and dysplastic lesions. The same study revealed that antibody titers for 9vHPV are 10 to 100 times higher than those generated by a natural infection. An intramuscular injection of 9vHPV is administered in the deltoid or anterolateral thigh region. For both adult and pediatric patients, a monotherapy consists of a 0.5 mL suspension. Depending on the patient's age at initial vaccination, a two- or three-dose schedule is followed for 9vHPV immunization [69].

9.5.2.1 USE IN SPECIFIC PATIENT POPULATION

1. **Renal Impairment:** The product labeling from the manufacturer does not provide any information regarding dose modification in individuals with renal impairment [70].
2. **Pregnancy:** According to the CDC and the American College of Obstetricians and Gynecologists (ACOG), the HPV vaccine is not recommended for use during pregnancy. It should be noted that if the vaccination is unintentionally administered to a pregnant woman, the safety data is reassuring. However, if a patient starts the vaccine series and subsequently becomes pregnant, it is advisable to delay further doses until after the pregnancy is completed [71].
3. **Considerations for Breastfeeding:** According to the CDC, immunizations given to nursing mothers do not affect the safety of breastfeeding for both mothers and infants. For example, a slightly higher proportion of breastfed infants who received the active quadrivalent HPV vaccine developed pneumonia within 30 days of maternal vaccination. However, the vaccine itself was not responsible for any adverse effects. Maternal vaccination does not prohibit breastfeeding, and healthcare providers may administer the HPV vaccine to nursing patients [72].

9.5.2.2 ADVERSE EFFECTS

Syncope, systemic events, and injection-site events are the most frequently reported side effects associated with 9vHPV. Reactions at the injection site, such as pain, swelling, erythema, and tenderness, were observed within five days of immunization. Systemic events, including headaches, pyrexia, lethargy, and nausea, were noted within 15 days. Additionally, syncope following the administration of the 9vHPV vaccination has been documented, which poses a significant risk of major secondary damage for patients [73].

9.5.2.3 CONTRAINDICATIONS

The only absolute contraindication to vaccination is an allergy to a known component of any vaccine. Therefore, 9vHPV is contraindicated in individuals with a history of hypersensitivity reactions to yeast, as it is a recombinant vaccine produced in Saccharomyces cerevisiae (brewer's yeast). The CDC advises against vaccination in cases of moderate or severe acute illness, and healthcare providers should postpone vaccinations until the symptoms of the acute illness subside. Additional contraindications include anaphylaxis to a previous dose of the quadrivalent vaccines for 9vHPV or 4vHPV [74].

9.5.2.4 TOXICITY

9vHPV is a vaccine with a high safety profile that is generally well-tolerated. Minor systemic adverse effects, such as headaches, pyrexia, tiredness, and nausea, have also been observed. Minor injection-site side effects include discomfort, edema, erythema, and soreness. The toxic effects of anaphylaxis and hypersensitivity reactions are more severe. For optimal response, IM epinephrine administration should be repeated every 5 to 15 minutes [75].

9.5.2.5 APPLICATION OF PREVENTIVE HPV VACCINES

This level of vaccination coverage is sufficient to achieve herd immunity in males within the general population, as well as a significant reduction in hrHPV prevalence and related cervical lesion rates. Importantly, assessments of global distribution of HPV vaccinations by location and economic level reveal that many disadvantaged populations of women worldwide receive minimal or no vaccinations [76]. One contributing factor is the relatively high cost of current vaccinations. Financing for vaccination programs competes with funding for screening-based strategies, and immunization offers little to no benefit to individuals who have already contracted HPV79. In fact, modeling suggests that, despite being less effective, the brief introduction of catch-up vaccination for older girls would delay the reduction of cervical cancer incidence by several years. Targeting adolescent girls before their first period was considered the most effective method for preventing cervical cancer, and this approach has been highly successful [77]. However, given the demonstrated long-lasting effects of HPV vaccination thus far, it may be conceivable to introduce a childhood vaccination schedule and even administer HPV vaccines in the same vial as other childhood vaccines as part of existing schedules to alleviate strain on infrastructure. To ensure that no child reaches sexual maturity without receiving the HPV vaccine, it would be necessary to gradually replace the current immunization regimen for teenagers. Despite the availability of licensed vaccines targeting specific hrHPVs, there is still a possibility that other hrHPV strains could cause disease. As a result, cervical screening, although expensive, is still performed even in individuals who have been vaccinated. The key performance indicators for cervical cytology, colposcopy referral criteria, colposcopy practice, and histology reporting have been modified due to the decreased incidence of disease in vaccinated women [78]. This impact could potentially be mitigated by extending the time between screenings and/or transitioning to HPV screening, including self-screening, as the primary method. However, implementing this dual strategy of promoting screening and immunization remains challenging, particularly in underdeveloped nations where logistics and costs are significant barriers. Nonetheless, childhood immunization has proven effective against various pathogens, and leveraging previous pediatric immunization regimens is a successful approach to enhance vaccination coverage and achieve substantial population benefits. The development of an effective multivalent formulation that can target the full range of HPV strains associated with illness remains a major challenge for L1 VLP technology. The 9vHPV vaccination protects against the seven most common types of cervical cancer. However, adding more L1 VLPs to protect against the other high-risk types, each of which only causes a very small percentage of malignancies, is likely to increase the price of the vaccine to an unaffordable level [79]. An alternative strategy is to find a single broadly protective antigen. The amino terminus of L2 contains numerous well-conserved protective epitopes, including residues 17 to 36, which are identified by the monoclonal antibody RG1. Early-stage therapy studies have investigated both strategies and found them to be immunogenic and well-tolerated. However, prophylactic efficacy was not tested in these short trials, and there was little evidence of therapeutic activity. Despite the implementation of preventative HPV vaccination programs being slow and uneven, there is still a need to screen for and treat HPV-associated cervical illness. The outlined issues need to be resolved for preventive HPV vaccination to be significantly improved worldwide [80].

9.5.3 GARDASIL-9 VACCINE

A purified main capsid L1 protein of HPV types, reassembled into VLPs, serves as the active ingredient in Gardasil9, a recombinant HPV 9-valent vaccination. Gardasil9 requires a potent adjuvant to generate high and long-lasting levels of Abs due to the use of pure recombinant proteins as antigens. Previously, it was understood that the adjuvant for Gardasil9, amorphous aluminum hydroxyphosphate sulfate (AAHS), necessitated one or more Toll-like receptor (TLR) agonists, such as the phospholipids found in the recombinant hepatitis B vaccine, RecombivaxHB®. However, neither the pure HPV L1 proteins nor Gardasil9 contain phospholipids [81]. Nevertheless, according to the FDA, Gardasil4 does contain unique recombinant DNA fragments specific to HPV L1, which may function as TLR-9 agonists in Gardasil9. Sanger sequencing revealed that HPV 18, 11, 16, and 6 L1 gene DNA were attached to insoluble AAHS nanoparticles (NPs), although their distribution within the same vaccine sample was irregular. These fragments exhibited non-B conformations, at least in part. Further research is necessary to optimize the HPV DNA molecules in Gardasil, which are well-known for their potent TLR-9 agonistic properties [82].

9.6 PROSEQUENCES AND CONSEQUENCES AND FUTURE CHALLENGES OF PREVENTIVE CANCER VACCINE

CV research is poised to make further groundbreaking discoveries, following the success of checkpoint inhibitors, adoptive cell therapy, and monoclonal Abs targeting tumor antigens. In cases where clinically advanced malignancies cannot be effectively managed by vaccination alone, the synergistic effects of combination therapy with other methods will be crucial. By using CVs to stimulate immune responses and subsequently amplifying them with inhibitors of negative regulatory immunologic signals, it is possible to enhance the efficacy of checkpoint blockade and other immunoregulatory mechanism inhibitors against "cold" tumors that are not naturally immunogenic. CVs have the potential to address the current challenges of low response rates and efficacy in existing cancer immunotherapy regimens [83]. Therefore, it is sensible to investigate CVs as a prophylactic or preventative measure, encompassing both primary and secondary prevention. For instance, intralesional administration of a DC vaccination targeting HER2/neu in ductal carcinoma *in situ* has shown promising results in reducing or eliminating HER2/neu expression. Research is also being conducted on preventative vaccines for high-risk patients with Lynch syndrome or colorectal cancer (CRC) with microsatellite instability (MSI) [84].

Immunotherapy offers enduring effectiveness and broad adaptability in the treatment of various tumor types. Its persistence lies in reactivating the body's immune system, enabling it to eliminate tumor cells over an extended period. The treatment can effectively halt tumor spread and recurrence, while also restoring and enhancing the body's immune system. It demonstrates the ability to fully detect, locate, and eliminate tumor cells with diligence. Furthermore, it can completely eradicate tiny lesions and residual tumor cells from the body. Compared to conventional treatments, immunotherapy has minimal adverse effects [85].

The treatment options are limited, and users are highly selective. Immunotherapy yields poor results when used to treat tumors of the "immune suppression type" and "immune exclusion type." The use of immunocheckpoint inhibitors carries the risk of unfavorable regulation, which may lead to autoimmune diseases and even death. After treatment, some patients may experience various unintended negative effects, including hyperprogressive illness, which could accelerate the patients' demise. Numerous factors influence the effectiveness of immunotherapy, and patients' chances of survival and prognosis remain unknown. The cost of treatment is high [14].

There are issues with the distribution of vaccinations, including the specific administration route. One example is the availability of CVs. Outside of cutting-edge research centers, it is challenging to promote the use of certain vaccines in the community due to their dependence on manufacturing or delivery facilities. The vaccination platform should be made more accessible to successfully increase the utilization of CVs. Unlike vaccinations against infectious organisms, which benefit from knowledge gained from therapeutically successful CVs, the optimal interval spacing, or doses of CVs are largely unknown [65].

A proven and vitally important method for treating cancer patients is cancer immunotherapy. Given the extensive clinical inquiry and research efforts devoted to expanding both endogenous and synthetic immunotherapy techniques, there is a need to concentrate on key issues and identify impediments to the development of clinical practice and fundamental knowledge. Here, we outline the top 10 obstacles to cancer immunotherapy, which range from a lack of trust in translating pre-clinical results to the need to determine the best immunotherapy combinations for any individual patient. It will take the combined efforts of fundamental scientists and medical professionals to overcome these obstacles, as well as the concentration of resources to hasten the knowledge of the intricate connections between cancer and the immune system and the creation of better cancer treatment alternatives [66].

The 10 important obstacles to cancer immunotherapy:

- Creation of pre-clinical simulations of human immunity.
- Identifying the key factors that influence cancer immunity.
- Recognizing the context of organ-specific tumor immunity.

- Understanding the distinctions between primary and secondary immune escape drivers at the molecular and cellular levels.
- Outlining the advantages of natural immunity against artificial immunity.
- Early clinical studies can efficiently and quickly evaluate combinations of cancer immunotherapy.
- Characterization of the impact of steroid and immunological suppression on autoimmune toxicities and cancer immunotherapy in full.
- Using composite biomarkers to enhance tailored methods.
- Improving regulatory endpoints for immunotherapy for cancer.
- Using combination regimens of multiple agents for cancer immunotherapy can potentially increase long-term survival [73].

KEYWORDS

- **cancer immunoprevention**
- **cancer immunotherapy**
- **cancer vaccine**
- **central nervous system**
- **circulating tumor cells**
- **pathogen recognition receptor**
- **sexually transmitted virus**
- **T-cells**
- **tumor specific antigens**

REFERENCES

1. Butenschoen, V. M., Nehiba, A., Meyer, B., & Wostrack, M., (2022). Neuropathic pain after spinal intradural benign tumor surgery: An underestimated complication? *Neurosurgical Review*, 1–7.
2. Steenland, K., & Winquist, A., (2021). PFAS and cancer, a scoping review of the epidemiologic evidence. *Environmental Research, 194*, 110690.
3. Eldridge, L., (2017). *Cancer Cells vs. Normal Cells: How Are They Different.* Verywell Health.
4. Fares, J., Fares, M. Y., Khachfe, H. H., Salhab, H. A., & Fares, Y., (2020). Molecular principles of metastasis: A hallmark of cancer revisited. *Signal Transduction and Targeted Therapy, 5*(1), 1–7.

5. Brennan, P., & Davey-Smith, G., (2022). Identifying novel causes of cancers to enhance cancer prevention: New strategies are needed. *JNCI: Journal of the National Cancer Institute, 114*(3), 353–360.
6. Debbouche, A., Nieto, J. J., & Torres, D. F., (2021). Focus point: Cancer and HIV/AIDS dynamics—From optimality to modelling. *The European Physical Journal Plus, 136*(2), 165.
7. Olcum, M., Tufekci, K. U., & Genc, S., (2022). MicroRNAs in genetic etiology of human diseases. In: *miRNomics 2022* (pp. 255–268).
8. World Health Organization (WHO). (2020). Global Health Estimates 2020: Deaths by Cause, Age, Sex, by Country and by Region, 2000–2019. WHO. Accessed December 11, 2020. who.int/data/gho/data/themes/mortality-and-global-health-estimates/ghe-leading-causes-of-death.
9. Chen, L., Spira, A., Ballinger, M., Kowanetz, M., Vansteenkiste, J., Mazieres, J., Park, K., et al., (2016). Atezolizumab versus docetaxel for patients with previously treated non-small-cell lung cancer (POPLAR): A multicentre, open-label, phase 2 randomized controlled trial. *The Lancet, 387*(10030), 1837–1846.
10. Sawicki, T., Ruszkowska, M., Danielewicz, A., Niedźwiedzka, E., Arłukowicz, T., & Przybyłowicz, K. E., (2021). A review of colorectal cancer in terms of epidemiology, risk factors, development, symptoms, and diagnosis. *Cancers, 13*(9), 2025.
11. Ye, T., Li, F., Ma, G., & Wei, W., (2021). Enhancing therapeutic performance of personalized cancer vaccine via delivery vectors. *Advanced Drug Delivery Reviews, 177*, 113927.
12. Wang, L., Pegram, M. D., & Wu, J. C., (2019). Induced pluripotent stem cells as a novel cancer vaccine. *Expert Opinion on Biological Therapy, 19*(11), 1191–1197.
13. Pappalardo, F., Chiacchio, F., & Motta, S., (2013). Cancer vaccines: State of the art of the computational modeling approaches. *BioMed Research International, 2013*.
14. Harrington, N., Chen, Y., O'Reilly, A. M., & Fang, C. Y., (2021). The role of trust in HPV vaccine uptake among racial and ethnic minorities in the United States: A narrative review. *AIMS Public Health, 8*(2), 352.
15. Lybaert, L., Vermaelen, K., De Geest, B. G., & Nuhn, L., (2018). Immunoengineering through cancer vaccines–A personalized and multi-step vaccine approach towards precise cancer immunity. *Journal of Controlled Release, 289*, 125–145.
16. Marhelava, K., Pilch, Z., Bajor, M., Graczyk-Jarzynka, A., & Zagozdzon, R., (2019). Targeting negative and positive immune checkpoints with monoclonal antibodies in therapy of cancer. *Cancers, 11*(11), 1756.
17. Benmebarek, M. R., Karches, C. H., Cadilha, B. L., Lesch, S., Endres, S., & Kobold, S., (2019). Killing mechanisms of chimeric antigen receptor (CAR) T cells. *International Journal of Molecular Sciences, 20*(6), 1283.
18. Davola, M. E., & Mossman, K. L., (2019). Oncolytic viruses: How "lytic" must they be for therapeutic efficacy? *Oncoimmunology, 8*(6), e1581528.
19. Wang, A., Xu, Y., Fei, Y., & Wang, M., (2020). The role of immunosuppressive agents in the management of severe and refractory immune-related adverse events. *Asia-Pacific Journal of Clinical Oncology, 16*(4), 201–210.
20. De Lombaerde, E., De Wever, O., & De Geest, B. G., (2021). Delivery routes matter: Safety and efficacy of intratumoral immunotherapy. *Biochimica et Biophysica Acta (BBA)-Reviews on Cancer, 1875*(2), 188526.

21. Donninger, H., Li, C., Eaton, J. W., & Yaddanapudi, K., (2021). Cancer vaccines: Promising therapeutics or an unattainable dream. *Vaccines, 9*(6), 668.
22. Kazemi, T., Younesi, V., Jadidi-Niaragh, F., & Yousefi, M., (2016). Immunotherapeutic approaches for cancer therapy: An updated review. *Artificial Cells, Nanomedicine, and Biotechnology, 44*(3), 769–779.
23. Jefferies, M., Rauff, B., Rashid, H., Lam, T., & Rafiq, S., (2018). Update on global epidemiology of viral hepatitis and preventive strategies. *World Journal of Clinical Cases, 6*(13), 589.
24. Zhao, Z., Zheng, L., Chen, W., Weng, W., Song, J., & Ji, J., (2019). Delivery strategies of cancer immunotherapy: Recent advances and future perspectives. *Journal of Hematology& Oncology, 12*(1), 1–4.
25. Zou, Z., Fairley, C. K., Ong, J. J., Hocking, J., Canfell, K., Ma, X., Chow, E. P., et al., (2020). Domestic HPV vaccine price and economic returns for cervical cancer prevention in China: A cost-effectiveness analysis. *The Lancet Global Health, 8*(10), e1335–344.
26. Berzofsky, J. A., Terabe, M., Trepel, J. B., Pastan, I., Stroncek, D. F., Morris, J. C., & Wood, L. V., (2018). Cancer vaccine strategies: Translation from mice to human clinical trials. *Cancer Immunology, Immunotherapy, 67*(12), 1863–1869.
27. Ogbu, B. N., Ilo, I. C., Makata, N. E., Agbapuonwu, N. E., Musa, K. M., & Uduji, O. F. (2019). *Factors Associated with the Utilization of Cervarix Vaccine Among Nursing and Midwifery Students in Anambra State, Nigeria. 8*(2), 1–6.
28. Lopes, A., Vandermeulen, G., & Préat, V., (2019). Cancer DNA vaccines: Current preclinical and clinical developments and future perspectives. *Journal of Experimental & Clinical Cancer Research, 38*(1), 1–24.
29. Clark, K. T., Trimble, C. L. (2020). Current status of therapeutic HPV vaccines. *Gynecol Oncol. 156*(2), 503–510. doi:10.1016/j.ygyno.2019.12.017.
30. Espina, C., Soerjomataram, I., Forman, D., & Martín-Moreno, J. M., (2018). Cancer prevention policy in the EU: Best practices are now well recognized; no reason for countries to lag behind. *Journal of Cancer Policy, 18*, 40–51.
31. Gallego, L. S., Dominguez, A., & Parmar, M., (2022). Human papilloma virus vaccine. In: *StatPearls [Internet]*. StatPearls Publishing.
32. Forbes, N. S., Coffin, R. S., Deng, L., Evgin, L., Fiering, S., Giacalone, M., Gravekamp, C., et al., (2018). White paper on microbial anti-cancer therapy and prevention. *Journal for Immunotherapy of Cancer, 6*(1), 1–24.
33. Lee, S. H., (2021). Toll-like receptor 9 agonists in HPV vaccine gardasil9. *International Journal of Vaccine Theory, Practice, and Research, 1*(2), 295–317.
34. Shemesh, C. S., Hsu, J. C., Hosseini, I., Shen, B. Q., Rotte, A., Twomey, P., Girish, S., & Wu, B., (2021). Personalized cancer vaccines: Clinical landscape, challenges, and opportunities. *Molecular Therapy, 29*(2), 555–570.
35. Black, E., & Richmond, R., (2018). Prevention of cervical cancer in sub-Saharan Africa: The advantages and challenges of HPV vaccination. *Vaccines, 6*(3), 61.
36. Canfell, K., (2019). Towards the global elimination of cervical cancer. *Papillomavirus Research, 8*, 100170.
37. Zhang, R., Billingsley, M. M., & Mitchell, M. J., (2018). Biomaterials for vaccine-based cancer immunotherapy. *Journal of Controlled Release, 292*, 256–276.
38. Maeng, H. M., & Berzofsky, J. A., (2019). Strategies for developing and optimizing cancer vaccines. *F1000Research, 8.*

39. International Agency for Research on Cancer, (2005). Cervix cancer screening. In: *Cervix Cancer Screening* (pp. 302–302).
40. Zucchetto, A., Ronco, G., Rossi, P. G., Zappa, M., Ferretti, S., Franzo, A., Falcini, F., et al., (2013). Screening patterns within organized programs and survival of Italian women with invasive cervical cancer. *Preventive Medicine, 57*(3), 220–226.
41. Bucchi, L., Baldacchini, F., Mancini, S., Ravaioli, A., Giuliani, O., Vattiato, R., Falcini, F., et al., (2019). Estimating the impact of an organized screening program on cervical cancer incidence: A 26-year study from northern Italy. *International Journal of Cancer, 144*(5), 1017–1026.
42. Altobelli, E., Rapacchietta, L., Profeta, V. F., & Fagnano, R., (2019). HPV-vaccination and cancer cervical screening in 53 WHO European Countries: An update on prevention programs according to income level. *Cancer Medicine, 8*(5), 2524–2534.
43. Rogovskaya, S. I., Shabalova, I. P., Mikheeva, I. V., Minkina, G. N., Podzolkova, N. M., Shipulina, O. Y., Sultanov, S. N., et al., (2013). Human papillomavirus prevalence and type-distribution, cervical cancer screening practices and current status of vaccination implementation in Russian Federation, the Western countries of the former Soviet Union, Caucasus region and Central Asia. *Vaccine, 31*, H46–58.
44. Engholm, G., Ferlay, J., Christensen, N., et al. (2010). NORDCAN--a Nordic tool for cancer information, planning, quality control and research. *Acta Oncol. 49*(5), 725–736. doi:10.3109/02841861003782017.
45. Dallo, F. J., & Kindratt, T. B., (2015). Disparities in vaccinations and cancer screening among US-and foreign-born Arab and European American non-Hispanic White women. *Women's Health Issues, 25*(1), 56–62.
46. Sudenga, S. L., Rositch, A. F., Otieno, W. A., & Smith, J. S., (2013). Knowledge, attitudes, practices, and perceived risk of cervical cancer among Kenyan women: Brief report. *International Journal of Gynecologic Cancer, 23*(5).
47. World Health Organization. (2013). *WHO Guidance Note: Comprehensive Cervical Cancer Prevention and Control: A Healthier Future for Girls and Women.* Web Link: https://www.who.int/publications/i/item/9789241505147.
48. Glick, S. B., Clarke, A. R., Blanchard, A., & Whitaker, A. K., (2012). Cervical cancer screening, diagnosis, and treatment interventions for racial and ethnic minorities: A systematic review. *Journal of General Internal Medicine, 27*(8), 1016–1032.
49. Sankaranarayanan, R., Anorlu, R., Sangwa-Lugoma, G., & Denny, L. A., (2013). Infrastructure requirements for human papillomavirus vaccination and cervical cancer screening in sub-Saharan Africa. *Vaccine, 31*, F47–52.
50. White, A., Thompson, T. D., White, M. C., Sabatino, S. A., De Moor, J., Doria-Rose, P. V., Geiger, A. M., & Richardson, L. C., (2017). Cancer screening test use – United States, 2015. *Morbidity and Mortality Weekly Report, 66*(8), 201.
51. Olson, B., Gribble, B., Dias, J., Curryer, C., Vo, K., Kowal, P., & Byles, J., (2016). Cervical cancer screening programs and guidelines in low-and middle-income countries. *International Journal of Gynecology& Obstetrics, 134*(3), 239–246.
52. Smith, R. A., Andrews, K. S., Brooks, D., Fedewa, S. A., Manassaram-Baptiste, D., Saslow, D., Brawley, O. W., & Wender, R. C., (2017). Cancer screening in the United States, 2017: A review of current American Cancer Society guidelines and current issues in cancer screening. *CA: A Cancer Journal for Clinicians, 67*(2), 100–121.

53. Centers for Disease Control and Prevention (CDC), (2013). Cervical cancer screening among women aged 18–30 years-United States, 2000–2010. *MMWR: Morbidity and Mortality Weekly Report, 61*(51, 52), 1038–1042.
54. Canfell, K., Sitas, F., & Beral, V., (2006). Cervical cancer in Australia and the United Kingdom: Comparison of screening policy and uptake, and cancer incidence and mortality. *Medical Journal of Australia, 185*(9), 482–486.
55. Gupta, R., Gupta, S., Mehrotra, R., & Sodhani, P., (2017). Cervical cancer screening in resource-constrained countries: Current status and future directions. *Asian Pacific Journal of Cancer Prevention: APJCP, 18*(6), 1461.
56. Gunderson, C. C., Nugent, E. K., McMeekin, D. S., & Moore, K. N., (2013). Distance traveled for treatment of cervical cancer: Who travels the farthest, and does it impact outcome? *International Journal of Gynecologic Cancer, 23*(6).
57. Phillips, A., Patel, C., Pillsbury, A., Brotherton, J., & Macartney, K., (2018). Safety of human papillomavirus vaccines: An updated review. *Drug Safety, 41*(4), 329–346.
58. Oh, N. L., Biddell, C. B., Rhodes, B. E., & Brewer, N. T., (2021). Provider communication and HPV vaccine uptake: A meta-analysis and systematic review. *Preventive Medicine, 148*, 106554.
59. Lee, G. H., & Lim, S. G., (2021). CpG-adjuvanted hepatitis B vaccine (HEPLISAV-B®) update. *Expert Review of Vaccines, 20*(5), 487–495.
60. Hagihara, K., Chan, S., Zhang, L., Oh, D. Y., Wei, X. X., Simko, J., & Fong, L., (2019). Neoadjuvant sipuleucel-T induces both Th1 activation and immune regulation in localized prostate cancer. *Oncoimmunology, 8*(1), e1486953.
61. Petersen, L. M., Fenton, J. M., Kennedy, L. S., LaRochelle, E. P., Bejarano, S., & Tsongalis, G. J., (2020). HPV, vaccines, and cervical cancer in a low-and middle-income country. *Current Problems in Cancer, 44*(6), 100605.
62. Khan, A. A., Liu, Z. K., & Xu, X., (2021). Recent advances in immunotherapy for hepatocellular carcinoma. *Hepatobiliary & Pancreatic Diseases International, 20*(6), 511–520.
63. Margolis, M. A., Brewer, N. T., Shah, P. D., Calo, W. A., & Gilkey, M. B., (2019). Stories about HPV vaccine in social media, traditional media, and conversations. *Preventive Medicine, 118*, 251–256.
64. Brinth, L. S., Pors, K., Theibel, A. C., & Mehlsen, J., (2015). Orthostatic intolerance and postural tachycardia syndrome as suspected adverse effects of vaccination against human papilloma virus. *Vaccine, 33*(22), 2602–2605.
65. Toh, Z. Q., Kosasih, J., Russell, F. M., Garland, S. M., Mulholland, E. K., & Licciardi, P. V., (2019). Recombinant human papillomavirus nonavalent vaccine in the prevention of cancers caused by human papillomavirus. *Infection and Drug Resistance, 12*, 1951.
66. Lipničanová, S., Chmelová, D., Ondrejovič, M., Frecer, V., & Miertuš, S., (2020). Diversity of sialidases found in the human body–A review. *International Journal of Biological Macromolecules, 148*, 857–868.
67. Roden, R., & Stern, P. L., (2018). Opportunities and challenges for human papillomavirus vaccination in cancer. *Nature Reviews Cancer, 18*(4), 240–254.
68. Ikeda, S., Ueda, Y., Yagi, A., Matsuzaki, S., Kobayashi, E., Kimura, T., Miyagi, E., et al., (2019). HPV vaccination in Japan: What is happening in Japan? *Expert Review of Vaccines, 18*(4), 323–325.
69. Diaz, A., Laufer, M. R., & Breech, L. L., (2006). American academy of pediatrics committee on adolescence. American College of Obstetricians and Gynecologists

Committee on Adolescent Health Care, Menstruation in girls and adolescents: Using the menstrual cycle as a vital sign. *Pediatrics, 118*(5), 2245–2250.
70. Levonorgestrel, I., (2019). *Drugs and Lactation Database (LactMed)[Internet]* (pp. 63–67). Bethesda (MD): National Library of Medicine (US).
71. Signorelli, C., Odone, A., Ciorba, V., Cella, P., Audisio, R. A., Lombardi, A., Mariani, L., et al., (2017). Human papillomavirus 9-valent vaccine for cancer prevention: A systematic review of the available evidence. *Epidemiology & Infection, 145*(10), 1962–1982.
72. Wise, L. D., Wolf, J. J., & Plitnick, L. M., (2018). Evaluation of a 9-valent HPV vaccine in Sprague-Dawley rats: Nonclinical studies assessing general, reproductive, and developmental toxicity. *Vaccine, 36*(43), 6401–6407.
73. Tsai, C. H., Wei, S. C., Lo, H. R., & Chao, Y. C., (2019). Baculovirus as versatile vectors for protein display and biotechnological applications. *Current Issues in Molecular Biology, 34*(1), 231–256.
74. Hegde, P. S., & Chen, D. S., (2020). Top 10 challenges in cancer immunotherapy. *Immunity, 52*(1), 17–35.
75. Tan, S., Li, D., & Zhu, X., (2020). Cancer immunotherapy: Pros, cons and beyond. *Biomedicine & Pharmacotherapy, 124*, 109821.
76. Sundstrom, B., Cartmell, K. B., White, A. A., Well, H., Pierce, J. Y., & Brandt, H. M., (2021). Correcting HPV vaccination misinformation online: Evaluating the HPV vaccination NOW social media campaign. *Vaccines, 9*(4), 352.
77. Gilkey, M. B., Grabert, B. K., Malo, T. L., Hall, M. E., & Brewer, N. T., (2020). Physicians' rhetorical strategies for motivating HPV vaccination. *Social Science & Medicine, 266*, 113441.
78. Kim, M., Lee, H., Kiang, P., Aronowitz, T., Sheldon, L. K., Shi, L., Kim, S., & Allison, J., (2019). HPV vaccination and Korean American college women: Cultural factors, knowledge, and attitudes in cervical cancer prevention. *Journal of Community Health, 44*(4), 646–655.
79. Chido-Amajuoyi, O. G., Jackson, I., Yu, R., & Shete, S., (2021). Declining awareness of HPV and HPV vaccine within the general US population. *Human Vaccines & Immunotherapeutics, 17*(2), 420–427.
80. Singh, J., Bowne, W. B., & Snook, A. E., (2021). Cancer vaccines and immunotherapy for tumor prevention and treatment. *Vaccines, 9*(11), 1298.
81. Cartmell, K. B., Mzik, C. R., Sundstrom, B. L., Luque, J. S., White, A., & Young-Pierce, J., (2019). HPV vaccination communication messages, messengers, and messaging strategies. *Journal of Cancer Education, 34*(5), 1014–1023.
82. Lahijani, A. Y., King, A. R., Gullatte, M. M., Hennink, M., & Bednarczyk, R. A., (2021). HPV vaccine promotion: The church as an agent of change. *Social Science & Medicine, 268*, 113375.
83. Dilley, S. E., Peral, S., Straughn, Jr. J. M., & Scarinci, I. C., (2018). The challenge of HPV vaccination uptake and opportunities for solutions: Lessons learned from Alabama. *Preventive Medicine, 113*, 124–131.
84. Bednarczyk, R. A., (2019). Addressing HPV vaccine myths: Practical information for healthcare providers. *Human Vaccines & Immunotherapeutics, 15*(7, 8), 1628–1638.
85. Leung, S. O., Akinwunmi, B., Elias, K. M., & Feldman, S., (2019). Educating healthcare providers to increase human papillomavirus (HPV) vaccination rates: A qualitative systematic review. *Vaccine: X, 3*, 100037.

CHAPTER 10

Providing Cognitive Platforms for the Effectiveness of Cancer Vaccines

ZAHRA ALIDOUSTI SHAHRAKI[1] and MOHSEN AGHABOZORGI NAFCHI[2]

[1]*Department of Computer Engineering, University of Isfahan, Isfahan, Iran*

[2]*Department of Computer Science and Engineering, Shiraz University, Shiraz, Iran*

ABSTRACT

Population growth in recent years has led to an increase in various diseases. The spread of these diseases, along with their unknown causes, has resulted in delays in providing effective treatments. Cancer, in particular, has become one of the most prevalent and painful diseases in recent years. The incidence of cancer varies based on factors such as gender, age, and geographic conditions. Therefore, the development of an intelligent system capable of diagnosing and preventing cancer can significantly reduce the occurrence of this disease and decrease mortality rates.

Furthermore, artificial intelligence (AI) and machine learning (ML) have made a significant impact on the field of medicine. By utilizing ML techniques, it becomes possible to diagnose specific symptoms experienced by individuals and predict the type of cancer they may be suffering from. Consequently, the accurate diagnosis of various cancerous tumors using ML algorithms can expedite patient treatment. In fact, cancerous tumors are examined through image processing, followed by detailed recognition using deep learning techniques. With the use of intelligent tools, such as smart watches, suspicious symptoms can be identified, enabling faster treatment for individuals. Smart applications are highly beneficial for all age groups,

Cancer Vaccination and Challenges, Volume 1: Strategies for Therapeutic Cancer Vaccine Development. Rishabha Malviya, Bhupendra Prajapati, and Sonali Sundram (Eds.)

particularly the elderly who may struggle with accurately diagnosing symptoms. However, despite these advancements, the importance of vaccination cannot be overlooked. Vaccination is a common method to prevent the spread of diseases, effectively reducing mortality rates. Encouraging the use of vaccines can prevent diseases and promote overall health within society. To prepare vaccines, it is crucial to explore the existing challenges and problems. This involves identifying disease symptoms in individuals and selecting vaccines that do not exacerbate other conditions. Smart medical applications can enhance the diagnosis of different disease symptoms, ensuring accurate identification of patients' problems and minimizing diagnostic errors. Additionally, the use of artificial intelligence technologies in vaccine design can streamline the production process, reducing the time required for each stage. In this chapter, we will examine various types of cancers, explore methods for designing and producing vaccines tailored to each cancer, and investigate the challenges and opportunities associated with utilizing artificial intelligence technologies for future cancer vaccine (CV) preparation.

10.1 INTRODUCTION

Cancer is a result of uncontrolled cell proliferation in various parts of the body. It is a potentially fatal disease if left untreated. Timely diagnosis and treatment of cancer are crucial aspects that doctors should address by utilizing technology. Nowadays, there are various and sometimes unknown symptoms that can lead to changes in the body's immune system and consequently result in cancer. Therefore, real-time diagnosis and the use of medication can aid in the prevention and complete cure of cancer.

Cancer is a well-known genetic disease. Individuals with a family history of irregular cell growth have a higher likelihood of developing cancer. Hence, it is important to accurately diagnose the health of individuals who are prone to cancer cell growth. Intelligent systems can assist in diagnosing and preventing this disease in such individuals. If the symptoms that may indicate cancer are evident, an intelligent system can help detect them. Developing smart applications and designing smartwatches can aid people in identifying symptoms. Every person experience changes in blood circulation speed, heart rate, or behavior. Therefore, an accurate diagnosis of these symptoms can benefit the health of anyone susceptible to the disease. Furthermore, vaccination can be an effective measure in preventing and treating numerous diseases. From childhood to old age, vaccination has successfully prevented the occurrence of many diseases based on each person's geographical location. It is worth noting that suspicious symptoms in the body can be identified, and artificial intelligence methods can assist in treating the disease. Designing smart vaccines has the potential to revolutionize medical science, although it presents significant challenges. Therefore, we will explore the obstacles involved in designing vaccines using artificial intelligence. We hope that the world of technology can take a significant stride towards improving human health.

10.2 FACTORS THAT CAN INFLUENCE VACCINATION

10.2.1 THE SIDE EFFECT OF VACCINATION

One of the important challenges in cancer vaccine (CV) production is the potential side effects on the body [1]. In recent years, concerns about the COVID-19 vaccine have led many people to forgo vaccination due to fear of potential issues. Therefore, it is crucial to develop methods for detecting any side effects that the vaccine may cause in individuals. Additionally, individuals with pre-existing medical conditions may be more susceptible to the vaccine's impact on their health. Therefore, implementing intelligent solutions such as designing applications to monitor an individual's health post-vaccination can play a vital role in ensuring people's safety. Symptoms such as shortness of breath, cough, or changes in skin color can indicate a negative effect of the vaccine. Therefore, an important challenge in vaccination is the accurate diagnosis of each individual's health symptoms using artificial intelligence. Additionally, it is crucial to investigate the precise monitoring of a person's health after vaccination, which can be achieved through the use of artificial intelligence networks.

10.2.2 HEALTHY LIFESTYLE

Healthy eating is an effective way to prevent the growth of cancer in humans [2]. Nutritionists and doctors today emphasize the importance of maintaining a healthy lifestyle. Both physical and mental well-being depend on proper nutrition. Artificial intelligence can play a positive role in promoting a healthy lifestyle. One aspect worth discussing is the impact of nutrition on vaccination. It can be argued that without proper nutrition, the body may not have sufficient health to respond effectively to vaccines, especially in the case of cancer recovery. In the following discussion, we explore the role of healthy nutrition in the effectiveness of vaccination and highlight how artificial intelligence can contribute to improving nutrition and the impact of vaccines on overall health.

In Ref. [3], the relationship between a healthy lifestyle and vaccination has been discussed. Individuals who maintain a healthy lifestyle are more conscious of their health and strive to improve their quality of life. Therefore, they utilize vaccination as a preventive measure against diseases. Adopting a nutritious diet and prioritizing mental well-being are among the positive aspects of cultivating a healthy life. It appears that individuals who opt for vaccination possess a greater understanding of scientific matters, which in turn contributes to their enhanced mental and physical health. Furthermore, the correlation between vaccination and a healthy lifestyle has been examined [4]. In fact, parents who lead fulfilling lives and endeavor to maintain a healthy lifestyle demonstrate attentiveness towards vaccination matters for both them and their children. It seems that individuals from families with higher educational attainment exhibit greater concern for vaccination issues pertaining to themselves and their children. Conversely, economically disadvantaged and socially marginalized families tend to overlook health matters and vaccination. Advertisements on social networks can serve as a means to inform these individuals and elevate their living standards by promoting health awareness.

10.2.3 SOCIAL NETWORKS

As mentioned in the previous section, social networks play a crucial role in raising awareness among individuals. Consequently, they can be utilized to promote vaccination and increase public knowledge. Presently, artificial intelligence has the capability to analyze users' personalities through social networks. Users actively engage in social network groups or channels that align with their interests. One potential solution to effectively guide users towards cancer prevention vaccination is through positive advertisements in news groups and social networks. Scientifically, this approach can provide assistance. It can be inferred that artificial intelligence indirectly plays a significant role in utilizing social networks to support and motivate individuals to get vaccinated, thereby enhancing their health literacy. In Ref. [5], the role of social networks in dispelling doubts about vaccination is discussed. The elderly population, who often have various health concerns, may be hesitant about vaccination. By improving their health literacy and encouraging vaccination, social networks can contribute to the well-being of the elderly. Different countries have specific social networks that are popular among their residents based on their interests. Therefore, awareness campaigns should be tailored to the level of user activity in social networks according to their geographic location.

In Ref. [6], the role of media and social networks in vaccination is discussed. According to surveys conducted on the Google search engine, it appears that people search in social spaces to learn more about vaccination and its effects on themselves. An interesting finding from the investigation is that labeling those who need to be vaccinated as "sick" has a negative impact on them. The term "patient" carries a negative connotation for them. Therefore, vaccination should be portrayed in a positive light, and vaccination advertisements on social networks should be crafted with positive encouragement.

In Ref. [7], the impact of social networks in providing information about COVID-19 vaccination is discussed. There have been numerous concerns regarding the vaccination of individuals against COVID-19. Social networks can play a crucial role in promoting vaccination through positive advertisements. People's confidence in vaccination can be assessed using deep learning algorithms. The obtained results are analyzed based on the volume of people's searches on social media. Consequently, if it is determined that a user is hesitant about vaccination, sending encouraging messages can help them overcome their concerns. It seems feasible to utilize data mining tools to analyze the data of each user across various social media platforms, enabling personalized advertisements based on their search activity.

10.2.4 COGNITIVE SCIENCE

The intellectual conditions of individuals are examined and investigated in the field of cognitive science. People's thoughts on societal issues, including mental and physical well-being, influence their life decisions. In this context, we will explore people's attitudes towards cancer vaccination. Investigating this matter is crucial to understand whether individuals interested in vaccination also prioritize other health-related concerns in their lives. The topic of instilling confidence in COVID-19 vaccination is discussed in Ref. [30]. It appears that people's attitude towards vaccination is linked to their trust in scientific information. The more individuals trust scientific sources, the more likely they are to comply with measures such as vaccination to safeguard their health. The issue of trust, whether complete or lacking, is a significant aspect in cognitive science, as discussed in Ref. [30]. Based on studies, individuals with a positive attitude towards health topics also exhibit a positive stance towards vaccination when necessary. However, it is important to note that one's personal attitude should not impede the well-being of others. Individuals who oppose vaccination should not impose their opinions on society, as this jeopardizes the health of others. To address this problem, it is essential to enhance public health by promoting positive thinking through intelligent applications. Designing smart applications tailored to individuals' attitudes has the potential to change people's perspectives and provide them with information about their health status. Informing individuals about their health and the importance of vaccinations can be achieved through the implementation of smart applications. This issue of promoting vaccination to enhance public health is a significant area of study in cognitive science.

Stress is among the factors that contribute to individuals experiencing fear and anxiety when facing diseases such as cancer, which can weaken their immune system [31]. This fear leads to more significant problems. In fact, lack of trust in reliable sources, non-compliance with medication and vaccinations, can hinder the patient's recovery. It appears that a smart user interface should be employed to address this issue. By designing a user interface, the factors of fear and stress in individuals can be reduced, thereby encouraging those in need to get vaccinated. Additionally, an intelligent user interface can assess the patient's health conditions based on relevant factors and propose preventive measures to inhibit cancer growth.

Predictive models can play a crucial role in preventing the development of cancer in the body. Convolutional neural networks can be utilized to classify symptoms of various diseases using prediction models [32]. If symptoms manifest in an individual's body, intelligent algorithms can trigger an alert. This approach can be a turning point in preventing the occurrence and progression of cancer, as well as in the treatment of diseases.

The implementation of ethical issues is crucial when utilizing medical intelligence tools [33]. While smart tools offer easy access to people's privacy, it is important to ensure that doctors in the healthcare field do not have access to data that violates patient privacy. Privacy is particularly important for cancer patients. As mentioned in previous articles, patients' immune systems may be compromised due to mental conditions such as fear and worry. Consequently, their immune system may not respond effectively to vaccinations. Violation of privacy is one of the factors that can cause discomfort in patients and potentially contribute to the growth of cancerous tumors, worsening the disease by reducing their immunity level.

Therefore, it is essential to investigate conditions that induce fear and worry in patients and diminish their immune response. Enhancing the patient's immune system can improve the effectiveness of vaccinations and other anticancer drugs on their body.

10.2.5 DECISION MAKING

Investigating the impact of people's feelings and emotions on their health is a complex issue [16]. The influence of emotions on people's decision-making can result in both correct and incorrect choices, which can disrupt an individual's life. Examining people's emotions at different times, taking into account geographical and environmental conditions, as well as their lifestyle, is a subject that requires careful investigation. In the realm of health, this aspect plays a particularly prominent role. As previously mentioned, emotional decisions that deviate from logic can cause physical and mental harm to individuals, with long-lasting effects on their lives. An important area of focus in this field is exploring the influence of emotions and individual decision-making on vaccination. A study has been conducted to examine the role of emotions in vaccination, as well as factors such as a person's nutrition. Negative emotions, such as embarrassment and shame, can have an adverse impact on people's decision-making. It is important to consider that individuals who make incorrect decisions may experience changes in their life circumstances, as well as a decline in their immune system. Consequently, vaccination may not have a positive effect on them. Therefore, it can be concluded that emotions and feelings indirectly affect the effectiveness of vaccination on patients' health.

The impact of emotions on cancer-related issues is significant [17]. In fact, various factors influence individuals' emotions and decision-making processes when dealing with cancer. The specific type of cancer a person has can lead to different decisions. Decisions are influenced by factors such as shame, embarrassment, fear, and more. Therefore, it is important to consider that a cancer patient may make different decisions based on their unique circumstances. If decisions are made impulsively, it can lead to problems. By utilizing machine learning algorithms, it is possible to predict the decisions individuals will make based on their specific conditions. If a decision does not contribute to the cancer recovery process, efforts should be made to provide appropriate solutions.

Recommender systems [18] can select optimal conditions based on individuals past experiences and provide intelligent recommendations tailored to their specific disease. These recommender systems utilize data mining techniques to analyze large amounts of patient data, which is recorded and used to generate intelligent recommendations. They play a crucial role in the field of healthcare. For example, they intelligently offer helpful suggestions in various situations based on the user information stored in the database. Recommender systems play a prominent role in addressing health and nutrition concerns. They can also assist individuals in overcoming fear and anger when making decisions. Overcoming negative factors can increase life expectancy based on environmental conditions, lifestyle, and enhance the immune system's ability to resist cancer and the growth of malignant cells. By utilizing an application designed with a smart recommender system, it can significantly contribute to improving overall health.

Multi-objective optimization [19] is an example of optimization that plays an important role in problem-solving. In multi-objective optimization, various variables that represent the desired goals in solving the problem are considered, and then the optimization algorithm is used to present the best solution for the problem. Additionally, recommender systems play a significant role in decision-making [23]. In fact, multi-objective problem-solving becomes easier to implement with the use of recommender systems. Recommender systems intelligently make the best decision by analyzing the current conditions and the importance of each goal in the problem [22]. Multi-objective recommender systems, which are based on decision-making systems, intelligently make the best decision by examining the problem's conditions [20]. By utilizing these systems, it becomes possible to consider different goals for the problem and then provide the best decision based on the conditions using machine learning algorithms [21].

Designing healthcare recommender systems can contribute to the improvement of healthcare [24]. By developing an application based on recommender systems, it is possible to enhance the well-being of individuals, particularly those who are battling cancer. The role of multi-objective recommender systems becomes highly significant in addressing health-related issues. Intelligent decision-making in critical situations is a crucial aspect that multi-objective decision-making systems can handle. These systems aim to provide the best recommendations with minimal errors by considering various aspects of human health. Vaccination is a critical topic that can be evaluated using multi-objective recommender systems. The patient's condition is intelligently monitored in real-time based on their medical history, physical health, and mental well-being. Additionally, the multi-objective decision-making system makes optimal decisions based on multiple criteria.

Machine learning techniques play a vital role in designing health recommender systems [25]. By utilizing clustering patterns and deep learning, important parameters related to individuals' health can be thoroughly examined. If there is a health issue, recommender applications using machine learning [26] can prescribe medication and provide immediate care by monitoring the patient's blood pressure and heart rate until the doctor arrives.

Designing recommender systems using natural language processing can also help cancer patients [27]. Psychologically, cancer patients require psychological support. Natural language processing can provide support to patients worldwide, regardless of their language. Additionally, these systems can assist in vaccinating cancer patients. Boosting morale in these patients is a crucial turning point that encourages them to get vaccinated.

Recommender systems have positive effects on breast cancer diagnosis [28], particularly in diagnosing cancer from mammography images. The BCRecommender system introduces a deep learning-based method for breast cancer diagnosis [29]. The key requirement for successful implementation in recommender systems is high accuracy in measuring criteria. It is also essential to have sufficient data for training the learning algorithm. Therefore, despite these criteria, the systems can play an effective role in cancer diagnosis and prediction. It appears that an efficient recommender system should be capable of providing suggestions for the use of food and drug systems based on available data, as well as offering instructions to prevent disease development based on each patient's conditions.

According to studies conducted in this field, recommender systems can utilize machine learning patterns to predict diseases and behaviors of each patient during their cancer period. These systems can assist in recommending drugs to cancer patients. Additionally, recommender systems play a significant role in vaccination. Learner recommender systems can be employed to prevent missed vaccinations and encourage cancer patients to get vaccinated. By expanding these systems, the accuracy of predictions can be enhanced, and errors caused by diagnosis and prediction can be minimized.

10.2.6 SMART TECHNOLOGIES

Anti-tumor vaccines can help to immunize the body against the spread of tumors. The neoantigen vaccine is one example of an immunogenic vaccine that aims to prevent the spread of cancerous tumors [34]. By utilizing the design of anti-cancer vaccines, it is possible to modify genes and eliminate defective or malignant genes, thereby reducing their ability to spread. Creating a protective environment within the body aids in reducing the disease. Administering vaccines within specific timeframes and at regular intervals helps to maintain a high level of immunity in the body. Artificial intelligence can assist in diagnosing the condition of each patient during the vaccination process. Developing a user interface tailored to each patient helps to minimize the spread of defective genes by maintaining a high level of immunity. Detection of defective genes can be achieved through a smart user interface, which also aids in assessing the patient's overall health by measuring blood sugar and blood fat levels. In the event of suspicious symptoms, medical tests can be conducted to examine the health of the genes. Bioinformatics plays a crucial role in cancer diagnosis and vaccine production for the prevention and treatment of cancer.

An important aspect in vaccine design is to consider gene mutations [34], which require thorough investigation. While vaccination can reduce the growth of cancer cells, it is possible for extreme mutations to occur in other cells, leading to increased growth. Therefore, it is essential to monitor the patient's condition in real-time using smart applications. During the vaccination process, the patient's condition should be closely monitored through smart applications, and any suspicious symptoms should be promptly reported to the doctors.

10.3 FACTORS THAT CAN REDUCE GROWTH OF CANCER

10.3.1 ARTIFICIAL INTELLIGENCE

There are various types of cancer treatment methods that play significant roles in reducing the incidence of cancer in patients [8]. However, due to the wide range of cancer incidence and the influence of factors such as environmental and geographical conditions on the speed of cancer growth, accurate confirmation is challenging. Artificial intelligence can assist in improving accuracy and reducing errors in cancer treatments to some extent. Smart applications are used to investigate the rate of cancer cell growth. If the patient experiences any changes in their body, these applications provide notifications to help prevent the growth rate of cells. Artificial intelligence algorithms analyze various factors such as the patient's nutrition and environmental conditions, including blood pressure and sugar levels. If necessary, the patient is provided with instructions through notifications from the smart applications.

In Ref. [9], the role of machine learning in predicting the occurrence of cancer is examined. Clustering algorithms are used to categorize patients based on their age, and the occurrence or growth of cancerous tumors is investigated. Artificial neural networks and decision trees are employed to detect the growth rate of cancer cells in patients. However, it should be noted that the growth of cancer cells varies among patients based on their individual life conditions. Therefore, similar data sets should be used when studying and checking using detection algorithms. The data sets should be based on similar criteria, such as patients who are close in age or have a similar lifestyle, to ensure better prediction through accurate clustering.

Cancer diagnosis and treatment methods can be performed using various artificial intelligence tools. Different approaches are described in the diagnosis and treatment of cancer [10]. Image processing can be utilized for cancer diagnosis. Artificial intelligence can be employed to analyze patterns of drug consumption or tumor mutations in individuals. Smart applications can assist doctors and patients in accurate cancer diagnosis, treatment, and disease prevention by segmenting and analyzing images of cancer tumors.

Deep learning techniques enable the diagnosis of uterine cancer from medical images. Image segmentation based on different criteria has facilitated cancer diagnosis for doctors. The obtained results have shown that the risk of cancer in patients can be reduced based on variables such as race, ethnicity, and disease severity [11]. Additionally, machine vision can aid in the diagnosis and prevention of prostate disease [12]. Artificial intelligence algorithms for target detection and clustering can effectively identify and examine areas prone to cancer.

One of the most crucial aspects of medical science is the early diagnosis of diseases, particularly cancer. Early detection of the disease helps prevent severe symptoms that can lead to fatality. In the case of cancer, patients may require long-term care, and the gradual progression of the disease can result in the gradual deterioration of various parts of the body over time. This gradual decline causes significant discomfort and pain for patients in the final months of their lives. Nowadays, artificial intelligence can assist in the early diagnosis of diseases. Regular screenings, especially for elderly patients or those with a genetic predisposition to the disease, can aid in timely diagnosis and treatment. If a person is at risk of developing cancer, certain medications can be administered to inhibit the growth of cancer cells. Predictive algorithms in deep learning can be employed for this purpose [13]. Additionally, it is possible to predict the necessary medications based on test results, blood pressure readings, and MRI and CT scans.

10.3.2 VACCINATION

Today, despite the various diseases that occur and cause the death of many people, the impact of vaccination in reducing their prevalence is undeniable. In recent years, vaccination against the COVID-19 disease has faced resistance from many individuals in most countries. However, post-vaccination research has revealed that vaccinated individuals are more resistant to contracting COVID-19 again compared to those who are unvaccinated. Additionally, the severity of illness and death among vaccinated individuals has significantly decreased. This demonstrates the proven effectiveness of vaccination over the years for individuals who prioritize medical science and their health. With the emergence of unknown diseases influenced by factors such as climate change, it is crucial to modernize vaccination methods. In other words, early disease diagnosis should be conducted using intelligent methods, and predictive algorithms in machine learning should be utilized to develop vaccines that can treat unknown diseases. By doing so, it is hopeful that artificial intelligence can identify future unknown diseases and provide appropriate vaccines based on environmental and biological changes occurring worldwide.

Vaccination can help decrease the occurrence of cancer cells in individuals with a history of genetic diseases or undergoing genetic changes in their bodies. Additionally, vaccination can help reduce abnormal cell growth caused by environmental factors and stress. In other words, vaccination can provide hope for individuals living in different communities.

As people become aware of the advancements in medical science, their hope for life, growth, and progress is ignited. I believe that many unknown diseases that disrupt people's lives today are a result of stress and changes in lifestyle. With the increasing incidence of these unknown diseases, artificial intelligence should be capable of predicting their occurrence. Hormonal changes or changes in body tissues may not exhibit symptoms, but the presence of mild symptoms such as unexplained headaches, fatigue, or sleepiness could indicate the initiation of cancer cell growth in the body.

By conducting the necessary tests and checking the test results using intelligent patterns, it can be determined whether the minor changes in a person's body will lead to the development of a disease. Predictive algorithms can provide more accurate predictions for disease diagnosis. These algorithms can predict the extent to which vaccination can enhance a patient's resistance to disease progression. The medications that can be used for vaccination against disease occurrence may vary based on individual environmental and biological conditions. Therefore, estimation algorithms can utilize artificial intelligence techniques to more accurately assess the impact of vaccination.

In-depth studies in the field of nutrition science indicate that promoting vaccination through nutrition science experts can play a more effective role [14]. Increasing people's awareness of health and nutrition issues, as well as improving their self-care literacy, can significantly enhance their motivation to get vaccinated for their own well-being. Data mining can be employed to achieve this objective. Segmenting individuals based on different criteria and conditions in their lives helps tailor information to each group's specific needs. By utilizing online questionnaires, it is possible to assess people's awareness of health and vaccination issues, and subsequently conduct semantic analysis on different societal groups. Through intelligent predictions, it becomes feasible to anticipate people's future needs and thereby encourage vaccination. An important aspect explored in Ref. [14] is the evaluation of individuals' satisfaction with vaccination clinics. Psychological factors appear to play a significant role in motivating people to get vaccinated. Enhancing the facilities provided at healthcare centers can elevate people's satisfaction levels and foster trust in healthcare services, thereby greatly benefiting individuals across various age groups.

Another crucial consideration is the impact of various factors on immunity during vaccination. Research in Ref. [15] indicates that factors such as nutrition, obesity, and chronic diseases experienced by individuals can influence the body's immune system. Consequently, vaccination may not effectively enhance the body's immunity level. This issue is particularly prevalent among the elderly and individuals who commonly suffer from conditions like high blood sugar due to their age. Artificial intelligence can indirectly identify and mitigate complications arising from old age or different diseases, thereby enhancing the effectiveness of vaccination for individuals with low immunity levels. Smart applications or programs can assist disease-prone individuals in improving their health and boosting their immunity through nutritional control, physical activities, and sports. Conducting diverse surveys can help ensure the safety of individuals with various conditions, including disabilities, physical and mental health issues, old age, and children, in order to safeguard their immune systems.

During the COVID-19 pandemic, the elderly population faced increased pressure to get vaccinated due to their various underlying health conditions. These individuals were understandably concerned about the potential impact of vaccination on their health, considering their medical history. It is important to take into account the potential side effects of vaccination based on each individual's medical background. Adverse reactions to vaccines in the elderly can even lead to fatal outcomes. Therefore, the use of smart applications can play a crucial role in monitoring the well-being of the elderly after vaccination, even during their sleep. This is an important aspect that should be taken into consideration.

It can be concluded that various factors, such as environmental conditions and genetics, can influence the immune system of individuals. Moreover, the immune system varies according to age. Vaccination should be administered based on each person's specific condition. By utilizing smart systems, it becomes evident that vaccination is necessary for everyone. Online applications can be used to assess the condition of each individual and monitor any potential side effects of vaccination. In case of any issues, online prescriptions can be provided to patients.

10.4 INTELLIGENTIZATION OF INJECTABLE VACCINES

mRNA is one of the primary components involved in the process of protein synthesis within the human body. It plays a crucial role in enhancing the efficiency of cellular development and lifespan. The design and production of mRNA vaccines are of utmost importance in bolstering the immune system [35]. By utilizing mRNA as a vaccine, we can effectively stimulate the immune response in the human body. The development of a sophisticated protein machine could serve as a groundbreaking advancement in the field of medical science. Through protein synthesis, the degradation of body cells can be minimized, resulting in a robust immune system. Previous mRNA-based vaccines have demonstrated resilience against viral infections. By incorporating artificial intelligence into the design of a novel mRNA vaccine, we can take a significant stride towards creating intelligent vaccines to combat viruses. In the future, such designs may even have the capability to monitor protein levels within the human body. However, a major challenge lies in the automatic multiplication of smart cells through injection into the human body. This endeavor is likely to encounter numerous obstacles, with patient resistance to smart cell injections being a significant concern. The apprehension and fear surrounding vaccinations cannot be disregarded. For instance, the COVID-19 vaccine faced various challenges, with some individuals harboring misconceptions about its potential to alter their genes or gender due to lack of knowledge. Designing smart cells that function similarly to intelligent vaccinations can yield numerous benefits. However, it will take time for individuals to develop complete confidence in this approach and embrace the concept of smart cell injections. Such vaccines have the potential to regulate genes and cells within the body, automatically injecting beneficial cells based on factors like aging or climate change. As mentioned, this idea presents significant challenges, but researchers are actively exploring and refining various concepts in this field.

One of the vaccines offered to immunize the body against cancer is the immunotherapy vaccine [36]. This vaccine identifies genes that cause tumor proliferation and simultaneously produces antigens. This vaccine, like the explanation that was given, faces challenges that must be investigated. One of the most important challenges that must be investigated using artificial intelligence algorithms is vaccine processing time. Responding to the vaccine at the time of need is an important point that must be implemented using response algorithms in the shortest possible time. These algorithms stop the growth of cancer cells by choosing the best path to reach the goal. The design of a Nano-vaccine can be effective in the implementation of this plan. As mentioned in Ref. [37], providing an optimal response in the shortest possible time during vaccination has the effects of vaccination on the patient's body to eradicate cancer cells. Neoantigens create immunity in the body. It seems that if new antigens can be designed automatically, it will activate immunization in the body spontaneously. Automatic design means designing a small structure in the body that automatically activates new genes when needed. This design makes cancer cells come under control and suppresses them at all times during reproduction.

One of the ways to deal with the growth of cancer cells is the use of multi-functional nanomaterial vaccines [38]. The results of these vaccines have been effective in preventing the growth of prostate cancer (PC) tumors. But they should be investigated and researched for other cancer cells. These nanomaterial vaccines can be effective in diagnosing and treating and even preventing cancer tumors. These vaccines can have a great transformation in the treatment of cancer. In fact, traditional methods such as surgery, chemotherapy, and radiotherapy are abandoned, and then multi-purpose vaccination replaces all treatments. With this method, the effect of intelligence platforms in the production of nanomaterial vaccines can be understood. The design of these vaccines using detailed studies in nanoparticles (NPs) and the deep learning application in the design of nanomaterial vaccines will greatly contribute to the development of science in this field of treatment.

Also, the Nano-engineered vaccines [39] can estimate and identify cancer cells using deep learning algorithms and take preventive measures to eliminate cancer cells. The estimation of cancer cells causes the immune system to regenerate immune cells. It can also strengthen weak immune cells. Machine learning patterns makes a great help in preventing the growth of cancer cells. It seems that the design of multi-purpose Nano-engineered vaccines can detect and completely suppress cancer cells in parallel regenerate weak immune cells. The multi-purpose action can strengthen the body's immune system against the growth of cancer cells in the shortest possible time.

In Ref. [40], the effect of immunotherapy in cancer treatment is discussed. This method is a new method to deal with cancer cells. The effect of immunotherapy along with smart methods creates a new branch in medical science, which is called immune engineering. In this branch, biological methods are used to stimulate the immune system and cause immunization to occur at the site of the tumor in the patient's body. This method can carry out immunization in a multi-target way, and different parts of the patient's body and cancer cells can be examined at the same time, then immunization can be done in several areas. Nanotechnology can play an effective role in immunizing the body using biological tools. In general, it can be concluded that nowadays smart solutions to diagnose and treat cancer cells cause vaccination and immunization in the patient's body. Smart vaccination can replace traditional vaccinations and create a great revolution in vaccination in medical science.

Non-small cell lung cancer (NSCLC) is an emerging model in the treatment and diagnosis of tumors and diseases [41]. This model facilitates the rapid identification of cancerous areas by detecting molecules associated with cancer cells. The swift impact of this method in disease diagnosis has heightened the prospects of patient survival. One notable advantage of this model is its ability to personalize vaccines based on predictive models. Personalizing the vaccine enables the identification of cancer cells specific to each individual's disease condition, thereby expediting the treatment process. This method also exemplifies intelligent vaccination, incorporating smart algorithms for swift disease prediction and diagnosis. Further exploration of these novel methods can provide assistance to a greater number of cancer patients. One possible application of this vaccination is the intelligentization of therapeutic molecules. The use of smart injectables in conjunction with smart vaccination can yield varying effects for each patient. This model holds the potential to enhance patient survival rates and foster greater motivation among cancer patients to receive vaccinations.

During cancer treatment, certain substances may suppress the body's immune system, allowing cancer cells to reactivate either voluntarily or involuntarily. The growth of cancer cells in an unfavorable environment reduces the effectiveness of vaccination. In Ref. [42], a series of anti-tumor effects have been introduced. Research and investigation on injectable substances indicate that the immune system should be examined at the precise time. Fluctuations in the immune system suggest significant growth of immunosuppressive factors. Therefore, vaccination timing should be carefully considered to suppress cancer cells.

The adjuvant enhances immunity and balance in vaccination [43]. This substance actually accelerates and prolongs the immune response to antigens [44]. It is prescribed in therapeutic vaccines to protect antigens. Protecting antigens in special conditions when the patient's immune system is weakened plays a crucial role in strengthening cells. This substance has shown effectiveness in treating lung cancer (LC) [45]. The injection of this vaccine can also impact the treatment of other cancers.

Immunotherapy based on nanoparticles (NPs) has a significant effect in treating malignant tumors. Combining this method with traditional cancer treatments like chemotherapy and radiation therapy can greatly aid in suppressing cancer cells [46]. This method provides innate immunity to cells and strengthens the body's immune response. The rapid advancements in NP-based immunotherapy have increased the life expectancy of cancer patients. Considering the role of life expectancy in overcoming diseases, it can be concluded that immunotherapy based on NPs can be a crucial step in achieving complete cancer treatment. By reducing stress and anxiety in patients, it can greatly enhance their recovery.

On the other hand, the design and production of DNA vaccines have been able to contribute to the treatment of cancer by enhancing the body's immune system [47]. DNA vaccines, which offer higher safety, greater stability, and easier production, can facilitate the process of eliminating cancer. These vaccines have shown promise in preventing LC. Numerous trials are currently underway to explore their potential in LC prevention, considering it is the most lethal form of the disease. However, it is important to note that the efficacy of a vaccine specific to a particular organ does not necessarily imply its effectiveness in eradicating cancer cells in other parts of the body. Extensive research is required to fully understand the use of vaccines in inhibiting cancer cells and enhancing the immune system.

Computational models play a crucial role in detecting the rate of cancer cell growth and tumor spread. In the case of LC, computational methods can assist in estimating the growth rate of tumors [48]. Accurate prediction of tumor growth rate enables real-time vaccination and helps prevent further fatalities. By employing artificial intelligence algorithms, the tumor growth rate can be estimated with minimal margin of error. Calculating the optimal timing for vaccination significantly contributes to boosting the body's immune system.

Furthermore, increasing the administration of antigens for immunization purposes contributes to reducing cancer mortality. Studies on small LC have demonstrated that machine learning techniques can be utilized to measure the amount of antigen introduced into the body [49]. Based on the rate of tumor growth, the quantity of antigen injection can be adjusted accordingly. Predictive techniques utilizing deep learning have proven highly effective in suppressing cancer cells and halting their growth.

It is important to note that the proliferation of cancerous tumors varies among individuals, resulting in unique characteristics for each type of cancer. This indicates that the type of cancer differs based on the genetic changes in each person. Consequently, the production of CVs should be tailored to each individual [50]. Therefore, mass production in vaccine manufacturing does not yield favorable results. Additionally, the administration of CVs should be timed and customized to the specific conditions of each person in order to be effective.

An issue that can be explored and considered as a challenge is the vaccination of millions of people worldwide during the COVID-19 pandemic. This raises questions about whether administering COVID-19 vaccines under different conditions in countries with distinct genetics and circumstances was the right approach. It also raises concerns about deaths caused by the coronavirus after vaccination due to the specific design of the vaccine. Although the spread of the coronavirus and related deaths have significantly decreased, this issue remains relevant in case another virus outbreak occurs. By conducting thorough research and analysis, accurate and effective planning can be made for the design of future vaccines.

However, the goal of vaccination is to reduce the growth of cancer tumors and mortality rates, taking into account the diverse genetics of individuals based on their geographical environments. The proliferation of cancer cells is influenced by the unique living conditions of each person. Paying attention to these factors can contribute to the development of a safe vaccine to combat cancer.

Recognizing cancer mutations using deep diagnostic algorithms brings about a significant transformation in the medical industry [51]. Cancer cells undergo successive mutations, which can lead to the rapid spread of cancer cells throughout the body's organs. In such cases, eradicating cancer cells throughout the entire body becomes a costly and time-consuming process, potentially risking the patient's life during treatment. To address this issue, monitoring the mutations occurring in cancer cells enables disease control. Deep learning algorithms and evolutionary algorithms estimate the growth rate of body cells, identifying the occurrence of cancer mutations at various stages. The calculation of mutations in cancer cells varies for each individual based on their immune system's level. By specifically investigating this matter for each patient according to their condition, it becomes more effective, enhancing the efficacy of vaccinations in targeting and destroying cancer cells during different mutation stages.

Considering the vast amount of information regarding vaccination conditions and the various mutations that occur over time, the utilization of data mining methods can aid in optimizing algorithms [52]. Information clustering methods enhance the efficiency of vaccination processes. Mining analysis has a positive impact on the vaccination process for cancer patients. Accurate identification of the cancer type and thorough examination of the patient's condition are essential. Therefore, data mining patterns contribute to improving vaccination conditions. Selecting the most suitable type of vaccination based on the patient's condition is a crucial aspect that can yield positive outcomes when combined with data mining and clustering algorithms.

In general, data mining algorithms play a significant role in the field of medical sciences and nanotechnology. The rise in the number of diseases due to population growth, genetic diversity, and cell mutations leads to the emergence of new, unknown diseases. To address this, smart applications based on deep learning can be employed to identify diseases and develop new vaccines tailored to these novel diseases. It appears that intelligent algorithms with innovative techniques need to be introduced periodically to identify diseases and design new vaccines. Simulating cancer cells using intelligent algorithms contributes to faster vaccination. Establishing a virtual vaccination method that can effectively target cancer cells would be a major advancement in vaccination practices. Virtual vaccination reduces errors and, consequently, achieves excellent results in combating mutated cancer cells for cancer patients.

10.5 ARTIFICIAL INTELLIGENCE IS A VIRTUAL VACCINATION AGAINST CANCER

Applications of artificial intelligence have the potential to revolutionize cancer treatment by creating virtual vaccinations. In other words, if artificial intelligence plays a role, vaccination can become a pivotal point in enhancing the body's immune system. This theory represents a significant stride towards transforming human health. Despite various concerns about the negative impacts of the smart revolution, the remarkable growth of smart applications continues to reshape people's lifestyles. It would be highly beneficial to utilize the advancements in science for positive purposes, improving both physical and mental well-being.

In today's world, the design of smart applications has significantly enhanced the quality of life. Smart devices designed for home environments alleviate physical fatigue among individuals. The expansion of social networks has fostered increased communication among people, enabling individuals to connect with peers based on their age and interests. This social connectivity contributes to improved mental health, reducing feelings of loneliness and isolation. The expansion of social networks plays a crucial role in enhancing people's mental well-being, which, in turn, positively impacts their physical health and reduces the risk of diseases. We should commend the role of smart tools and applications that leverage artificial intelligence techniques in elevating people's quality of life.

Artificial intelligence has played an indirect yet significant role in enhancing the mental and physical well-being of individuals worldwide. It is worth noting the concept of artificial intelligence as a "virtual vaccination" in promoting societal health. This term combines the medical understanding of vaccination with its psychological implications for mental health.

I firmly believe that vaccination goes beyond the mere administration of chemicals to combat cancer cells. Vaccination can yield remarkable outcomes from a psychological standpoint. Patients who have lived with cancer often find hope in their life expectancy and maintain a positive mindset, which empowers them to fight the disease through exercise and an improved quality of life. Therefore, I reiterate that the most valuable insights and ideas are those that approach the treatment and health of individuals from a comprehensive perspective. It is essential to engage in a thorough and diverse discussion on the various aspects of artificial intelligence's role in vaccination.

10.6 BLOCKCHAIN TECHNOLOGY AND CANCER VACCINATION

Blockchain plays an effective role in storing and retrieving information in medical science. It is also crucial to safeguard the privacy of patients. Therefore, tools should be employed to prevent information from being compromised. Blockchain technology has the capability to encrypt patients' information, including their medical history, medications, and other health-related issues, within a secure set of blocks [53]. Blockchain can also be utilized in the context of vaccination. During the COVID-19 pandemic, certain countries successfully leveraged blockchain capabilities in vaccination efforts to curb the spread of the virus. These countries employed blockchain to verify COVID-19 test results through blockchain activation. The health centers of each country should have control over the timeline of corona patients, their medical history, and the duration of their treatment. This close monitoring is essential to effectively manage the spread of the disease within society. The utilization of blockchain in verifying COVID-19 tests aims to prevent the falsification of test results. It can be argued that countries with high death rates and inadequate control over the spread of COVID-19 have demonstrated poor performance in utilizing smart tools such as blockchain.

A specialized architectural structure has been implemented to store patient information during the COVID-19 pandemic [54]. Patient data, medical center details, disease records, and vaccination schedules are stored separately in each node. This architecture enhances the speed of accessing information and facilitates efficient vaccination scheduling. Gathering patient information, including their living conditions and genetic structure, is a crucial approach in preventing diseases like cancer. This comprehensive data enables medical centers to take precise actions for cancer prevention. Analyzing individual patient records aids in determining the optimal vaccination timing. By leveraging blockchain technology, an intelligent architecture can efficiently process a vast amount of patient information at a national level. The blockchain-based smart software can identify individuals who require vaccination based on their specific conditions. Implementing blockchain clustering can further expedite access to health information with enhanced accuracy. It appears that blockchain clustering holds great potential in analyzing patients' health data, as swift access to such information promotes patients' recovery and prolongs their lifespan.

Preventing drug counterfeiting can have a significant impact on the economic processes of every country. Those who engage in the falsification of medicine drive up the cost of purchasing medication, depriving those in need of access. To address this issue, the utilization of blockchain technology can enhance security in the production and consumption of medicine within medical centers and clinics [55]. Given that vaccinations are crucial in preventing the spread of cancer, particularly during critical times, blockchain can play a pivotal role in safeguarding patient information and ensuring the provision of vaccines, especially in underdeveloped and impoverished nations. The cryptographic storage of vaccine types and the number of individuals requiring vaccination, based on the living conditions in cities and provinces, is a crucial aspect that can be achieved through blockchain algorithms.

Many patients who require cancer treatment through vaccination often experience mental weakness, feelings of isolation, and sadness. To enhance their sense of freshness and vitality during the treatment period, it is important to provide them with a positive experience. When patients are aware that their treatment process is being supervised and secured by medical centers, they will feel more psychologically secure, alleviating feelings of loneliness and isolation. Addressing this issue is of utmost importance. Therefore, one of the primary responsibilities of medical centers is to maintain comprehensive patient records within the healthcare system. By visiting a medical center, patients can easily receive treatment and communicate with their doctors. Through the use of blockchain technology, doctors can securely access patients' medical records, including vaccination history, physical and mental conditions, at any medical center [56]. This technology eliminates the need for patients to repeatedly explain their medical history, which can be tiresome and discouraging. Additionally, it reduces the duration of treatment. A blockchain-based platform design enables seamless access to patients' records across different medical centers, fostering a positive doctor-patient relationship throughout the treatment period.

One of the challenges during the COVID-19 pandemic is the issuance of counterfeit vaccination certificates. This poses a threat to effective crisis management and hampers control over the movement of infected individuals. Consequently, the spread of COVID-19 becomes uncontrollable on a global scale. To mitigate the spread of the virus and other potential future outbreaks that may impact global health, a secure blockchain-based system should be developed. This system ensures the safe storage of individuals' information within the community and enables effective management of disease records [57]. Setting the time of vaccination for cancer patients accurately through the blockchain treatment system ensures that their treatment is carried out regularly, reducing mortality rates. Previous studies have emphasized the importance of reducing stress and increasing the sense of security for cancer patients during their treatment. Blockchain-based systems provide this sense of security to patients under the care of medical staff, while doctors in the community ensure their well-being. This feeling of security boosts their strength, maintains their morale, and psychologically strengthens their immune system. The ultimate goal of designing and implementing AI-based applications and software is to support the physical and mental health of patients. Additionally, fostering a friendly relationship between doctors and patients encourages patients to easily seek treatment and vaccinations.

The design of the UAIVS system has been able to create high security in maintaining the privacy of patients in vaccinations [58]. This system is highly cost-effective and scalable. The cost of adding patient information data to this system is affordable and the process is efficient. These capabilities make this system unique. In the Fog system, the design and implementation of an intelligent blockchain-based system have been discussed [59]. In this system, patient information is stored in each block, ensuring privacy protection. Medical records such as vaccinations and treatment time are stored in separate blocks. This system offers high processing speed and quick access to information. A public block is used for communication between blocks, enabling faster access to public data. As mentioned, privacy protection is highly secure in the architecture of this system.

In Ref. [60], an international authentication system for issuing vaccination certificates is introduced. This system protects data using encrypted algorithms and ensures the confidentiality of patient information. The architecture of this system aims to validate vaccine passports and prevent the falsification of vaccination documents. It is possible that this encryption algorithm can be utilized to enhance security in other centers where data security is required.

According to studies conducted on the role of the blockchain system in cancer vaccination, it can be concluded that the blockchain architecture plays an important role in maintaining patient information within a country's healthcare organization. By keeping track of patients' vaccination information and their history of cancer treatment, it becomes possible to carry out vaccinations accurately and with fewer errors. Additionally, due to the large volume of data in medical centers, a blockchain system can increase the speed of data processing and reduce the time required to access information. The combination of encryption algorithms and data mining in a blockchain-based system can create a secure database for storing patient information. Undoubtedly, implementing this system can improve the speed of handling cancer patients and reduce mortality rates. These measures are aimed at maintaining the health of patients in healthcare centers. It is hoped that in the future, with the advancement of technology, blockchain-based applications with even more advanced architectures will be utilized to preserve information.

10.7 CONCLUSION

One of the factors that has always caused death throughout human life has been various diseases. If a disease is not prevented and treated, it causes cells to multiply, eventually covering the entire body and leading to death. Cancer is one of the dangerous diseases that can cause death. According to conducted studies, the prevalence of cancer in the human body is not specific to age or gender. Even young people can be affected by genetic and environmental factors, such as stress on their immune system, which triggers the growth and proliferation of malignant cells in their body. Based on these studies, the best approach is to reduce stress and increase peace and well-being in life. In any case, in order to reduce the number of deaths caused by cancer, various factors should be investigated. If a person has a genetic disease, they can prevent and reduce the growth of cancer by undergoing various tests and taking medication. Each individual should strive for their health and longevity against diseases, based on their living conditions and environment.

The field of medicine is vast, and the impact of artificial intelligence on cancer vaccination is a significant topic in medical science. The most valuable theories are those that challenge different aspects of cancer treatment. This chapter aims to discuss the role of artificial intelligence in cancer vaccination from both medical and psychological perspectives. Extensive research in the field of psychology has indirectly contributed to this issue. Overall, artificial intelligence has influenced people's lives in various ways, indirectly contributing to improving their quality of life. Consequently, an improvement in people's quality of life leads to comprehensive physical and mental well-being. The purpose of this chapter is to examine various cognitive factors that influence cancer vaccination and discuss intelligent approaches in the development of cancer drugs. By challenging the presented methods, we aim to create a comprehensive platform that guides social thinking towards vaccination, ultimately enhancing people's health. An interesting aspect of this chapter is the investigation of the role of medical smart tools and the effectiveness of cognitive methods in mental health, which can imperceptibly eliminate cancer cells. We hope that the ideas presented in this chapter can contribute to the design of effective platforms for cancer treatment and vaccination. Undoubtedly, vaccination will reduce the mortality rate and improve living conditions for people of all age groups.

KEYWORDS

- **artificial intelligence**
- **blockchain**
- **cancer**
- **cognitive science**
- **decision making**
- **machine learning**
- **social networking**
- **vaccination**

REFERENCES

1. Hwang, J. S., Kim, S. G., Shin, T. H., Jang, Y. E., Kwon, D. H., & Lee, G., (2022). Development of anticancer peptides using artificial intelligence and combinational therapy for cancer therapeutics. *Pharmaceutics, 14*(5), 997.
2. Kassam, S., & Freeman, L., (2022). The role of nutrition in cancer prevention-should you listen to your doctor or influencer? *American Journal of Lifestyle Medicine.* 215598276221112331.
3. Barut, S., Baransel, E. S., & Akarsu, A. C., (2022). Do healthy lifestyle behaviors affect COVID-19 vaccination attitudes in generation Z? *Medicine, 11*(1), 53–58.
4. Wiley, K. E., Leask, J., Attwell, K., Helps, C., Degeling, C., Ward, P., & Carter, S. M., (2020). Parenting and the vaccine refusal process: A new explanation of the relationship between lifestyle and vaccination trajectories. *Social Science & Medicine, 263*, 113259.
5. Fernandez-Villa, J. M., Galnares-Carcoba, V., Solis-Mata, J. S., Flores-Ruiz, C., & Perez-Zepeda, M. U., (2022). Social networks and COVID-19 vaccination hesitancy in Mexican older adults. Proceedings of scientific research Universidad Anáhuac. *Multidisciplinary Journal of Healthcare, 2*(3), 11–19.
6. Stahl, J. P., Cohen, R., Denis, F., Gaudelus, J., Martinot, A., Lery, T., & Lepetit, H., (2016). The impact of the web and social networks on vaccination. New challenges and opportunities offered to fight against vaccine hesitancy. *Medecine et Maladies Infectieuses, 46*(3), 117–122.
7. Bernados, Jr. S., & Ocampo, L., (2022). How do people decide on getting vaccinated? Evaluating the COVID-19 vaccination program through the lens of social capital theory. *Social Sciences, 11*(4), 145.
8. Zhou, X., Qu, M., Tebon, P., Jiang, X., Wang, C., Xue, Y., Zhu, J., et al., (2020). Screening cancer immunotherapy: When engineering approaches meet artificial intelligence. *Advanced Science, 7*(19), 2001447.

9. Themis, P., Konstantinos, P., Michalis, V., & Dimitrios, I., (2015). Machine learning applications in cancer prognosis and prediction. *Computational and Structural Biotechnology Journal, 2015.*
10. Liang, G., Fan, W., Luo, H., & Zhu, X., (2020). The emerging roles of artificial intelligence in cancer drug development and precision therapy. *Biomedicine & Pharmacotherapy, 128*, 110255.
11. Wagner, V., Piver, R., Levine, M., Backes, F., Cohn, D., Copeland, L., Cosgrove, C., et al., (2022). Using machine learning to predict venous thromboembolism in patients with uterine cancer (081). *Gynecologic Oncology, 166*, S55, 56.
12. Salman, M. E., Çakar, G. Ç., Azimjonov, J., Kösem, M., & Cedimoğlu, İ. H., (2022). Automated prostate cancer grading and diagnosis system using deep learning-based Yolo object detection algorithm. *Expert Systems with Applications, 201*, 117148.
13. Hunter, B., Hindocha, S., & Lee, R. W., (2022). The role of artificial intelligence in early cancer diagnosis. *Cancers, 14*(6), 1524.
14. Yao, M., Gu, X., Mo, Y., Xia, C., & Tang, L., (2022). The role of health education in vaccination nursing. *Journal of Healthcare Engineering, 2022.*
15. Bianchi, F. P., & Tafuri, S., (2022). Vaccination of elderly people affected by chronic diseases: A challenge for public health. *Vaccines, 10*(5), 641.
16. Mostafapourdehcheshmeh, M. (2022). *Exploring the Role of Emotions on Health Decisions: Case Studies in Cancer Screening, Vaccination, and Nutrition.* https://uwspace.uwaterloo.ca/handle/10012/18125.
17. Mazzocco, K., Masiero, M., Carriero, M. C., & Pravettoni, G., (2019). The role of emotions in cancer patients' decision-making. *Ecancermedicalscience, 13.*
18. Sanchez-Bocanegra, C. L., Sanchez-Laguna, F., & Sevillano, J. L., (2015). Introduction on health recommender systems. In: *Data Mining in Clinical Medicine* (pp. 131–146). Humana Press, New York, NY.
19. Deb, K., (2014). Multi-objective optimization. In: *Search Methodologies* (pp. 403–449). Springer, Boston, MA.
20. Pang, J., Guo, J., & Zhang, W., (2019). Using multi-objective optimization to solve the long tail problem in recommender system. In: *Pacific-Asia Conference on Knowledge Discovery and Data Mining* (pp. 302–313). Springer, Cham.
21. Ahmed, S. T., Sankar, S., & Sandhya, M., (2021). Multi-objective optimal medical data informatics standardization and processing technique for telemedicine via machine learning approach. *Journal of Ambient Intelligence and Humanized Computing, 12*(5), 5349–5358.
22. Zheng, Y., & Wang, D. X., (2022). A survey of recommender systems with multi-objective optimization. *Neurocomputing, 474*, 141–153.
23. Jameson, A., Willemsen, M. C., & Felfernig, A., (2022). Individual and group decision making and recommender systems. In: *Recommender Systems Handbook* (pp. 789–832). Springer, New York, NY.
24. Etemadi, M., Abkenar, S. B., Ahmadzadeh, A., Kashani, M. H., Asghari, P., Akbari, M., & Mahdipour, E., (2022). A systematic review of healthcare recommender systems: Open issues, challenges, and techniques. *Expert Systems with Applications*, 118823.
25. Shaikh, S. G., Suresh, K. B., & Narang, G., (2022). Recommender system for health care analysis using machine learning technique: A review. *Theoretical Issues in Ergonomics Science*, 1–30.

26. Rustam, F., Imtiaz, Z., Mehmood, A., Rupapara, V., Choi, G. S., Din, S., & Ashraf, I., (2022). Automated disease diagnosis and precaution recommender system using supervised machine learning. *Multimedia Tools and Applications*, 1–24.
27. Leung, Y. W., Park, B., Heo, R., Adikari, A., Chackochan, S., Wong, J., Alie, E., et al., (2022). Providing care beyond therapy sessions with a natural language processing–based recommender system that identifies cancer patients who experience psychosocial challenges and provides self-care support: Pilot study. *JMIR Cancer, 8*(3), e35893.
28. Rani, S., Kaur, M., & Kumar, M., (2022). Recommender system: Prediction/diagnosis of breast cancer using hybrid machine learning algorithm. *Multimedia Tools and Applications, 81*(7), 9939–9948.
29. Bhargava, H., Makeri, Y. A., Gyamenah, P., Gupta, S., Vyas, G., Sharma, A., & Chatterjee, S. (2022). *BCRecommender System for Breast Cancer Diagnosis Using Machine Learning Approaches*. https://doi.org/10.21203/rs.3.rs-1985404/v1.
30. Newman, D., Lewandowsky, S., & Mayo, R., (2022). Believing in nothing and believing in everything: The underlying cognitive paradox of anti-COVID-19 vaccine attitudes. *Personality and Individual Differences, 189*, 111522.
31. Acar-Burkay, S., & Cristian, D. C., (2022). Cognitive underpinnings of COVID-19 vaccine hesitancy. *Social Science & Medicine, 301*, 114911.
32. Bhinder, B., Gilvary, C., Madhukar, N. S., & Elemento, O., (2021). Artificial intelligence in cancer research and precision medicine. *Cancer Discovery, 11*(4), 900–915.
33. Mishra, S., Abbas, M., Jindal, K., Narayan, J., & Dwivedy, S. K., (2022). Artificial intelligence-based technological advancements in clinical healthcare applications: A systematic review. *Revolutions in Product Design for Healthcare, 2022*, 207–227.
34. Liu, Z., Lv, J., Dang, Q., Liu, L., Weng, S., Wang, L., Zhou, Z., Kong, Y., Li, H., Han, Y., & Han, X., (2022). Engineering neoantigen vaccines to improve cancer personalized immunotherapy. *International Journal of Biological Sciences, 18*(15), 5607–5623.
35. Samantaray, M. U., & Khuntia, R. P., (2022). mRNA vaccines against emerging infectious diseases; A challenging approach of novel vaccine discovery. *Int. J. Med. Phar. Drug Re., 2022*, 52.
36. Xie, X., Song, T., Feng, Y., Zhang, H., Yang, G., Wu, C., You, F., Liu, Y., & Yang, H., (2022). Nanotechnology-based multifunctional vaccines for cancer immunotherapy. *Chemical Engineering Journal,* 135505.
37. Le, I., Dhandayuthapani, S., Chacon, J., Eiring, A. M., & Gadad, S. S., (2022). Harnessing the immune system with cancer vaccines: From prevention to therapeutics. *Vaccines, 10*(5), 816.
38. Fu, B., Huang, X., Deng, J., Gu, D., Mei, Q., Deng, M., Tang, S., & Lü, M., (2018). Application of multifunctional nanomaterials in cancer vaccines. *Oncology Reports, 39*(3), 893–900.
39. Yin, W. M., Li, Y. W., Gu, Y. Q., & Luo, M., (2020). Nanoengineered targeting strategy for cancer immunotherapy. *Acta Pharmacologica Sinica, 41*(7), 902–910.
40. Shams, F., Golchin, A., Azari, A., Mohammadi, A. L., Zarein, F., Khosravi, A., & Ardeshirylajimi, A., (2021). Nanotechnology-based products for cancer immunotherapy. *Molecular Biology Reports*, 1–24.
41. Rodriguez, P. C., & Sanchez, B., (2013). Challenges and opportunities for cancer vaccines in the current NSCLC clinical scenario. *Current Topics in Medicinal Chemistry, 13*(20), 2551–2561.

42. Wu, H., Li, H., Liu, Y., Liang, J., Liu, Q., Xu, Z., Chen, Z., et al., (2022). Blockading a new NSCLC immunosuppressive target by pluripotent autologous tumor vaccines magnifies sequential immunotherapy. *Bioactive Materials, 13*, 223–238.
43. Broich, K., (2009). Committee for medicinal products for human use (CHMP) assessment on efficacy of antidepressants. *European Neuropsychopharmacology, 19*(5), 305–308.
44. Sasaki, S., & Okuda, K., (2000). The use of conventional immunologic adjuvants in DNA vaccine preparations. In: *DNA Vaccines* (pp. 241–249). Humana Press.
45. Dondulkar, A., Akojwar, N., Katta, C., Khatri, D. K., Mehra, N. K., Singh, S. B., & Madan, J., (2022). Inhalable polymeric micro and nano-immunoadjuvants for developing therapeutic vaccines in the treatment of non-small cell lung cancer. *Current Pharmaceutical Design, 28*(5), 395–409.
46. Banstola, A., Jeong, J. H., & Yook, S., (2020). Immunoadjuvants for cancer immunotherapy: A review of recent developments. *Acta Biomaterialia, 114*, 16–30.
47. Huang, T., Liu, L., Lv, Z., Zhao, K., Yi, Q., & Zhang, J., (2022). Recent advances in DNA vaccines against lung cancer: A mini review. *Vaccines, 10*(10), 1586.
48. Nimsarkar, P., Gulhane, P., & Singh, S., (2022). Understanding the molecular kinetics in NSCLC through computational method. In: *Systems Biomedicine Approaches in Cancer Research* (pp. 129–163). Springer, Singapore.
49. Ni, C., (2022). Personal neoantigens from patients with NSCLC induce efficient antitumor responses. *Unveiling the Impact of Local or Systemic Therapeutic Strategies on the Tumor Microenvironment.*
50. Sahin, U., & Türeci, Ö., (2018). Personalized vaccines for cancer immunotherapy. *Science, 359*(6382), 1355–1360.
51. Lang, F., Schrörs, B., Löwer, M., Türeci, Ö., & Sahin, U., (2022). Identification of neoantigens for individualized therapeutic cancer vaccines. *Nature Reviews Drug Discovery, 21*(4), 261–282.
52. Chen, C., Yaari, Z., Apfelbaum, E., Grodzinski, P., Shamay, Y., & Heller, D. A., (2022). Merging data curation and machine learning to improve nanomedicines. *Advanced Drug Delivery Reviews*, 114172.
53. Carniel, A., De Melo, B. J., & Hirata, C. M., (2022). A blockchain-based approach for COVID-19 vaccine lifecycle. In: *International Conference on Enterprise Information Systems* (pp. 71–85). Springer, Cham.
54. Arumugam, S. K., & Sharma, A. M., (2022). Blockchain: Opportunities in the healthcare sector and its uses in COVID-19. In: *Lessons from COVID-19* (pp. 61–94). Academic Press.
55. Omidian, H., & Omidi, Y., (2022). Drug counterfeiting and decentralized whistleblowers. *Drug Discovery Today.*
56. Chen, J., Chen, X., & Chen, C. L., (2022). A traceable blockchain-based vaccination record storage and sharing system. *Journal of Healthcare Engineering, 2022.*
57. Kobbaey, T., Bilquise, G., Alqatawna, J. F., & Dashti, O., (2022). A blockchain-based vaccination model for COVID-19 and other infectious diseases. In: *2022 8th International Conference on Information Technology Trends (ITT)* (pp. 189–195). IEEE.
58. Sharma, M., Kothapalli, K., & Gujar, S., (2022). BlockVac: A universally acceptable and ideal vaccination system on blockchain. In: *2022 IEEE International Conference on Blockchain (Blockchain)* (pp. 320–325). IEEE.

59. Costa, H. J., Da Costa, C. A., Righi, R. D., Antunes, R. S., Santana, J. F., & Leithardt, V. R., (2022). A fog and blockchain software architecture for a global scale vaccination strategy. *IEEE Access, 10*, 44290–44304.
60. Lee, H. A., Wu, W. C., Kung, H. H., Udayasankaran, J. G., Wei, Y. C., Kijsanayotin, B., Marcelo, A. B., & Hsu, C. Y., (2022). Design of a vaccine passport validation system using blockchain-based architecture: Development study. *JMIR Public Health and Surveillance, 8*(4), e32411.

CHAPTER 11

Survival Data and Therapeutic Outcomes of Cancer Vaccine

PATEL PRIYA,[1] PRAJAPATI BHUPENDRA,[2] and PATEL GAYATRI[1]

[1]*Department of Pharmaceutics and Pharmaceutical Technology, Ramanbhai Patel College of Pharmacy, Charotar University of Science and Technology (CHARUSAT), CHARUSAT Campus, Changa, Gujarat, India*

[2]*Department of Pharmaceutics and Pharmaceutical Technology, Shree S.K. Patel College of Pharmaceutical Education and Research, Ganpat University, Kherva, Gujarat, India*

ABSTRACT

Patients with cancer receive therapeutic cancer vaccines (CV) with the aim of enhancing their own immune responses to destroy cancer cells. Unlike preventive vaccines, which are given to healthy individuals, therapeutic vaccines target and kill cancer cells while sparing healthy cells. Theoretically, these vaccines could be used to halt the growth of advanced cancers and relapsed tumors that are resistant to conventional medical treatments such as surgery, radiation therapy, and chemotherapy. Tumor immunotherapy research has made significant advancements in the past two decades, and numerous experiments are currently being evaluated in clinical settings.

There are also vaccines available that can protect healthy individuals from certain virus-related cancers. Similar to flu or chickenpox immunizations, these vaccines provide protection against specific viruses. However, it is important to receive these vaccines before contracting the virus in order to benefit from their immunization.

According to data from GLOBOCAN 2018, there were 7,84,821 cancer-related deaths, 2,258,208 individuals living with cancer, and 1,157,294 new

Cancer Vaccination and Challenges, Volume 1: Strategies for Therapeutic Cancer Vaccine Development. Rishabha Malviya, Bhupendra Prajapati, and Sonali Sundram (Eds.)

cases of cancer in India in 2018 (within five years of diagnosis). The type of cancer, stage of the disease, and time since diagnosis all impact the chances of survival and prognosis.

In this chapter, we have discussed the survival rate and prognosis associated with CVs. Additionally, we have introduced the benefits and challenges involved in the development of a new generation of cancer-preventing vaccines.

11.1 INTRODUCTION

Current malignancies can be treated using active, passive, or immunomodulatory immunotherapies [1]. Passive immunotherapy refers to the management of exogenously formed substances, such as lymphocytes or antibodies (mAbs), instead of active immunotherapies [2], which enhance the patient's own immune system's ability to combat an immune response, mediate an immune response to identify tumor-associated antigens (TAAs), and eliminate cancerous cells [3]. Immunomodulatory medications aim to enhance overall immunological responsiveness to boost anticancer immune responses, rather than targeting specific antigens. Prostate cancer (PC), lung cancer (LC), and colorectal cancers (CRC) are more prevalent in men than in women, who are more susceptible to breast, lung, and CRCs. All four cancer types have shown a decline in incidence and fatality rates in both genders, with the possible exception of LC in women. This is likely due to advancements in treatments, medication combinations, and early detection. However, despite these positive developments, cancer remains a significant global public health issue that necessitates the development of cutting-edge methods for diagnosis and treatment to improve patient health outcomes [4]. The absence of tumor-specific antigens and the weakened immune responses against TAAs, which the immune system typically recognizes as self-antigens, are two of the main obstacles to the development of an effective cancer vaccine (CV). To address the compromised immune response to Tumor Associated Antigens (TAAs), various therapeutic techniques have been proposed in recent years. These include vaccines derived from cells, DNA, RNA, proteins, or peptides, as well as vaccines based on vectors [5]. All of these therapeutic approaches involve the activation of antigen-presenting cells (APCs) and the induction of a cytotoxic T lymphocyte-mediated immune response that is specific to the antigen. Among the most effective APCs are dendritic cells (DCs), and numerous strategies have been employed to enhance their ability to stimulate T lymphocytes [6].

Despite recent achievements, several clinical studies have faced criticism due to the lack of immunological evidence supporting the significant improvements in overall survival (OS) and time to progression observed. Therefore, it is crucial to investigate the precise components of each immune response in the context of anticancer vaccination. This exploration should focus on the specificity of the response and the identification of potential escape mechanisms, in order to develop novel and potent anticancer vaccines. A deeper understanding of how different modalities impact the cytotoxic T lymphocytes (CTL) response is essential [7].

Therapeutic cancer vaccinations aim to educate the body's defenses to recognize and selectively target tumor cells for elimination. Two potential targets for immune response are TAAs or tumor-specific antigens (TSAs). Due to their specificity, TSAs are considered ideal for cancer immunotherapy. The majority of TSAs consist of mutant proteins that arise from somatic mutations in the original protein sequence. The fact that several of these proteins have been shown to be crucial for carcinogenesis and cancer development makes targeting TSAs an important benefit [8]. However, targeting TSAs has a significant disadvantage in that most of the mutations found are specific to each tumor, which may necessitate the creation of individualized immunotherapy for specific individuals. On the other hand, TAAs are typically expressed on tumors with the same histology and are present in malignancies of many origins. Targeting TAAs has significant drawbacks due to their limited immunogenicity, resulting from the immune system's early development of tolerance for self-antigens [9]. Further investigation is needed to determine whether combining vaccinations with low doses of chemotherapeutic drugs or radiation therapy, administered in the proper sequence, could enhance vaccine-induced immunity. The driving force behind this is the encouragement from preclinical and clinical data.

11.2 OVERVIEW OF THE CV AND ITS ROLE

The development of therapeutic CVs has seen a significant increase over the past decade. The utilization of innovative antigen delivery technology, understanding of the native immune response, and the discovery of various TAAs have all contributed to the advancement of vaccine design. Therapeutic CVs aim to prevent cancer, treat minimally recurrent disease, establish long-lasting antitumor memory, and minimize undesired side effects. However, there are obstacles to overcome in achieving these goals, such as immunosuppression and immunoresistance induced by tumors [10].

When selecting a vaccine platform, several factors need to be considered. The time required for personalized vaccine production is crucial in choosing the appropriate vaccination platform. Nucleic acid vaccines (NAVs) have shown to be a cost-effective treatment for many metastatic diseases. Combination therapy can also be employed in vaccine development to prevent further disease progression and create an immunological environment that supports a successful immune response. Additionally, the method of administration and the number of doses should be taken into account when selecting a vaccine platform.

The optimization and design of antigens, along with the platform selection, play a critical role in vaccine development. Tumor antigen coupling binding vectors, such as tetanus endotoxin and diphtheria toxoid, can enhance the immunogenicity of an antigen. Vaccines containing virus-like particles (VLPs) or other antigens with protein structural optimization have shown to improve the immune response. Furthermore, bioinformatics and deep sequencing methods are essential in guiding vaccine design [11].

11.3 TYPES OF VACCINES

There are four different types of CVs which are discussed in subsections.

11.3.1 VACCINES BASED ON CELLS

CVs initially take the form of cell-based vaccinations. In order to stimulate a broader immune response to the antigen, cell-based CVs are often created from whole cells or cell fragments that closely resemble tumor antigens. The DC vaccine is an essential component of cell-based vaccinations. Clinical tests of personalized DC-based neoantigen CVs have shown promising anti-tumor results. The development of DC vaccines is still hampered by the lengthy procedure and high cost [12]. Examples of current ongoing trials are represented in Figure 11.1.

11.3.2 VACCINES BASED ON PEPTIDE

The specific tumor antigen site is significantly stimulated by peptide-based subunit vaccinations, which also include chemical and biosynthetic preparations of predicted or known specific cancer antigens. Adjuvants and

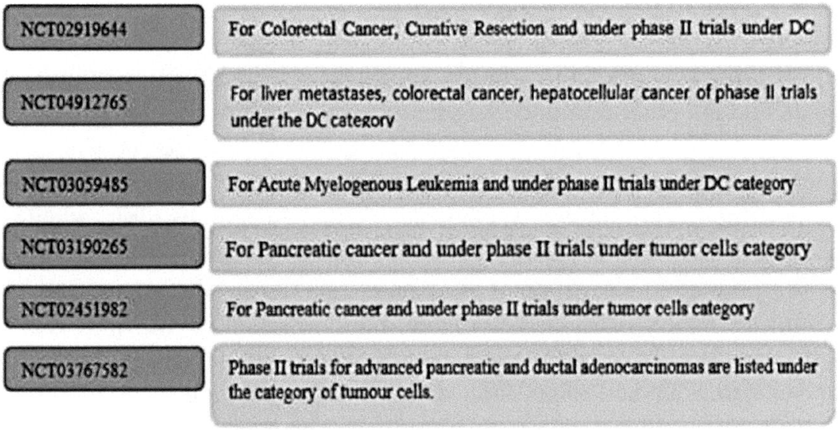

FIGURE 11.1 Examples of current ongoing trials on cell-based vaccine.

peptide-based subunit vaccines can effectively elicit a humoral immune response, making them ideal for the prevention and treatment of viral infectious diseases [13].

Key examples of current ongoing trials of vaccine based on peptide shown in Figure 11.2.

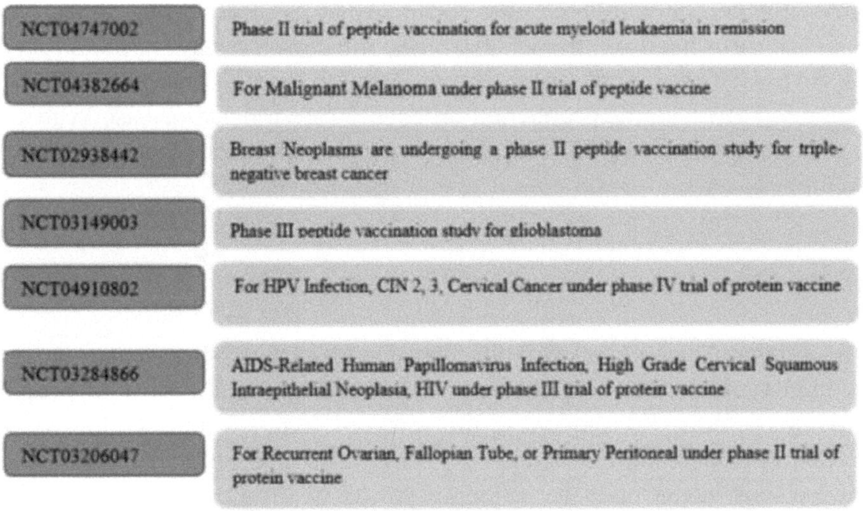

FIGURE 11.2 Examples of current ongoing trials of vaccine based on peptide.

11.3.3 VACCINES BASED ON VIRUS

The key advantage of virus-based vaccines is their ability to stimulate the innate and adaptive immune systems to work together, resulting in an effective and long-lasting immune response. Oncolytic viruses (OVs), a novel therapy, selectively target and kill tumor cells while also promoting anti-tumor responses. Examples of OVs include herpes simplex virus, adenovirus, measles, and more. T-VEC, a first-generation recombinant herpes simplex virus product, has shown great promise in this field. Adenovirus, in addition to herpes simplex virus, is another commonly used oncolytic virus. It is often utilized as a vector for delivering specific genes. Adenovirus is advantageous for gene transfer and expression of tumor antigens due to its well-defined gene structure and ease of use. Adenoviruses can be rapidly produced in large quantities and have a wide range of host cell tropism. Furthermore, adenoviruses naturally cause mucosal infections, making them a promising candidate for vaccine delivery. Both oncolytic and non-replicating adenovirus-based vaccines have shown promise in preclinical and clinical trials [14].

Viruses can be genetically modified to include sequences encoding tumor antigens since they are naturally immunogenic. Adenovirus, along with other recombinant viruses, can spread infection across immune cells. The modified virus vaccinations can significantly increase the immune system's exposure to tumor antigens and produce anti-tumor immunity. Additionally, the oncolytic virus can be employed as a vector. Apart from providing cancer antigens, the virus itself has the ability to lyse the tumor, release tumor antigens, improve vaccination effectiveness, and develop long-lasting immunological memory. However, the production procedure for viral vector-based vaccines is complicated. Figure 11.3 describes examples of current ongoing trials of vaccines based on viruses.

11.3.4 VACCINE BASED ON NUCLEIC ACID

NAVs deliver genetic information that encodes cancer antigens to the host through typical physiological mechanisms, enabling the host to produce antigen proteins. Subsequently, the released tumor antigens may initiate immunological reactions that can eliminate cancer cells. DNA is more stable and persists in the body for a longer period of time compared to mRNA, primarily due to the extensive utilization of RNA enzymes and the structural differences between the two. Consequently, the initial nucleic acid

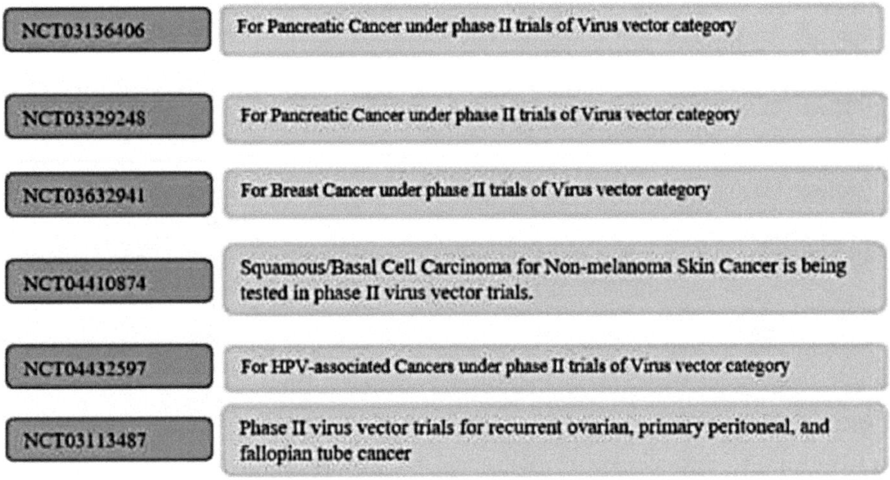

FIGURE 11.3 Examples of current ongoing trials of vaccine based on virus.

immunizations primarily focused on DNA vaccines. Unlike mRNA, which can rapidly translate and express antigens in the cytoplasm, DNA molecules must first enter the cell nucleus to initiate transcription. As a result, the generation of mRNA antigens is rapid and efficient. In comparison to mRNA vaccines, DNA vaccines elicit a lower immune response due to the requirement of an additional step for entry into the cell nucleus. Moreover, plasmid DNA has the capability to produce multiple copies of mRNA once it enters the nucleus, resulting in the production of a greater number of antigens compared to a single mRNA molecule. Furthermore, while DNA vaccines may carry a risk of insertion mutations, mRNA vaccinations do not pose a risk of insertion and integration into the genome [15].

One effective vaccination platform is the NAV (non-replicating viral vector). The NAV significantly induces MHC I-mediated CD8+ T cell responses, making it an appealing platform for creating vaccines. NAVs allow for the delivery of numerous antigens simultaneously, supporting both cellular and humoral protection. Additionally, NAVs can encode full-length tumor antigens, enabling antigen-presenting cells (APCs) to cross-present a large number of epitopes or present multiple antigens at once. They are particularly suitable for the production of specialized neoantigen vaccines due to the quick and straightforward creation of NAVs [16]. Figure 11.4 describes the examples of current ongoing trials of vaccine based on nucleic acid.

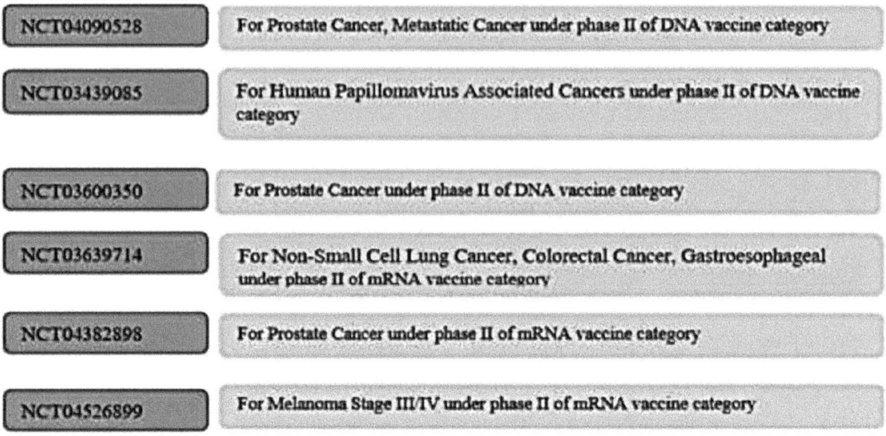

FIGURE 11.4 Examples of current ongoing trials of vaccine based on virus.

11.4 GLOBAL SURVIVAL DATA OF CANCER

Forecasts from the World Health Organization (WHO) indicate that, with an expected 10 million fatalities in 2020, cancer will be the leading cause of death globally [17]. In 2020, the highest rates of new cancer cases were reported in the following categories: 2.26 million breast cancer cases, 2.21 million LC cases, 1.93 million colon and rectum cancer cases each, 1.20 million non-melanoma skin cancer cases, 1.41 million PC cases, and 1.09 million stomach cancer cases. In terms of cancer-related deaths in 2020, LC accounted for 1.80 million cases, followed by colon and rectum cancer (9,16,000), liver cancer (8,30,000), stomach cancer (7,69,000), and breast cancer (7,69,000) [18].

11.4.1 CANCER INCIDENCE WORLDWIDE IN MEN

In 2020, LC was the most prevalent type, accounting for 15.4% of all new cancer cases reported in men globally. The top three malignancies, which made up 41.9% of all cancers, were LC, PC, and CRCs (excluding non-melanoma skin cancer). Stomach and liver cancers were two other prevalent tumors that contributed to more than 5% [19].

11.4.2 THE PREVALENCE OF CANCER IN WOMEN WORLDWIDE

Breast cancer was the most prevalent type of cancer in women globally in 2020, accounting for 25.8% of all new cases. The top three cancers in women, which accounted for 44.5% of all cancers (excluding non-melanoma skin cancer), were LC, breast cancer, and CRCs. Cervical cancer was the fourth most prevalent malignancy in women, with 6.9% of all new cases identified in 2020 [20].

11.5 PROGNOSIS OF CANCER

Our arsenal against cancer now encompasses a diverse array of medications and equipment, spanning from laboratory research to bedside application. Despite promising prospects, many cancer treatments still exhibit inadequate long-term response rates. Recent research indicates that individuals with similar types of tumors may respond differently to various treatment methods, with varying degrees of sensitivity [21]. Furthermore, clinical trials utilizing targeted drugs and gene microarray-based disease profiling have explicitly demonstrated the underlying genetic and phenotypic heterogeneity of human cancer. It is a shared goal to transition a cancer patient's treatment from an ineffective approach to a more effective one. Therefore, it is crucial, from a therapeutic standpoint, to identify which cancer patients can benefit from a particular therapy and which ones cannot. Early prognosis during cancer therapy is particularly desirable due to the rapid progression of the disease. To monitor macroscopic changes in tumor size during cancer treatment *in vivo*, a range of molecular imaging techniques, such as magnetic resonance imaging (MRI) and X-ray computed tomography (CT) imaging, have been employed. Numerous validation studies have been conducted to evaluate the treatment response of solid tumors to conventional cytotoxic chemotherapy. However, the situation becomes more complex when it comes to recently developed therapeutic methods such as gene therapy, immunotherapy, and antivascular therapy. In the early stages of cancer treatment, tumor necrosis and cavitation often occur before changes in tumor size, making it challenging to modify treatment regimens. Some studies have even indicated that alterations in tumor metabolism frequently occur before changes in tumor volume. These findings support the notion that assessing size change alone is not a sufficiently sensitive indicator of tumor response, but rather a quantitative measurement of tumor metabolism is needed. Additionally, relying on changes in tumor size to gauge treatment response often leads to inaccurate results. Functional imaging techniques that focus on various physiological characteristics of tumors have gained popularity for evaluating the early effects of tumor therapy. By capturing these early changes, we can better identify the appropriate therapeutic strategy for each patient in a timely manner, thus improving early treatment planning. The advantages of using this approach are virtually limitless in terms of enhancing patient well-being, from reducing physical burdens to improving therapeutic outcomes [22].

In this regard, several functional molecular imaging methods have been developed, including single-photon emission computed tomography, positron emission tomography (PET), MRI, ultrasonic imaging, optical imaging, and single-photon emission computerized tomography (SPECT). Functional molecular imaging tools enable intuitive observation of tumors' underlying physiology, metabolism, biomarker distribution, and blood perfusion. In other words, before observing changes in tumor growth, the information obtained from functional molecular imaging is utilized to determine how medications enter their target area, impact the tumor microenvironment (TME), and identify the presence of treatment resistance [23].

11.6 FACTORS GOVERNING PROGNOSIS

1. **Glandular Involvement and Distant Metastasis, Whether Present or Absent:** One could argue that the degree of distant metastasis and glandular involvement is inversely correlated with life expectancy. The corollary of this theory states that the essential anatomic position of both the original growth and the metastasis determines how close to death the disease will be [24].
2. **Fixation of Growth:** The complexity of surgery, the risk of an immediate operation, and the likelihood of a full recovery all increase with a cancer's degree of attachment to nearby structures [25].
3. **Location:** There is a consensus that early treatment, especially when the patient is compelled to undergo testing due to a tumor, hemorrhage, discomfort, or obstruction, offers the best prognosis for tumors found in easily accessible organs during a physical examination. Therefore, it can be stated that benign or malignant tumors that are more likely to cause early pain, hemorrhage, mechanical blockage, or a clearly visible mass are preferable. In general, it can be said that a growth has a better prognosis if it is easier to diagnose and treat and if it presents incapacitating symptoms early [26].

4. **Renal and Cardiac Function:** The heart and kidneys play a similar role in cancer as they do in overall health. The prognosis is worse, and the risk of immediate treatment is higher when cardiac and renal efficiency are diminished [27].
5. **Anemia:** This illness appears to be linked to two distinct conditions: dietary restrictions, a single bleed, or persistent or frequent leaking from the growth. Although it is difficult to determine whether it is not one of the other two, there may be a third criterion. This refers to anemia caused by a probable cancer toxin, or a disease associated with the breakdown of the tumor. In general, it can be said that the risk of immediate therapeutic failure and life expectancy increase with the severity of anemia. However, the prognosis following the removal of the growth may not necessarily change if the anemia is due to a single or recent series of hemorrhages from the growth [28].
6. **Size of Growth:** It has been demonstrated that the size of the growth and the presence of glandular involvement have a clear relationship that holds prognostic significance. It is reasonable to state that the likelihood of glandular involvement increases with the size of the primary growth, resulting in a worse prognosis [29].
7. **Age:** The prognosis appears to be somewhat influenced by age, according to the prevailing clinical opinion. Consequently, clinical experience shows that cancer in a young individual often leads to a quicker death compared to an older person. However, due to the numerous other variables at play, statistical analysis of this issue is highly challenging. In addition to general characteristics such as pulmonary, cardiac, and renal efficiency, there may still be differences in size, location, fixation, duration, growth pattern, and glandular involvement even when age is taken into account [29].
8. **Direction of Growth:** Whether the majority of the tumor is located within the lumen or surface of the organ, or if it has penetrated the organ's wall and is spreading to nearby organs or large cavities, such as the thoracic cavity in cases of breast cancer or the peritoneal cavity in cases of intestinal malignancies. Some intestinal tumors, particularly those in the stomach and colon, only minimally or not at all invade the biological wall as they actively grow into the lumen. Others infiltrate the wall, serosa, and adjacent structures instead of

growing towards the lumen. The latter type has the worst prognosis of the two. It may also be significantly smaller than those that grow towards the lumen and is more likely to involve glands [30].

9. **Weight Loss:** This in cancer patients can be attributed to various factors, including food quality and quantity, mechanical obstructions, discomfort, anxiety, and possibly infections and absorption of decomposing tissues. On its own, weight loss may not be very significant, but when combined with other prognostic markers, it adds more weight to the prognosis. It is important to note that underweight individuals often pose a lower risk for surgical complications compared to overweight individuals [31].

10. **Cellular Differentiation, Lymphocytic Infiltration, Fibrosis, and Hyalinization:** Differentiation, which forms the basis of many tumor classification methods, is crucial, although perhaps not as crucial as many physicians believe. The immediate clinical interest is weaker compared to research interest. It is widely accepted that the level of differentiation of tumor cells and their growth rate are somehow connected, but several other factors also influence the clinical prognosis [32].

11. **Duration of Disease:** The duration of the disease, as described in most written accounts of cancer, is often inaccurate and misleading. Over time, specific indications and symptoms that are not necessarily related to the type or grade of cancer become more significant. Factors such as size, anatomical location, mechanical obstruction, and bleeding all contribute to the signs and symptoms of cancer. Even pain, unless caused by a mechanical constraint, is usually not a reliable indicator. None of these symptoms can be used to diagnose cancer or predict the effectiveness of treatment. These symptoms and signs may have been present long before the formation of cancer and are common in many benign conditions. In other words, in the majority of cases, it is often challenging to determine or estimate the duration of the actual malignant development. It is not particularly useful as a sole prognostic indicator. Inoperable tumors often have visible histories, especially in the stomach and colon. The duration of the illness is generally only meaningful when considered in conjunction with other factors [33].

11.7 THERAPEUTIC CV

Therapeutic CVs function by precisely stimulating the immune system using tumor antigens to generate an anticancer response. However, therapeutic cancer vaccinations are still considered a strategy without proven clinical benefits. When compared to recently established immunotherapies such as immune checkpoint blockade (ICB) or Chimeric antigen receptor (CAR) T-cell treatments, therapeutic vaccinations have a limited track record of producing clinical responses in patients with advanced cancer.CVs, a form of immunotherapy, aim to treat cancer by enhancing the body's inherent anti-cancer defenses. Unlike cancer prevention vaccines, cancer treatment vaccines are administered to individuals who already have cancer, focusing on targeting cancer cells rather than the disease itself.

The concept of treatment vaccines is based on TAAs, which are believed to be present in cancer cells but absent or present at low levels in healthy cells. Therapeutic immunizations can train the immune system to recognize these antigens, mount a response against them, and eliminate cancer cells that harbor them [5].

There are three basic methods for creating CVs [34]:

- Your own tumor cells may be used to create them. In other words, they are made to specifically provoke an immune response against traits that are unique to your cancer.
- These vaccines can be developed using TAAs present on the cancer cells of multiple patients with a specific type of cancer. Once administered, patients whose tumors express the antigen may experience an immune response. This type of vaccine is currently in the testing phase.
- DCs, a type of immune cell, can be utilized for their production. The patient's immune system responds to an antigen on tumor cells as a result of DC vaccinations. Some individuals with advanced PC are treated with the DC vaccine Sipuleucel-T, which has been approved by the FDA.

Oncolytic viral treatment, sometimes referred to as a CV, is a distinct form of cancer therapy. It employs an oncolytic virus, a type of virus that selectively targets and eliminates cancer cells while sparing healthy ones. The first oncolytic virus treatment to receive FDA approval is a medication called talimogene laherparepvec (T-VEC, or Imlygic®). It is based on herpes simplex virus type 1 (HSV-1). Although this virus can infect both cancer cells and healthy cells, healthy cells can eliminate the virus while cancer cells cannot [35]. A tumor is directly injected with T-VEC. The virus continues to replicate itself, causing cancer cells to undergo apoptosis. Dying cells release fresh viruses and other compounds, which may trigger an immune response against cancer cells throughout the body.

11.8 TARGETED ANTIGENS

In the context of developing anti-cancer vaccines, the selection of the tumor antigen to target has garnered significant interest. Despite the extensive exploration of various cancer vaccination techniques, targeted antigens remain a focal point of discussion. These antigens can be broadly categorized into two main groups. The first group, known as TAAs, includes differentiation antigens, cancer testis antigens (such as MAGE-3 and NY-ESO-1), overexpressed antigens (such as Her2/neu, survivin, and MUC-1), and others. TAAs are expressed by both tumor cells and normal cells (such as Mart1, PSA, and PAP). Although normal cells also express TAAs to some extent, their immunogenicity can elicit specific T-cell responses. However, TAAs are subject to a certain level of self-tolerance. On the other hand, TSAs (Tumor-Specific Antigens) evade central thymic tolerance. TSAs include oncogenic viral proteins found in virus-induced malignancies, as well as neoantigens formed by non-silent somatic mutations of normal proteins. These neoantigens are recognized as foreign by the immune system. Neoantigen-based vaccine techniques have shown promising results in early (Phase I) human clinical trials and have demonstrated specific anti-tumor immune responses in various animal models [36].

11.9 THERAPEUTIC VACCINE STRATEGIES

There are numerous forms of therapeutic vaccination, such as those using whole tumor cells, tumor-cell lysates, or cancer cells that have had their genes altered. It is also typical to utilize vaccinations made of proteins or peptides, RNA or DNA, DC-based vaccines loaded with DNA, RNA, or peptides, viral vectors designed to express cancer antigens, and other vaccines. The development of vaccine platforms has also made it possible to use viral vectors with modified viruses engineered to express both specific antigens and immunomodulatory chemicals. Numerous studies have investigated autologous DCs pulsed with tumor antigens as potential CVs. In a phase

III clinical trial, peptide-pulsed DCs were examined as a first-line therapy for patients with metastatic melanoma, but DC immunization was found to be ineffective compared to chemotherapy. A DC-based vaccination is currently being tested in glioblastoma patients as part of a randomized phase II/III clinical trial (NCT03548571). Many factors, including the delivery of vaccines using DCs, the preservation of DC viability and maturation, and the consistency of *ex vivo* creation, can be limiting [37].

The development of vaccine platforms has also enabled the use of viral vectors containing modified viruses that express specific antigens and immunomodulatory chemicals. Research has been conducted using the modified vaccinia virus (MVA) Ankara strain. In a phase II clinical trial, patients with metastatic renal cell cancer (mRCC) were administered the TG4010 vaccine. This vaccine contains the immunostimulatory cytokine IL-2 and MVA that expresses the tumor antigen MUC-1 (mRCC). The induction of the MUC-1 immune response was observed, and the safety of the vaccine was confirmed. However, the vaccination did not provide any clinical advantage. These modified viral techniques have also been employed in the production of other vaccines, such as TroVax. TroVax is an MVA that expresses the fetal oncogene 5T4 (MVA-5T4). It was investigated in a phase III clinical trial for the treatment of kidney carcinoma, but no therapeutic benefit was observed [38]. Consequently, there are various types of therapeutic vaccination approaches. Furthermore, research on vehicle delivery and adjuvants is increasingly emphasized to enhance vaccine efficacy.

11.9.1 THERAPEUTIC VACCINES FOR PC

In April 2010, the FDA approved sipuleucel T (brand name Provenge®, Dendreon Corporation, Seattle, USA) as the only FDA-licensed therapeutic cancer vaccination for the treatment of metastatic castration-resistant PC, marking a significant development for active immunotherapy (mCRPC). Three immunological co-stimulatory molecules, commonly referred to as PSA-TRICOM and produced by Bavarian Nordic in Kvistgaard, Denmark, are present on lymphocytes: intracellular adhesion molecule 1 (ICAM 1), lymphocyte function-associated antigen 3, and CD80 (B7) [39]. These molecules are markers of T lymphocyte activation (LFA 3). Following boosting with a vaccinia-based vector, priming with a fowlpox-based vector is administered, both of which are administered with GM CSF. Two allogeneic PC cell lines that have undergone radiation treatment and are constitutively expressing GM CSF are used in the active immunotherapy GVAX vaccine for PC (GVAX-PCa). Despite initial indications of clinical safety and efficacy in patients with mCRPC, two phase III trials were prematurely stopped due to lack of therapeutic benefit and increased mortality [40].

11.9.2 THERAPEUTIC VACCINES FOR BREAST CANCER

In clinical trials, several HER2-derived peptides have been assessed in combination with GM CSF, cyclophosphamide, or poly-ICLC, among other immunostimulatory drugs. Active immunotherapy targeted against HER2 may only be truly successful in women with breast tumors expressing HER2 at modest levels. A subset of patients with low levels of HER2 expression who received GM CSF plus the HER2-derived peptide E75 (also known as NeuVaxTM) demonstrated longer disease-free life (DFS) [41]. A phase III clinical trial is currently investigating the use of E75 plus GM CSF to determine its efficacy in preventing the recurrence of early-stage, node-positive breast cancer in patients with low-to-intermediate risk. One of the most recent peptide vaccine treatments for breast cancer focuses on the WT1 (Wilms tumor protein) antigen, a transcription factor involved in cell proliferation, differentiation, and death. Early studies have shown that injection of WT1 emulsified with MontanideTM ISA 51 adjuvant can induce tumor regression and WT1-specific CTL expansion in patients with breast cancer. In a phase III trial involving 1,028 women with metastatic breast cancer, sialyl-Tn, an epitope identified in various glycoproteins conjugated with KLH, was investigated, but it was found to have no effect on survival. However, for a pre-stratified sample of these women who also received concurrent endocrine treatment, the time to progression and OS did improve significantly. One non-peptide technique involved administering vaccines engineered to express mucin 1 (MUC1) and IL2 to two out of 31 patients with metastatic breast cancer, resulting in partial tumor reduction [42].

11.9.3 THERAPEUTIC VACCINES FOR LC

The vaccine GSK1572932A combines the immunological adjuvant AS15 with the melanoma-associated antigen 3 peptide. Although none of the improvements were statistically significant, the results of a phase II trial involving 182 patients with resected stage IB-II MAGE A3-positive NSCLC showed a trend towards improved outcomes with GSK1572932A compared to placebo [43]. GV1001, a 16-amino-acid peptide vaccine that matches the active site of human telomerase reverse transcriptase (hTERT), along with GM CSF immunization, induced specific immune responses in up to 80% of patients with unresectable stage III NSCLC. The CIMAvax EGF vaccine aims to generate an antibody-mediated immune response rather than a cell-mediated immunological response. Recombinant human EGF, a carrier protein, as well as additional components, are combined with the adjuvant MontanideTM ISA 51. The median OS in a phase II study including 80 patients with stage IIIB-IV NSCLC was comparable to that attained with the best supportive treatment alone. The survival rate was higher in patients who had a robust immune response to the CIMAvax EGF immunization. A TGF-2 antisense gene has been inserted into four NSCLC cell lines as part of the Belagenpumatucel L vaccination (LucanixTM, NovaTx Corporation, San Diego, USA). A larger dose of belagenpumatucel L increased survival in phase II research. Tergenpumatucel L, which contains three LC cell lines transfected with a murine 1,3-galatosyltransferase gene, was associated with a median survival time of 11.3 months in a phase II study including 28 patients with carcinoma or recurrent NSCLC. Stimulation of antigen-specific T cells' IFN release was linked to improved OS [44].

11.9.4 THERAPEUTIC VACCINES FOR MELANOMA

A sequence-optimized peptide of the MontanideTM ISA 51 adjuvant, systemic IL-2, and gp100-specific immunization were utilized to enhance clinical outcomes in a phase III trial involving 185 patients with locally advanced stage III and stage IV melanoma. Out of the nine patients with advanced cancer who received adoptive cell therapy using autologous engineered T cells transduced with an anti-MAGE A3 TCR, five exhibited significant clinical remission. If there are no adverse reactions to the immunizations, a tolerance mechanism that prevents the presence of high-avidity TCRs, such as those used in the adoptive T cell therapy (ACT) study, may develop. Advanced melanoma patients who received DC therapy in combination with KLH and a cocktail of melanoma-associated antigens experienced a longer median survival time compared to non-responders. The phase III study of an allogeneic whole-tumor-cell vaccine given with Bacillus Calmette-Guérin as an adjuvant was abruptly terminated due to lack of efficacy after resection of metastatic melanoma. In a metastatic melanoma trial comparing whole tumor cell immunization with a DC-based strategy, vaccination with the autologous DC vaccine resulted in significantly longer lifespans than vaccination with the irradiation, entire tumor cell vaccine [45].

11.9.5 THERAPEUTIC VACCINES FOR PANCREATIC CANCER

The therapeutic targeting of pancreatic tumors with tumor-specific KRAS mutations has been successful in 90% of cases. In a phase I-II trial, 38 patients with advanced-stage pancreatic cancer were administered an immunization consisting of synthetic mutant RAS peptides and GM CSF. It was observed that responders had longer life durations compared to non-responders [46].

11.9.6 THERAPEUTIC VACCINE FOR CC

Regarding therapeutic vaccines for CRC, it has been demonstrated that Ad5 [E1, E2b]-CEA (6D), a novel adenoviral gene delivery system encoding the CEA antigen, enhances cell-mediated immunity in 61% of patients with advanced-stage CRC. Whole cancer cell immunotherapies have also shown promise in CRC. Phase III trials with OncoVAX® (Vaccinogen Inc., Maryland, USA), an irradiated, autologous tumor-cell vaccine with BCG adjuvant, revealed improvements in recurrence-free and OS for patients with stage I to stage IV colon cancer (CC). However, these improvements were primarily observed in patients with stage II disease [47].

11.10 CONCLUSION

The immune system is undoubtedly the most effective tool for long-term cancer control due to its ability to selectively target cancer cells and adapt to a changing tumor, as well as its inherent memory function. Neoantigens now provide the elusive antigen target for successful cancer vaccination. These neoantigens are distinct from each other and highly specific to the tumor. To fully unleash a patient's immune response and train it specifically against their cancer, constantly developing immune-modulating medicines such as CPB should be made available. These therapies include personalized vaccines that target these antigens alongside cutting-edge immune adjuvants. Active immunotherapy is becoming a significant addition to traditional cancer treatments, but there are still many crucial questions to address. It is essential to identify the ideal antigen, adjuvant, and delivery system combinations, as well as develop effective strategies to combat tumor-associated immunosuppression. Effective and long-lasting anti-cancer effects undoubtedly require combinations of complementary immunotherapies, and both investigator-initiated and industry-sponsored clinical development programs must take this into account. A significant amount of work is needed to discover novel predictive biomarkers and prospectively validate them in actual clinical settings.

There are numerous potential biomarkers. While measuring levels of circulating proteins, DNA, tumor cells, and immune cells before therapy is relatively straightforward, levels in the TME are likely more accurate predictors of response. Examining immunological responses both pre- and post-therapy could be beneficial in predicting long-term outcomes. Therapeutic CVs present an appealing alternative to immunotherapy due to their potential for safety, specificity, and long-lasting effects, potentially even leading to cures through the development of immunological memory. Unfortunately, previous attempts to create potent therapeutic CVs have been largely unsuccessful. However, recent research on CVs has shown promising clinical findings, thanks to the lessons learned from past failures. Three key insights have driven this breakthrough: the significance of immunogenic antigens, the importance of highly effective immunization vectors, and a better understanding of how to counteract tumor-mediated immunosuppression. With the help of new techniques, more immunogenic TAAs can be generated, and tumor-specific neoantigens can be identified. Improved vaccine technologies currently under investigation include enhanced co-stimulatory components, multi-antigen vaccinations, and the activation of both CD8 and CD4 T cell responses. Additionally, combinations with checkpoint modulators and other emerging immunosuppressive drugs show promise, although further research is needed to determine the most effective combinations and optimal dosing schedules for each component. We believe that simultaneous advancements in all three areas will be crucial for the success of CVs.

KEYWORDS

- **immunization**
- **immunological response**
- **lung cancer**
- **modified vaccinia virus**
- **positron-emission tomography**
- **targeted antigens**
- **therapeutic cancer vaccines**
- **tumor-specific antigens**

REFERENCES

1. Kennedy, L. B., & Salama, A. K. S., (2020). A review of cancer immunotherapy toxicity. *CA Cancer J. Clin., 70*(2), 86–104.
2. Kaur, K., & Khatik, L. G., (2020). Cancer immunotherapy: An effective tool in cancer control and treatment. *Current Cancer Therapy Reviews, 16*(1), 62–69.
3. Cox, M. C., Lapenta, C., & Santini, S. M., (2020). Advances and perspectives of dendritic cell-based active immunotherapies in follicular lymphoma. *Cancer Immunol. Immunother., 69*(6), 913–925.
4. Ferlay, J., et al., (2021). *Cancer Statistics for the Year 2020: An Overview, 149*(4), 778–789.
5. Hollingsworth, R. E., & Jansen, K., (2019). Turning the corner on therapeutic cancer vaccines. *NPJ Vaccines, 4*, 7.
6. Pan, R. Y., et al., (2018). Recent development and clinical application of cancer vaccine: Targeting neoantigens. *J. Immunol. Res., 2018*, 4325874.
7. Lee, Y. T., Tan, Y. J., & Oon, C. E., (2018). Molecular targeted therapy: Treating cancer with specificity. *Eur. J. Pharmacol., 834*, 188–196.
8. Saxena, M., et al., (2021). Therapeutic cancer vaccines. *Nat. Rev. Cancer, 21*(6), 360–378.
9. Lu, L., et al., (2021). Targeting tumor-associated antigens in hepatocellular carcinoma for immunotherapy: Past pitfalls and future strategies. *Hepatology, 73*(2), 821–832.
10. Tran, T., et al., (2019). Therapeutic cancer vaccine: Building the future from lessons of the past. *Semin. Immunopathol., 41*(1), 69–85.
11. Vergati, M., et al., (2010). Strategies for cancer vaccine development. *J. Biomed. Biotechnol., 2010*.
12. Copier, J., & Dalgleish, A., (2006). Overview of tumor cell-based vaccines. *Int. Rev. Immunol., 25*(5, 6), 297–319.
13. Malonis, R. J., Lai, J. R., & Vergnolle, O., (2020). Peptide-based vaccines: Current progress and future challenges. *Chem. Rev., 120*(6), 3210–3229.

14. Ura, T., Okuda, K., & Shimada, M., (2014). Developments in viral vector-based vaccines. *Vaccines (Basel), 2*(3), 624–641.
15. Leitner, W. W., Ying, H., & Restifo, N. P., (1999). DNA and RNA-based vaccines: Principles, progress and prospects. *Vaccine, 18*(9, 10), 765–777.
16. Ho, W., et al., (2021). Next-generation vaccines: Nanoparticle-mediated DNA and mRNA delivery. *Adv. Healthc. Mater., 10*(8), e2001812.
17. Jiang, Y., et al., (2021). Global pattern and trends of colorectal cancer survival: A systematic review of population-based registration data. *Cancer Biol. Med., 19*(2), 175–186.
18. Sung, H., et al., (2021). Global cancer statistics 2020: GLOBOCAN estimates of incidence and mortality worldwide for 36 cancers in 185 countries. *CA Cancer J. Clin., 71*(3), 209–249.
19. Ferlay, J., et al., (2019). Estimating the global cancer incidence and mortality in 2018: GLOBOCAN sources and methods. *Int. J. Cancer, 144*(8), 1941–1953.
20. Mattiuzzi, C., & Lippi, G., (2019). Current cancer epidemiology. *J. Epidemiol. Glob. Health, 9*(4), 217–222.
21. Hagerty, R. G., et al., (2005). Communicating prognosis in cancer care: A systematic review of the literature. *Ann. Oncol., 16*(7), 1005–1053.
22. Wajant, H., Pfizenmaier, K., & Scheurich, P., (2003). Tumor necrosis factor signaling. *Cell Death Differ., 10*(1), 45–65.
23. Weissleder, R., (2006). Molecular imaging in cancer. *Science, 312*(5777), 1168–1171.
24. Debnath, J., & Brugge, J. S., (2005). Modeling glandular epithelial cancers in three-dimensional cultures. *Nat. Rev. Cancer, 5*(9), 675–688.
25. Adam, M. N. M., et al., (2022). Evaluation of the effect of time of fixation and microwave treatment on quality of fatty tissue fixation in breast cancer specimens. *Science Progress and Research (SPR), 2*(1), 402–408. https://do.org/10.52152/Apr/2022.166.
26. Wang, Y., et al., (2019). HMGA1 in cancer: Cancer classification by location. *J. Cell Mol. Med., 23*(4), 2293–2302.
27. Su, Y., et al., (2020). Renal clearable hafnium-doped carbon dots for CT/fluorescence imaging of orthotopic liver cancer. *Biomaterials, 255*, 120110.
28. Busti, F., et al., (2018). Anemia and iron deficiency in cancer patients: Role of iron replacement therapy. *Pharmaceuticals (Basel), 11*(4).
29. Nunney, L., (2018). Size matters: Height, cell number and a person's risk of cancer. *Proc. Biol. Sci., 285*(1889).
30. Clinton, S. K., Giovannucci, E. L., & Hursting, S. D., (2020). The world cancer research Fund/American Institute for Cancer Research third expert report on diet, nutrition, physical activity, and cancer: Impact and future directions. *J. Nutr., 150*(4), 663–671.
31. Álvaro, S. E., et al., (2021). Impact of weight loss on cancer patients' quality of life at the beginning of the chemotherapy. *Support Care Cancer, 29*(2), 627–634.
32. Maccarty, W. C., (1922). Longevity in cancer: A study of 293 cases. *Ann. Surg., 76*(1), 9–12.
33. Boyne, D. J., et al., (2019). Association between adjuvant chemotherapy duration and survival among patients with stage II and III colon cancer: A systematic review and meta-analysis. *JAMA Netw. Open, 2*(5), e194154.
34. Josefsberg, J. O., & Buckland, B., (2012). Vaccine process technology. *Biotechnol. Bioeng., 109*(6), 1443–1460.

35. Fukuhara, H., Ino, Y., & Todo, T., (2016). Oncolytic virus therapy: A new era of cancer treatment at dawn. *Cancer Sci., 107*(10), 1373–1379.
36. Tacken, P. J., Torensma, R., & Figdor, C. G., (2006). Targeting antigens to dendritic cells *in vivo*. *Immunobiology, 211*(6–8), 599–608.
37. Melero, I., et al., (2014). Therapeutic vaccines for cancer: An overview of clinical trials. *Nat. Rev. Clin. Oncol., 11*(9), 509–524.
38. Banchereau, J., & Palucka, A. K., (2005). Dendritic cells as therapeutic vaccines against cancer. *Nat. Rev. Immunol., 5*(4), 296–306.
39. Parsons, J. K., et al., (2018). A Randomized, double-blind, phase II trial of PSA-TRICOM (PROSTVAC) in patients with localized prostate cancer: The immunotherapy to prevent progression on active surveillance study. *Eur. Urol. Focus, 4*(5), 636–638.
40. Obradovic, A. Z., et al., (2020). T-cell infiltration and adaptive Treg resistance in response to androgen deprivation with or without vaccination in localized prostate cancer. *Clin. Cancer Res., 26*(13), 3182–3192.
41. Giacobino, C., et al., (2021). Extracellular vesicles and their current role in cancer immunotherapy. *Cancers (Basel), 13*(9).
42. Su, X., et al., (2021). Breast cancer-derived GM-CSF regulates arginase 1 in myeloid cells to promote an immunosuppressive microenvironment. *J. Clin. Invest., 131*(20).
43. Quaratino, S., Forssmann, U., & Marschner, J. P., (2017). New approaches in immunotherapy for the treatment of lung cancer. *Curr. Top. Microbiol. Immunol., 405*, 1–31.
44. Rodríguez, P. C., et al., (2010). Clinical development and perspectives of CIMAvax EGF, Cuban vaccine for non-small-cell lung cancer therapy. *MEDICC Rev., 12*(1), 17–23.
45. Saxena, M., et al., (2022). Abstract CT108: Immunogenicity of poly-ICLC matured dendritic cells as an adjuvant for NY-ESO-1 and melan-A-MART-1 peptide vaccination compared to Montanide® ISA-51 VG, in study subjects with melanoma in complete clinical remission but at high risk of disease recurrence. *Cancer Research, 82*(12_Supplement), CT108–CT108.
46. Earl, J., et al., (2021). Somatic mutation profiling in the liquid biopsy and clinical analysis of hereditary and familial pancreatic cancer cases reveals KRAS negativity and a longer overall survival. *Cancers (Basel), 13*(7).
47. Jia, W., et al., (2022). Colorectal cancer vaccines: The current scenario and future prospects. *Front. Immunol., 13*, 942235.

CHAPTER 12

Regulatory Considerations for Cancer Vaccines Development

SHIVANI N. JANI,[1] HITESH B. KATARIYA,[2] and BHUPENDRA G. PRAJAPATI[3]

[1]*Department of Pharmaceutical Quality Assurance, SMT. S. M. Shah Pharmacy College, Gujarat Technological University, Amsaran, Gujarat, India*

[2]*Department of Pharmaceutics, SMT. S. M. Shah Pharmacy College, Gujarat Technological University, Amsaran, Gujarat, India*

[3]*Department of Pharmaceutics, Ganpat University, Mehsana, Gujarat, India*

ABSTRACT

This chapter offers a much-needed summary on the regulatory aspects of vaccine development and marketing approval, more specifically focused on the Cancer vaccine. Focusing on the clinical evaluation of the vaccines, this chapter also draws attention to the documentation required for vaccines development. Also, the chapter describes various guidelines for the same, and step wise procedure to follow for the clinical evaluation identifying the safety, immunogenicity and efficacy of the new vaccines as well as documentation or other supporting material required for changing any information of the prescribed vaccines. Ambiguity arises in scientific advices and application dossiers for the vaccine development and marketing, but this chapter describes points to consider while planning for the same. Moreover, case studies and examples are discussed at the end to give more detailed explanation of the topic.

Cancer Vaccination and Challenges, Volume 1: Strategies for Therapeutic Cancer Vaccine Development. Rishabha Malviya, Bhupendra Prajapati, and Sonali Sundram (Eds.)

12.1 INTRODUCTION

A category of diseases where aberrant cell multiplication occurs in conditions referred to as "cancer" and has the potential to spread to other body parts and invade them. Cancer is a serious health concern in all countries around the world. Ongoing research in this field has recently led to notable breakthroughs that support the theory of a non-toxic cancer treatment, in which the natural immune response of the individual destroys cancerous growth. With the introduction of new technology and cutting-edge diagnostics, the novelty and variety of techniques in this field are more diverse than ever.

A vaccine is a biological preparation that stimulates the body's immune system to fight against other diseases. Generally administered intravenously, current developments have led to the discovery of new routes of administration, such as oral or nasal spray. The main aim of tumor vaccines is to provoke an immune response to cancerous tissue. The clinical evaluation of vaccines is a key point during their development stage, as vaccines are an imperative part of the prevention and control cycle of infectious diseases. The guidelines provided by various regulatory authorities for clinical trials of vaccines focus on safety, potency, effectiveness, and immunogenicity of the newly developed vaccines. They also support any necessary changes in documentation, labeling, or manufacturing processes for already authorized vaccines available for sale. The guidelines for vaccine development and marketing have been amended in the last few years on various points, for example, some vaccines include allergens from numerous pathogens or even from different subtypes of a single pathogen. There were several issues that needed to be addressed for the clinical trials of these types of vaccines. Some tricky cases arose where the earlier guidelines did not provide sufficient knowledge on topics such as delivering vaccines to expectant mothers to protect their babies during the first few months of life, factors to consider when conducting vaccination efficacy trials, and vaccines designed for heterologous prime-boost protocols. Taking into account the challenges faced in the dossier application, the guidelines for vaccine development were revised. This included the addition of new points, such as a detailed description of the variables to consider when designing and analyzing findings from similar virological experiments. For example, when determining the limits of noninferiority and evaluating the therapeutic effects that do not meet pre-established standard norms, it is crucial to consider elements such as severity, mortality threat, potential to prevent infectious diseases, and the reliability of the assays used to measure the immune response [1]. All of these factors have contributed to the development of the current regulatory guidelines for vaccine development.

12.2 BACKGROUND

St. Peregrine's tumor [2] and Coley's toxins [3], respectively, provide the earliest evidence for the human body's capacity to eliminate tumor tissue in response to an external infection or when a pro-inflammatory drug has been administered. A systemic inflammatory process was necessary for Coley's toxin which was a combination of dead bacteria injected directly into the tumor, to show its effectiveness. Despite the effectiveness of this strategy, Coley's poisons were not widely adopted. A published study of abrupt regression of malignant melanoma after a febrile reaction to tetanus-diphtheria-pertussis vaccine treatment has just been described, which evokes memories of Coley's toxin in modern times [4]. Although the fundamental cause of this impact is yet unknown, it appears plausible that infections or inflammatory reactions may expose aberrant self-antigens. MUC1 mucin, a membrane-bound glycoprotein that is highly expressed on the epithelial layer of glandular cells, serves as an intriguing illustration of this [5]. MUC1 is released on cells undergoing inflammation, in pre-malignant lesions, and on cancerous cells in an improperly glycosylated form [6]. A lower prevalence of ovarian cancer has also been linked to the existence of antibodies to the aberrant MUC1 after mumps infection. This suggests that immunity to aberrant self-antigens produced during infection may eventually protect against precancerous tumors that express the same improperly glycosylated MUC1.

Although the development of tumors in immunosuppressed people is consistent with the idea of immune surveillance, [7] it's indeed likely that the focus of this surveillance is mostly on the causing viruses. It becomes clear that viruses can persist and cause cellular change as a result of the diminishing anti-viral immunity.

12.3 REGULATORY CONSIDERATIONS

Prior to granting a marketing authorization, it is crucial from a regulatory standpoint to guarantee the quality, stability, and efficacy of medications. Patients and healthcare practitioners can rest easy knowing that newly licensed medications have a favorable benefit: risk ratio in the target group thanks towards this regulatory check. To make sure that the benefit: risk balancing remains favorable after licensing, it is crucial to continue collecting clinical testing results post authorization, which may include (depending on the amount of data submitted at the time of application) comprise registry data or new clinical trial data. Such information also enables the detection of extremely uncommon adverse events that were not visible in the preliminary clinical data. Cancer vaccines as a category of treatment are subject to the same regulatory standards for a favorable benefit: risk balance, an appropriate customized risk mitigation plan, and pharmacovigilance. Guidelines covering a variety of quality, non-clinical, and clinical matters are periodically released by the regulatory bodies for Europe, the EMA, and the United States, the FDA. The (ICH) publishes standards that have been agreed upon by the US, Japan, and the EU. These ICH recommendations include regulatory, multidisciplinary, and quality, safety, and efficacy issues. As a result, the ICH recommendations and adherence thereto serve as a roadmap for drug development across several geographies. A section on cancer vaccines has been included to the EMA guideline, while the FDA has published regulatory rules pertinent for the therapeutic development of cancer vaccines. These materials also include references to additional regulatory requirements that are pertinent to clinical oncology research. Along with this, all the regulated countries have their own guidelines for cancer vaccine development [8].

The guidelines of USFDA specifically mentions the details of the products which can be included under this vaccine development. For e.g., they have defined Therapeutic cancer vaccine as those items which are designed to trigger particular reactions to a tumor antigen and are used to treat individuals who have already been given a cancer diagnosis. The guidelines of USFDA are governed by the Center for Biologics Evaluation and Research (CBER). These cancer vaccines work therapeutically by inducing or amplifying the host immune reaction to the pathogen in vivo. The cancer vaccine guidelines are specific and do not apply to vaccinations used to treat or prevent infectious diseases, to products meant to elicit or enhance non-specific immune responses, or to those meant to lower the cancer incidence in people who have never had the disease before. Additionally, adoptive immunomodulatory products like T cell or Natural killer (NK) cell products that may specifically attack the tumor to mediate their therapeutic impact will not be covered by the cancer regulatory guidelines [9]. Because the methods of action of cancer vaccines and adoptive immunomodulatory agents differ and so these guidelines do not apply to the formulation of those medicines.

Similarly, the guidance document of EMA describes that it is applicable to vaccines made up of many antigenic components designed to prevent a variety of contagious diseases as well as vaccines aimed towards several subgroups of a single pathogen that are used to cause infectious diseases. It applies to vaccinations that include-

a. Living organisms that are natively non-virulent to living beings or have been modified or genetically engineered to lessen their pathogenicity;

b. Microbes that have been inactivated by physical or chemical techniques;
c. In order to boost their immunogenicity, antigenic compounds obtained from pathogens or released by them may be employed in their natural condition, detoxified through chemical or physical processes, or clustered, polymerized, or tethered to a carrier;
d. Compounds that are antigenic and are created chemically or by genetic engineering;
e. Vaccinations using live bacteria or viruses that contain foreign antigens;
f. Naked DNA, such as plasmids that have been modified to express particular antigens.

The recommendations cover vaccination effectiveness, immunogenicity, and/or safety trials. It takes into account the circumstances in which a pre-licensure proof of vaccine efficacy might or might not be necessary, the planning of pre-licensure studies to gauge vaccine efficacy, and the evaluation of vaccine efficacy in the aftermath of approval. It also takes into account any evidence that may be offered by nonclinical studies to demonstrate vaccine efficacy, but it excludes other kinds of nonclinical research that are covered by other vaccination-related guidelines. Because of its in-depth discussion in a different CHMP guidance, vaccine pharmacovigilance is not a part of EMA guidelines for vaccine development [10].

In India, under the Central Drug Standard Control Organization (CDSCO), there's no specific definition given in the Guidance for Industry on Submission of Clinical Trial Application for Evaluating Safety and Efficacy, but it has a very good point that is, it directly mentions the list of documents needed to supply to the authority for approval of product along with the format in which the submission is required [11].

In Canada, Health Canada takes care of all these things, more specifically, Health Canada's Biologics and Genetic Therapies Directorate checks the vaccines marketing approval and its guidelines. The document says that its goal is to create uniform guidelines for the submission of license requests for vaccines intended for use in humans. The licensing procedure will be facilitated and, ultimately, vaccination availability will increase if information requirements are uniform across nations. It is anticipated that the Region will gain from having a common agenda by using its financial and technological resources more effectively [12].

In different countries, the guidelines are divided into various parts like, USFDA guidelines is divided into detailed description of early and late phase considerations for clinical trials. The Indian guidelines of CDSCO are divided into sections like: - general Information, chemistry manufacturing control, non-clinical data, proposed phase I/II studies, proposed phase III studies. Health Canada's guidelines for vaccines has adopted the modules followed by ICH CTD, i.e., five modules; administrative information of that particular country as well as prescribing information, summary of common technical documents, quality reports, non-clinical as well as clinical study reports [13]. And the guidelines for EMA are divided into various parts like immunogenicity, efficacy, effectiveness, safety, and topics on special population like pregnant women, geriatric patients and immunodeficient population [14]. Figure 1.1 shows different stages of development of vaccine where most important steps are discussed below in the chapter.

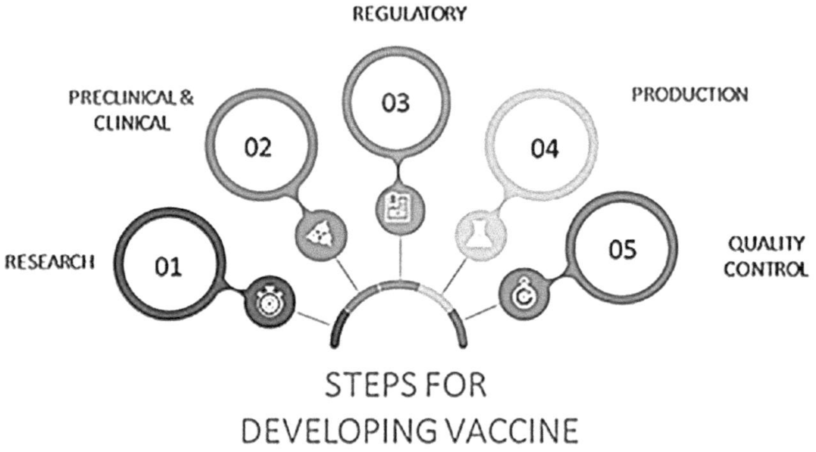

FIGURE 12.1 Steps for vaccine development

Source: Adapted from Ref. [68]

12.3.1 QUALITY PERSPECTIVE

Although there are guidelines currently offered for several product types (e.g., gene therapy, cell therapy) on the EMA webpage, it is not practicable to account for all circumstances in any one guideline due to the complexity, variety, and rapid increase in our understanding of tumor cell science and the immune response, as well as the novel strategies for therapy, such as sophisticated medicinal products. This is so because regulatory experience guides the writing of guidelines. It should be emphasized, nevertheless, that adhering to the rules as they stand is not required if it cannot be done. It is advisable to contact with policymakers to explain these differences and the specifics of the drug discovery program in situations where pharma companies offer well-justified perspectives, especially when these views are based on fresh scientific understanding and/or diseases. The investigators must have a strong explanation and, ideally, preliminary good quality findings to back their strategy in order to properly justify variations. Evaluation of the efficacy of cancer vaccines may be more difficult, especially when dealing with ATMPs as opposed to, say, peptides. It can be difficult to evaluate the quality characteristics of, say, a DC-based vaccination in relation to a pharmacodynamic (PD) marker or an efficacy objective. As a result, release standards for some vaccinations must be decided on a particular instance basis. Ideally, these standards should be connected to the product's mechanism of action [15].

The guidelines of various countries have different requirements for quality section for e.g., Health Canada have a separate module for quality which has requirements like the quality data presented under Module 3 must be current, thorough, sufficiently detailed, pertinent, and substantial to support the acceptance of a vaccination submission. This module will help in the easier preparation of the Quality Overall Summary (QOS) and will speed up the submission approval process if it is fully completed.

The antigen or vaccine component is referred to as the drug substance. The whole module for drug substance for the first pharmaceutical product should be followed by the full module of drug substance for the second drug substance module 3.2.S, and finally by the whole module 3.2.P of drug product for a vaccination containing more than one drug substance. To differentiate the data for each drug substance properly, the name of the drug product should be mentioned in the title of all relevant sections and subsections [16].

The complete Module 3.2.P Drug Product for one constituent or dosage should be accompanied by the whole Module 3.2.P Drug Product for the subsequent constituent or dosage if the vaccine has more than one dosage form or is delivered in various components, such as lyophilized granules with an infusion diluent. The titles of the respective Module 3 sections should include the name of the constituent or dosage form. The whole submission will be in the form of Common technical document [17].

There is no requirement of eCTD format by Health Canada. The quality module requires a lot of details like general information on drug substance and product, which involves name, structure, properties of drug substance and product; name of the manufacturer, detailed description of the manufacturing process, which includes culturing and harvesting of cells, detoxification and purification process, conjugation process, filling process, storage conditions, etc. Then control of materials will include details on strains used, embryonated eggs, seed banks, etc. After that there will be description about the process development and validation [18].

The requirements of CDSCO for vaccines in quality module are same as those for Health Canada, just at a point of post-approval stability protocol they differ.

12.3.2 NON-CLINICAL PERSPECTIVE

Based on the cancer vaccine and tumor target, non-clinical models have varying degrees of utility. Peptides, enzymes, cells such DCs, T cells, tumor cells and mixtures of vaccines with vectors, adjuvants, cytokines and ICI treatments are just a few of the numerous items and approaches that have been investigated. There are numerous potential target antigens in cancer. The concept proof and mechanism of action might very well, in some cases, be challenging or unlikely to illustrate in the non-clinical designs due to the absence of appropriate animal models for a number of these methodologies as well as the unique variables that are implicated when, for example, autologous therapies are explored. When developing autologous customized vaccinations, this is very troublesome. Products like human cells are not suitable for pharmacological testing in animals because preclinical testing must make use of the product that will be consumed by humans. Before organizing these investigations, the usefulness of animal models (such as chimeric animals, humanized mice, naked mice, must be taken into account on a situation basis. The benefit of providing evidence of concept if this is achievable, in situations where an appropriate murine model may be employed, lends support for the drug trials. Even in such situations, however, there are still some restrictions. These are related in addition to the distinctions between patient and mouse studies; often, young mice are employed when a treatment-resistant inducible tumor is being treated. As in case of prophylactic agents, patients already have a tumor that has been present for extended periods of time, as well as some immunodeficiency and sensitivity to the tumor; this is due to both tumor specific inhibition in the tumor environment as well as to previous chemotherapy and radiotherapy. Moreover, even when non-human animals are used, cross variations in the innate immunity can have a big impact. Non-human animal use is discouraged since, although being more similar to humans, there are still significant differences, a few of which were only discovered following severe adverse effects in a clinical study [19]. It should be done in vitro testing on living tissue to look at the spread of a potential TAA. Where appropriate, comprehensive in vitro testing on living cells to confirm the anticipated effects of altering the candidate antigen must be carried out. To assess antigen expression, tumor tissue and tumor cell lines can be utilized. Examining the quantity and level of malignant cells (such as DCs, MDSC, CD4C T, TIMs, Treg, and CD8C T cells) may also help to establish the evidence of concept as our understanding of the impact of the tumor environment on invading cells has grown. Additionally, if tumor sequencing testing is feasible after immunization, the results could bolster proof of concept even more [20].

A chapter on therapeutic cancer vaccines is included in the EMA guideline for anticancer products. The EMA guidelines specify that "non-clinical in vivo and in vitro solid evidence studies should be submitted to support the intended starting dose and timeframe in phase I investigations," although there is an exception in situations when there isn't an appropriate animal model. In some circumstances, using in vitro experiments with living cells as evidence may be appropriate. This warning recognizes the challenge of offering a typical non-clinical strategy in support of numerous cancer vaccine kinds. Companies use a variety of methods to generate cancer vaccines, and this variation presents a significant difficulty for the regulatory bodies when debating individual projects and taking necessary non-clinical research into account. An adjustable, case-by-case, scientific approach to pre - clinical safety assessment has been accepted by all regions of ICH (the United States, European Union and Japan,) in order to aid clinical advancement and marketing authorization [18]. According to ICH S6, the standardized guidance providing suggestions on the preclinical studies of biotechnology products in overall states. Although monoclonal antibodies employed as immunogens and cancer vaccines not designed for the treating of infectious diseases are not included in the vaccine guidance, according to WHO and CHMP regulations on vaccines, the suggestions made in these publications may still be taken into consideration as pertinent for the formulation of cancer vaccines. This viewpoint is supported by the CHMP guidelines on vaccine adjuvants, which states that "the guideline should, in principle, indeed be relevant to non-clinical aspects and quality of therapeutic vaccines." For non-clinical development, based on the analytical approach taken, there are many more guidelines like; CPMP note for guidance on gene transfer products which is recommended for DNA based cancer vaccines. For medications comprising monoclonal antibodies, tissue-based medicines, cell and gene therapies, the FDA offers particular guidelines. The WHO has published a policy on preclinical studies of DNA vaccines for informal discussion. The CHMP guidance on the pre-clinical assessment of anticancer medicines is primarily relevant to cytotoxic/cytostatic medications, which has little borne on cancer vaccines. However, certain broad ideas from the non-clinical formation of cancer vaccines might be pertinent [21].

The following goals are part of the non-clinical examination for cancer vaccines, which has roughly the same outcomes as testing for all medications: establish "confirmation" of the effectiveness in a simulation using animals, (b) specify the pharmacological or target organs systems that could be hazardous, (c) offer quantitative data on safety, such as the recommended starting dose and the location and timing of clinical trials. EMA and USFDA have the same goals on this. The amount and quality of knowledge about the target, natural and aberrant pathophysiology of the target, the metabolic routes implicated in its action mechanism, its tissue diffusion, the amount of in vitro analysis, and the targeted patient group, illness severity, dosing regime, and treatment duration are all factors that influence how much non-clinical research will be done. This is true for most new medicines as well. Cancer vaccines are no different from other new medicinal items in that evidence proving their quality, safety, and efficacy must be supplied in order to receive marketing approval. When a batch of products is released, they are thoroughly characterized and the production system is subjected to tighter regulation and validation. A product's safety is typically evaluated through the execution of non-clinical safety investigations and evaluation of safety goals in clinical trials. However, at the initial (preclinical) phase of development, results are gathered in animal efficacy designs to aid in the assessment of potential effective doses and dosing regimens for use in subsequent clinical trials. Efficacy eventually needs to be proven in the corresponding human patient group [22]. Therefore, non-clinical investigations on cancer vaccines in the framework of this study include both safety and efficacy studies carried out on animals. The following specific difficulties for cancer vaccine's must be solved in conjunction to these broad standards. The development of the non-clinical approach must take into account the following issues relating to this "vaccine-principle": In most cases, a cancer vaccine is made up of a vaccine antigen, deposited on, or mixed in, a strong adjuvant. In keeping with "conventional" vaccines, the injected product's effect incites an immune reaction without being directly related to the clinical effect. The biologically active principle assumed to be in charge of the therapeutic outcome is this immunomodulatory reaction to the vaccine. Programs for testing must take into account needs brought on by the type of vaccine antigen (such as a DNA construct or just a monoclonal antibody), the adjuvant utilized, and the truth that cancer vaccines are typically generated using biotechnology technologies [23].

CDSCO contains separate module on non-clinical studies, which includes details like testing strategy of non-clinical studies, pharmacokinetics, toxicology and pharmacology of vaccines. The toxicological sub-section requires information on general as well as special toxicology. General toxicological study includes details on the animal model used and justification for the same, the tolerance level, the parameters which were monitored and dose of vaccine. Special toxicological study includes study on special population, and also studies on carcinogenicity and genotoxicity. Immunogenicity of product and adjuvant should be shown in detail [24].

The guidelines of Health Canada have module 4 which is specifically dedicated to non-clinical study reports. Here the details need to be submitted in CTD form rather than in eCTD format. The study reports contain first detail of pharmacological reports which contains primary pharmacodynamics reports. Pharmacodynamics research for a vaccination product is often carried out to assess the vaccine's toxicity or, in the case of accessible animal models, its ability to give protection. In order to support the requirement for the adjuvant, a pharmacodynamics assessment may include the pharmacology of an adjuvanted vaccine. After that, secondary pharmacodynamics is not needed in the case of vaccines. Then safety pharmacology is performed and its results are tabulated. A safety pharmacology study's goal is to find out how the potential vaccination would affect key bodily processes [25]. The NRA may occasionally advise safety pharmacology studies for vaccinations even though they are not often required. For instance, these studies should be implemented into the safety evaluation program if records from nonclinical and/or human clinical studies indicate that the adjuvanted vaccine might well affect physiological functions (such as the renal functions, central nervous, body temperature, respiratory and cardiovascular systems) other than the innate immunity. Then goes the pharmacokinetics data, where they are not generally performed for vaccines but for the vaccines containing live recombinant vectors or new adjuvants, biodistribution studies can be done. If pharmacokinetic studies are done, then the report must be submitted for the following things: Analytical method and validation of it, Absorption, Distribution, Metabolism, Excretion and Drug interactions. Then for toxicology studies, the reports have to be submitted in the form of Single-dose toxicity, repeat-dose toxicity, genotoxicity (not done for vaccines generally) [26].

12.3.2.1 SELECTION OF ANIMAL MODEL

The (very specific) underlying mode of action and specificities of each agent determine the species of animals to be used for the therapeutical and/or toxicological study of cancer vaccines. The utilization of more than one experimental animal during the entire preclinical procedure may be necessary due to the intricacy of vaccine research. In general, screening process for animal models can vary depending on whether a substance's effectiveness or toxicity is being assessed. Nonhuman primate species may be useful for toxicology testing due to the closeness of their body's immune system and the often-similar manifestation sequence of the antigen, just as special mice species may be utilized for effectiveness testing but are typically not suited for toxicity testing [27]. Effectiveness and toxicological assessment of cancer vaccines demand an intact immune system, as well as a medication that is biologically active inside the chosen species, as crucial entry requirements for animal model(s). Non-clinical evaluation of cancer vaccines is similar to evaluation of biotechnological goods in that it assumes knowledge of the therapy's ostensible mechanism of action.

12.3.2.2 TISSUE CROSS-REACTIVITY

Studies of tissue cross-reactivity may be important during the non-clinical progression of cancer vaccines in two ways: first, as a requirement for choosing the right species of animals (evaluation of various organisms' tissue cross-reactivity), as well as foremost, to look at potential undesirable incidents that result from the adhesion of a vaccination allergen or the stimulated immune reaction on non-target specific tissue. Assessment of tissue interaction is crucial when the vaccination protein responds with tissues immediately, such as when it does with development receptors, growth regulators, or antibodies, but it is of little utility for other vaccines (e.g., DNA plasmid vectors). Immunohistochemical cross-reactivity tests that show the impact of an injected antibody and the elicited immune reaction on any unintended reaction and/or toxicity toward living tissue different than the designated target may be expected from EU and US agencies. Elevated concentrations of interaction in human tissue locations that are not therapeutic point has the requirement for more thorough histopathological analysis in toxicity investigations. Cross-reactivity investigations on antibody molecules (fragments or intact) that are supplied in deposited form (for instance, deposited onto aluminium hydroxide) ought to have low potential for cell binding, thus these studies may not be very useful [28].

12.3.2.3 SELECTION OF DOSE LEVEL AND SCHEDULE OF VACCINATION FOR NON-CLINICAL STUDIES

Due to the following factors, dose range determining toxicity tests often carried out in animals on traditional medications may not be highly relevant for cancer vaccines. Cancer vaccine antigens are typically given at very low dosages, significantly lower than the amounts anticipated to cause side effects; A tenfold rise in dose may not inevitably translate into a tenfold rise in responsiveness because it is well recognized that the immune response may function as an "on-off button"; When it comes to vaccines, the clinical impact is sometimes not caused by the medicine itself but instead by the immune response that is elicited against the manufactured antigen [29]. Therefore, obtaining the best immune response might serve as the basis for dose ranges and dosing regimen in non-clinical research. In a similar vein, the vaccination frequency should be chosen to provide the best induction. To sustain a beneficial immune response, periodic vaccine boosters may be necessary; The availability of the product may restrict the practicality of doing such experiments at large doses [30].

12.3.2.4 PHARMACODYNAMICS

Beginning with the subjective and numerical characterization of the generated immune response toward the cancer vaccine, which symbolizes the ideal reaction anticipated to correspond with clinical success, animal efficacy evaluation of cancer vaccines is conducted. Since the emphasis of cancer vaccines is the possible impact of antigenicity on the non-clinical safety data (i.e., through the potential creation of "neutralizing" antibodies that might impact the results of the toxicity testing), proposed readings pertaining the immunogenicity for biotechnology goods as highlighted in ICH S6 are pretty much irrelevant. Research on immunogenicity consequently form element of the "Primary Pharmacology" evaluation of "conventional" vaccines since, like "traditional" vaccinations, the antigenicity of cancer vaccines is not a criterion of harm but an indication of efficacy [31].

Similar to other medicines, initially in cancer vaccines' development a comprehensive pharmacodynamics (PD) review is conducted to help choose the best clinical dosage and treatment plan. Additionally, biomarkers generated from these investigations might be approved auxiliary markers that connect to the anticipated clinical result or toxicity, speeding up the process of developing drugs in general. The creation of an efficacy assay may also be supported by research on pharmacodynamic factors. Instead of being used alone, cancer vaccines may be used with chemotherapy, radiation, or other biosimilars in combination treatment. Therefore, interaction experiments in an animal paradigm that is applicable for both the targeted vaccination therapy and the conventional medication should be taken into account in pharmacodynamic research, with a focus on potential impacts on immunogenicity. Compounds containing rodent or recombinant antibodies that might be delivered before to or following a commercially available or under development product that also contains a rodent or recombinant antibody should receive special consideration [32].

12.3.2.5 TOXICOLOGY PROGRAM

According to ICH S6, the main objectives of preclinical safety assessment are a) to discover the relevant target tissues for toxicity and the cyclability of such harm, as well as b) a starting safe dose and successive dose progression methods in humans. Since vaccine harm might result from the immunogen, from adjuvants or immunomodulators added to the vaccine, from the evoked immune reaction, or from a mixture of the three, testing for safety and toxicity of cancer vaccines can be challenging (as with "conventional" vaccinations). Despite the fact that cancer vaccines are believed to be better from other cancer medications, proving the efficacy and safety of vaccines depends on a methodical, step-by-step process in which businesses and authorities handle each product separately based upon the kind of vaccine and its believed mode of action. The choice of an acceptable experimental animals for toxicity evaluations is frequently problematic for the research design [33]. By comparing cancer vaccine to small molecule medications, a modified sequence of investigations is typically needed for safety/toxicity profile. Contrary to small molecule therapeutic drugs, the major part of cancer vaccine screening is frequently finished before the start of a Phase I trial in living beings. In comparison, non-clinical experiments to discuss severe and reproductive toxic effects and carcinogenicity are typically conducted concurrently with the clinical studies for miniscule molecule therapeutic drugs. Despite being used at low quantities; adjuvants can need to undergo their individual toxicological testing. The possible toxicity of "novel" adjuvants with natural or synthetic sources in specific has to be assessed. Testing is typically not necessary for adjuvants that have been around for a while, such as calcium phosphate, aluminium phosphate and aluminium hydroxide. Various studies need to be reported under this i.e., Single dose and repeated dose toxicological studies, adjuvants, points related to DNA vaccines, pharmacokinetics and toxicokinetics, immunotoxicity studies, local tolerance reports, carcinogenicity experiments, fertility performance and developmental toxicity [34].

12.3.3 CLINICAL PERSPECTIVE

The need for demonstrating the action mechanism of the vaccine will significantly depend on human in vivo tests results due to the anticipated constraints in the non-clinical strategy for several cancer vaccine items. In contradiction to mice cancer models, the immunological status of cancer patients will differ at baseline, and a decline in overall immune function is linked to increased age and prior therapy. Through a process known as immunoediting, the tumor changes to avoid detection and elimination by the innate immunity while immunosurveillance is in operation. Additionally, the tumor microenvironment suppresses the immune system, which is evident even at the earliest stages of tumorigenesis. These elements may affect a patient's capacity to react to a vaccination. The clinical program should take baseline immunological status measurement into account (e.g., TAA-specific T cells, serum Igs, MDSC, CD4C, DC, Tregs, CD8C). These factors could affect prognosis or anticipate how a patient would react to a vaccine (for example, by employing a pharmacodynamic (PD) read-out of impact in preliminary clinical trials). The layout of pivotal clinical studies can be influenced by the discovery of a link between baseline immunological status and vaccination response in early phase research [35].

Periodic in vitro examinations (blood/tumor) can be used to monitor the effectiveness of treatment and, if practical, can offer highly helpful data during the early stages of clinical growth. More compelling data could be obtained through direct ex-vivo study of peripheral plasma (e.g., number of

Tregs, TAA-specific Ig level and TAA-specific T cells,) than from in-vitro proliferated cells retrieved from peripheral blood. In the lack of repeated tumor biopsies, it can be a technically practical strategy to look for a PD impact, even though it is recognized that the peripheral blood will not always represent alterations in the immune system inside the tumor. If it is possible, it must be done to check for TAA-specific T lymphocytes in peripheral blood. Additionally, some interpretation of a PD impact will confirm the action mechanism, reinforce the evidence of concept, and help with dose determination [36].

Determining the ideal treatment duration could also benefit from the use of such information. Utilizing all accessible clinical data is advised because the information backing the concept proof may be restricted in regards to the proof provided by non-clinical data. There is data that vaccination alone can have a counterintuitive impact on the cells which infiltrate tumors, increasing the tumor-specific immunodeficiency. This is a major problem for cancer vaccines. This suggests that vaccination might not necessarily be a secure method. This discovery supports the advantages of establishing evidence of concept in individuals and/or establishing a post-vaccination PD readout. Such human data have to be accessible before constructing crucial studies. Prophylactic cancer vaccines are a more recent strategy that might require less complexity in regards of combating immune suppression. Furthermore, it is intended to focus on individuals who do not yet have cancer but are at an elevated danger of developing one, like those who have precancerous tumors [37]. So, in terms of patient selection and selecting workable objectives for the key clinical trials, a preventative cancer vaccine strategy will necessitate a different methodology than a therapeutic cancer vaccine. Alternative efficacy indicators will need to be created and validated in this area. A two-way conversation with regulators is advised due to the paucity of regulatory rules and the minimal knowledge of this field. Clinical dose discovery investigations and serial immune response monitoring are typically necessary for developing therapeutic cancer vaccines. The development of new cancer vaccine relies heavily on assays; hence the guidelines also call for comprehensive descriptions of the analytical tests. Even if it is not always possible, periodic tumor biopsies are crucial since the results may show anti-tumor action. In that situation, evidence from preliminary clinical trials with a small group of patients who underwent repeated tumor biopsies might offer proof of theory. It is likely that visualizing results on tumors that cannot be safely sampled could potentially give proof for such a reaction, even if it is not clearly mentioned in the guideline. Given that patients with late-stage cancer and massive tumors, who would have a short life span, will have immune depletion, choosing the right treatment groups for the critical studies is challenging. Therefore, the EMA recommendation proposes that individuals with a reduced or minimal illness severity could be examined. Although patients with early disease (e.g., recently diagnosed instances) and who have not received extensive pretreatment may have a greater chance of effectiveness from an immune activation standpoint, picking one such group where alternative therapy available will require to be validated [38].

When such a situation arises, engagement with regulators to obtain agreement for the clinical advancement strategy is encouraged. This is especially the case when there is a strong justification that favors such a strategy and strong evidence of concept. The recommendations for learning of efficacy stipulate that "updated criteria characterizing progression is approved if adequately justified" and permit a delayed reaction for cancer vaccinations. This is consistent with the updated Response Evaluation Criteria in Solid Tumors (RECIST) guidelines, which emphasize that an efficient immune response may take some time to establish and that this might cause a slower tumor reaction than cytotoxic drugs. However, the EMA recommends overall survival (OS) as the primary effectiveness endpoint. Although overall survival is a defined target, provided a cancer vaccine has demonstrated good evidence of concept and improves progression-free survival (PFS) compared to a suitable comparator, this would be acceptable. This is especially important for cancers in which it would not be viable to conduct pre-licensing trials for the lengthy periods of time necessary to gather adequate data on overall survival. However, in these situations, further OS post-licensing data will be very crucial to confirm that there is no indication for any OS-harmful effects [39].

It is advised to conduct a double-blind experiment to evaluate PFS. The experiment should use a randomized efficacy assessment if double-blinding is not feasible. It is recommended that radiological evaluations be evaluated centrally by evaluators who have been blinded if they constitute the primary read-out of performance [40].

The construction of accompanying diagnostics is covered in greater detail in the FDA Guidance for Industry on Clinical Considerations for Therapeutic Cancer Vaccines8. But just like the EU recommendations, there are still questions about which patient group to utilize and how to pick a workable objective. The efficacy outcome of relapse will require extensive follow-up if the patient cohort chosen has little (or no) remnant disease. Figure 1.2 shows flowchart on how to design the clinical trial which is used in the approval of cancer vaccines [41].

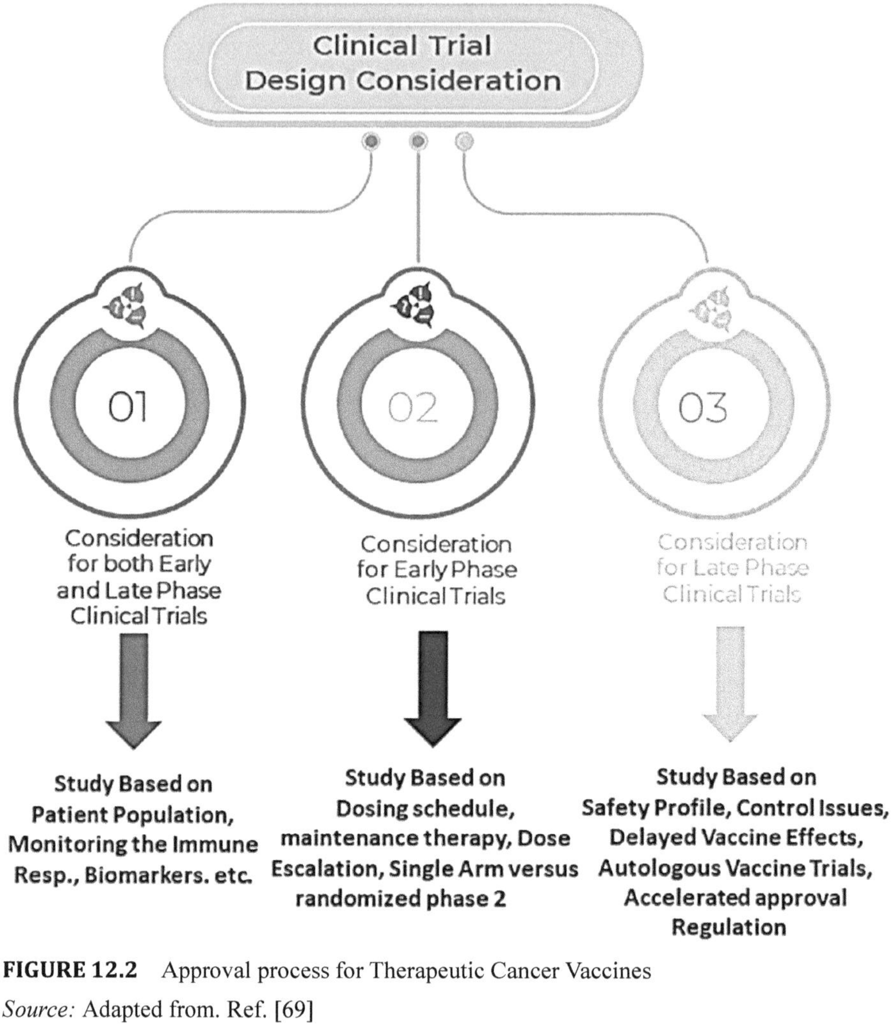

FIGURE 12.2 Approval process for Therapeutic Cancer Vaccines

Source: Adapted from. Ref. [69]

The FDA views immune response tracking as experimental, and it considers such assessments relevant for evidence of concept, dose determination, and potential association with clinical effectiveness. The FDA offers some advice on excipients and multi-antigen treatment in addition to supporting the use of investigational biomarkers as solid evidence tools. The recommendation in the guideline is that the primary clinical objectives be clinically significant and reviewed with the FDA due to the great variety of methodologies used for therapeutic vaccinations [43].

The core standards for quality, safety, and efficiency remain the same for every product, and it is evident from a reading of the EMA and FDA regulations that the process to obtain marketing authorization will be case-by-case. Discussion with regulatory agencies to actively sought agreement on the grounds of a well explained rationale and scheduled clinical design program is recommended in cases where the regulations that are currently available do not involve the stance taken in the development of a new drug, as is anticipated for numerous curative cancer vaccines as well as for every preventative cancer vaccine. (35) There are currently no regulatory recommendations available for cancer vaccinations used as preventative measures. Because the participants in these trials will be fit individuals as opposed to those getting therapeutic cancer vaccines, it is crucial to consider this form of preventive vaccination differently. Because of this, patient health would be a greater issue here than it would be with a therapeutic vaccination given to patients with advanced cancer [44].

Another factor to take into account is that a gene therapy medication will have greater safety ambiguity than, say, a peptide vaccination. Drug manufacturers should view the lack of regulatory guidelines for preventive cancer vaccines less as a hindrance and rather as a potential to impact regulatory opinion. Regulators appreciate such early communication, and these discussions may take the shape of a conference with the EMA's Innovation Task Force (ITF) [45].

Although regulatory authorization is a must, it does not ensure effectiveness in terms of patient accessibility and payment in the long run. Considerations for this early commercialization are crucial. In order to make sure that the endpoints specified by authorities will be agreeable to HTAs and payers, communication with Health Technology Assessment (HTA) Authorities is also advised. The overarching message from numerous seminars around the globe, while not indication-specific, is unmistakable: whatever the program, it is best to get regulatory and HTA/payer feedback early on. A method like this is anticipated to lower the possibility of failure during the reimbursement phase following licensing. The UK offers simultaneous technical advice conferences with NICE and the MHRA as well as similar joint advising protocols at the national scale [46].

CDSCO has Module 5 which is designated as reports of clinical studies. It includes detailed data on Phase-1,2 and 3 trials. Phase 3 studies also contain bridging studies in it. Then it includes details on any special considerations needed for vaccines and details on adjuvants. Phase 4 studies or plan of pharmacovigilance if prepared can also be submitted. For mixed vaccines or already approved vaccines but manufactured by new manufacturer requires a non-inferiority study also. And at last co-administration with other vaccines create any significant change or not is also been shown [47].

Similarly, Health Canada also has Module 5 as clinical study reports. Human clinical trials are often divided into three distinct stages: phase I, phase II, and phase three. In certain nations, formal regulatory clearance is needed before beginning any of these studies. Depending on the nation, this permission may take several forms, such as an Investigational New Drug Application (IND) in the US or a Clinical Trial Certificate in the UK. In addition, clinical studies in all nations must also receive ethical approval [48].

All research involving human beings must undergo a thorough ethical evaluation in conformity with the Helsinki Declaration. Studies carried out in Canada require a Clinical Trial Application (CTA). Sponsors are invited to ask Health Canada for advice on technical, clinical, quality, and related regulatory hurdles at a suitable phase of product development before filing a CTA by scheduling a pre-CTA meeting. Safety is the main focus of phase I investigations [49].

Phase I clinical investigations test a vaccine for the first time in small groups (for instance, 20 healthy individuals) to determine its safety, tolerability, as well as, if necessary, clinical testing and immunological characteristics. Phase II studies contain a greater number of participants and are designed to gather preliminary data on a vaccine's general safety as well as its capacity to have the targeted impact (often a particular immune response) with in chosen population. Comprehensive Phase III investigations are necessary in order to completely evaluate the protective effectiveness and toxicology of a vaccine. Phase III clinical studies are the key studies that will determine if a license will be granted, and it will be necessary to gather enough information to show that a new drug is both safe and efficient for the intended use. A vaccine must ideally have been fully characterized before the start of Phase III of development, and the entire manufacturing technique, requirements, and batch discharge testing protocols ought to have been created [50].

To support quality modifications made after Phase III investigations, more data may be needed. On the grounds of the results from Phase III study, a licensing application may be made to an NRA, and if accepted, the vaccine will subsequently be marketed in that nation. The type of vaccination and the antigenic substance must be taken into consideration while designing the clinical development program's framework. For instance, the clinical assessment of a vaccination that exclusively contains innovative antigen(s) may naturally differ greatly than that of a vaccine which also comprises one or more antigens that have already undergone clinical testing. These variables also impact the necessity of clinical protection studies, their viability, and whether it is reasonable to base an approval solely on immunogenicity evidence. In such cases, it is the applicant's responsibility to defend the clinical development program's structure and content. Pre-submission consultations with regulatory bodies may help to make sure that the resulting data package's content is probably to be accepted [51].

The format for submission will be CTD format. Various reports are required to be submitted under Module 5, for e.g., Biopharmaceutical studies, which are generally not done for vaccines but if immunogenicity study reports are submitted, then bioanalytical assay needs to be performed. Then bioavailability studies, comparison of bioequivalence and bioavailability, in-vitro in-vivo correlation studies reports need to be submitted. Human pharmacokinetic studies are not performed for vaccines, reports of human pharmacodynamic studies need to be submitted. The immunogenicity studies for vaccines are often carried out to validate the dosing, dosage schedule, and composition of the finished product. Then the reports on safety and efficacy are also to be submitted. Documentation of all clinical tests done by the sponsor (or otherwise available) to evaluate the effectiveness and security of the vaccine, along with all finished and all ongoing trials involving the vaccine in planned and non-proposed indications, must be given [52].

This needs to incorporate the findings of the immunogenicity tests done to back up the dosage choice, timing, and composition of the finished product. There is a chance that the receiver or contacts will acquire clinically severe illnesses after receiving live attenuated vaccines (bacterial or viral, particularly vaccine vectors). It is recommended to submit clinical study findings with details on shedding, reversion traits, and transmission to individuals [53].

Along with general vaccines, the trend of personalized vaccines is increased. Figure 1.3 shows the detailed pathway of how to develop personalized vaccines, and what is the future in it. In Europe, EMA has given a separate guideline for clinical evaluation of vaccines. It contains in detail description of immunogenicity, efficacy, safety, effectiveness and trials on special population. USFDA has its own guidelines for clinical studies design considerations. It consists of 2 sub-sections; early phase and late phase clinical trials. Following are some clinical factors that apply to both initial and later phase clinical studies: population of the patient, immune response monitoring, biomarkers as the proof of efficacy, adjuvants that activate the immune system, multi-antigen vaccines, immediately following

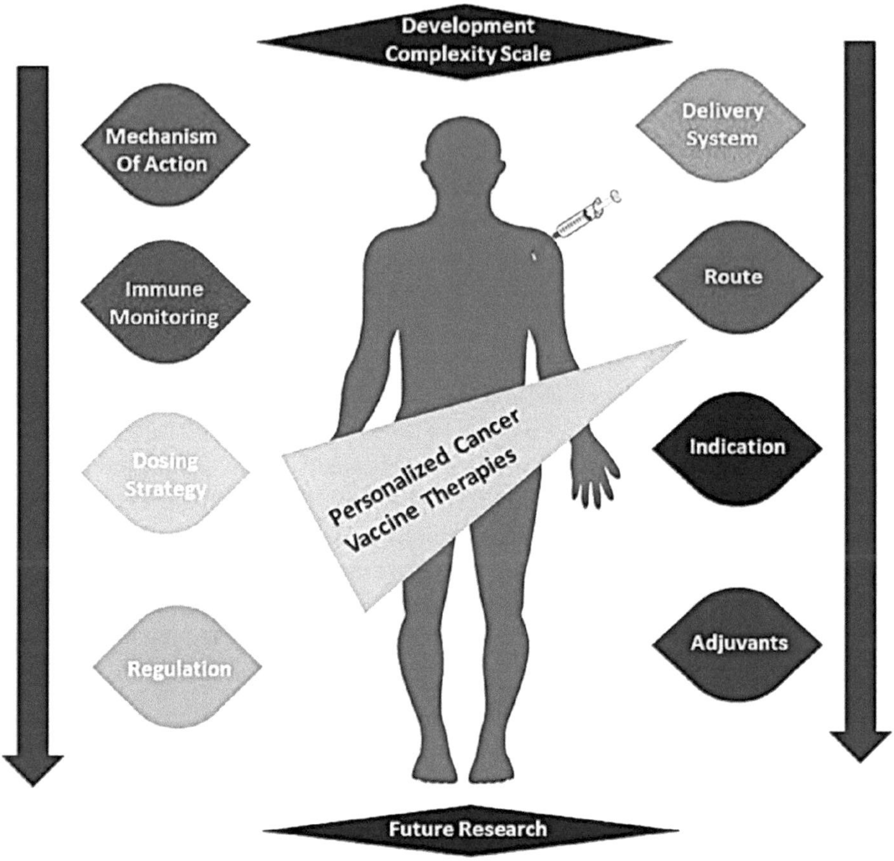

FIGURE 12.3 Development pathway of personalized cancer vaccine therapies

Source: Adapted from Ref. [70]

or soon after the initial dose of cancer vaccinations disease progression or recurrence, subsequent and concomitant therapies. The standard methodology for clinical testing of a chemotherapeutic drug entails preliminary trials in participants with progressed metastatic diseases and various tumor types to establish the maximum tolerable dose (MTD) and ideal schedule as well as evaluation of clinical activity via tumor reduction (objective tumor responses) [54]. Therefore, tumor reduction that becomes visible within the initial 8 weeks of exploratory therapy can be used to assess the MTD and clinical efficacy with brief monitoring period in clinical studies of conventional cytotoxic therapy. The agent is then tested for efficiency and safety in a metastatic scenario of a single tumor form in a large, typically randomized, and controlled environment. The same medication may be developed and evaluated in individuals with low disease burden or no signs of residual disease after its effectiveness and safety in the context of metastatic disease are established. In contrast, it typically takes 2–3 months for an anti-tumor immune reaction to emerge, which is necessary for a cancer vaccination to be active or successful. In phase 1 clinical studies, cytotoxic drugs are typically examined in populations with a varied mix of tumor forms at different clinical phases [55].

The main objective of these phase 1 trials is frequently to ascertain the MTD and the low toxicity of the tested drugs; as a result, it is understood in these trials that every particular drug may have a variable impact on various tumor types. Agents are examined in phase 2 clinical studies with a somewhat homogenous patient group and a specific tumor type if they are determined to have a tolerable level of toxicity. It is also worth giving thought to designing an assay or numerous different methods to quantify the target antigen utterance in tumor cells of specific patients and utilizing that data in purposeful sampling or response monitoring when the suggested method of action includes a particular antigen or numerous different therapeutic targets [47]. The Center for Devices and Radiological Health generally oversees the regulation of these assays (CDRH). As a result, sponsors of cancer vaccines who are considering using an assay in the labelling of the cancer vaccine or funders of these assays who also are intending to create the diagnostic test for use with a particular cancer vaccine could perhaps request a discussion with both the pertinent product review office (CBER) as well as the pertinent device review division (CDRH). This is especially crucial in cases when using the test proves to be vital for the safe and efficient use of the medicinal substance (which is referred as companion diagnostics) [56].

To identify and quantify the constituents of the immune responses, many surveillance assays may be required because a therapeutically efficient anti-tumor response comprises a multi-component mechanism. It is vital to create assays that gauge the immune response(s) regarded as the most significant and pertinent elements of the anti-tumor reaction. In an effort to track the postulated immunologically-mediated anti-tumor reaction, minimum two immunological assays must be utilized. Specific factors should be included in the assay standardization process to account for regional immune response variability. The clinical protocol should include a detailed description of the assay parameters, including the assay situations, responsivity and selectivity of the assay, any in vitro acceleration steps involved, negative and positive limits, cutoff values for identifying negative and positive lab tests from patient samples, and the numerical quantitative methodologies to be utilized for the test outcomes, before the clinical trials are launched [57].

The FDA promotes the creation of exploratory biomarkers to provide evidence of concept and advance our knowledge of how certain drugs work. The creation of biomarkers as substitutes for effectiveness, however, is outside the purview of this advice. Adjuvants may be used in cancer vaccine preparations along with vaccination antigens to enhance or focus a particular immune reaction to an antigen. The prospective hazard of the adjuvant individually and of the experimental clinical vaccine-adjuvant mixture should be evaluated in preclinical research as necessary before the clinical injection of the vaccine-adjuvant mixture [58].

These preclinical investigations should be designed to closely resemble the intended clinical vaccination schedule and administration method. The general conditions for adding such adjuvants to biological products with a license are outlined in 21 CFR 610.15. According to 21 CFR 610.15(a), one of these requirements is the submission of proof that administering the recommended adjuvant to a pharmaceutical won't compromise its efficacy or safety. Numerous tumor-associated antigens may be present in cancer vaccine preparations in order to stimulate a variety of immune responses unique to the tumor and maybe thwart any tumor escape pathways. On a case-by-case scenario, it may not always be necessary to independently assess the safety and efficacy of every element of a multi-antigen vaccination. [51] The table 1.1, 1.2 and 1.3 gives us details of the cancer vaccines which are approved by FDA and which are still under process [59].

TABLE 12.1 FDA Approved Vaccines for Cancer Prevention [60]

Serial No.	Type of the cancer	Vaccine name
1.	Human Papillomavirus related, cervical, vaginal, anal, neck and head, vulvar and penile cancers	Cervarix
2.	Human Papillomavirus related, cervical, vaginal, anal, neck and head, vulvar and penile cancers	Gardasil-4
3.	Human Papillomavirus related, cervical, vaginal, anal, neck and head, vulvar and penile cancers	Gardasil-9
4.	HBV-related hepatocellular carcinoma	Hepatitis B (HBV) vaccine (HEPLISAV-B)

TABLE 12.2 FDA Approved Therapeutic Vaccines for Cancer [60]

Serial No.	Type of the cancer treated	Vaccine name
1.	Bladder cancer in early stages (Through local instillation into the bladder)	Bacillus Calmette–Guérin (BCG)
2.	Prostate cancer	Sipuleucel-T (Provenge)

TABLE 12.3 List of Cancer Vaccines Currently Under Clinical Trial Phase [60]

Serial No.	Condition/Name	Interventions	Clinical Phase	CT number
1.	Mammaglobin-A DNA vaccine for Breast cancer patients receiving neoadjuvant endocrine therapy receive the mammaglobin-A DNA vaccine.	Mammaglobin-A DNA Vaccine	1	NCT02204098
2.	Immunotherapy in Patients with Non-Metastatic, Node-Positive, HER2 Negative Breast Cancer in Remission Can Help Prevent Cancer Recurrence	pUMVC3-IGFBP2-HER2-IGF1R Plasmid DNA Vaccine Sargramostim	1	NCT02780401
3.	Concurrent VRP-HER2 Vaccination and Pembrolizumab Evaluation in Breast Cancer Patients	VRP-HER2 Pembrolizumab	2	NCT03632941
4.	Neoadjuvant Therapy for HER2+ Breast Cancer: HER2 Directed Dendritic Cell Vaccine	Dendritic Cell Vaccine (DC1) Neoadjuvant Chemotherapy	1	NCT03387553
5.	Patients With Stage III–IV HER2-Negative Breast Cancer Receiving Vaccine Therapy	CD105/Yb-1/SOX2/CDH3/MDM2-polyepitope Plasmid DNA Vaccine	1	NCT02157051
6.	Breast Cancer: HER2 Pulsed DC Vaccine to Prevent Recurrence of Invasive Breast Cancer	HER2 pulsed Dendritic Cell Vaccine	1	NCT02063724
7.	Before Surgery for the Treatment of In Situ HER2-Expressing Ductal Carcinoma, a Vaccine (H2NVAC)	Granulocyte-Macrophage Colony-Stimulating Factor Multi-epitope HER2 Peptide Vaccine H2NVAC	1	NCT04144023

Source: Adapted from "Cancer vaccines, treatment of the future: With emphasis on her2-positive breast cancer," by Pallerla, S., Abdul, A.U., Comeau, J., Jois, S., 2021, *International Journal of Molecular Sciences, 22*(2), p. 779 (https://www.ncbi.nlm.nih.gov/pmc/articles/PMC7828795/).

12.4 CASE STUDIES

Cancer treatments are somewhat tricky, whether in case of vaccines or in case of tablets or any other dosage forms. This portion of chapter explains few cases where difference was seen in the results of phase 2 and phase 3 of clinical trials. And due to these challenges in cancer vaccine development, it becomes difficult to develop one and to apply for a marketing authorization application for the same [61].

12.4.1 BRIVANIB

Goal: To treat hepatocellular cancer.

Sponsor: Bristol-Myers Squibb.

Difficulty revealed at the time of Phase-3 clinical studies: Lack of efficacy.

Different endpoints in phase 3 trials: Brivanib showed identified surprising toxicities in phase 3 trials despite showing promising anti-tumor efficacy in phase 2 trials, but failed to increase cumulative lifespan of patients relative to approved treatment.

The majority of primary liver cancers, or hepatocellular carcinomas (HCC), affect four out of every five malignancies that originate in the liver. Based on the phase and degree of cirrhosis, therapy alternatives for liver cancer comprise operation to excise the tumor, ablation to cut off the tumor's blood supply, radiotherapy, and transplantation. Sorafenib is the only medication that has been approved by the FDA; it slows tumor development and increases survival by blocking specific signals involved in cell development or function. Individuals who are not suitable for locally tailored therapy are typically given sorafenib. Brivanib was created to treat individuals that do not react to sorafenib or who experience severe negative effects from the medication. In addition to the growth factors that sorafenib targets, brivanib blocks a new growth factor [62].

In a phase 2 study, brivanib was administered daily to 55 participants with severe HCC in the first-line scenario. Based on the reported data, one patient who received brivanib experienced a complete response, three experienced a partial response, and 24 had unchanged disease checked by utilizing computerized tomography (CT)/magnetic resonance imaging (MRI) measures of tumor volume. Following the failure of sorafenib treatment or the discontinuation of sorafenib due to unacceptable side effects, a subsequent group of 46 patients was given brivanib. According to the study reports, 19 patients had unchanged disease after treatment, and two participants had a moderate response utilizing the same CT/MRI tumor measurement parameters. Together, the studies demonstrated the anticancer effectiveness of brivanib, and over half of the individuals were categorized as having stable illness after treatment. The research also discovered that individuals with advanced HCC had a controllable safety profile.

Numerous phase 3 RCTs conducted to focus on brivanib's effects proved statistically substantial anticancer activity, however they did not uncover any proof that the drug's use increases the survival rates of HCC patients. In a phase 3 research comparing brivanib and sorafenib, almost 1,100 participants with severe HCC who had not previously received pharmacological treatment were randomly assigned to get either sorafenib or brivanib. The major objective of the study—i.e., non-inferiority of sustenance not achieved because the overall survival median in the brivanib group was 9.5 months and in the sorafenib group was 9.9 months. Because patients on brivanib experienced significantly greater incidences of poor appetite, tiredness, hypertension, vomiting, and lower serum sodium levels than those taking sorafenib, the authors came to the conclusion that brivanib was "relatively poorly tolerated" than sorafenib. The individuals who got brivanib showed a more marked deterioration in functional ability and social functioning, according to the authors.

395 individuals with severe HCC who had previously taken sorafenib were randomly assigned to receive brivanib or a placebo in a different phase 3 study. In this trial, there was no statistically significant difference between participants who took brivanib and those who received a placebo in terms of overall survival. In a separate phase 3 trial, Asian participants with severe HCC who had previously failed sorafenib treatment were examined to see if brivanib may prolong survival relative to placebo. However, this study was stopped by its sponsors, so no findings are published.

In a fourth phase 3 research, patients with HCC who received brivanib in addition to chemoembolization were contrasted with those who received only chemoembolization. The other phase 3 trials described above did not demonstrate an improvement in the overall longevity of individuals with HCC, hence this experiment was stopped early. This study's conclusion revealed that brivanib did not increase overall survival (26.4 versus 26.1 months) [62].

12.4.2 FIGITUMUMAB

Goal: Advanced non-small cell lung cancer adjuvant therapy (NSCLC).

Sponsor: Pfizer.

Difficulty revealed at the time of Phase-3 clinical studies: Lack of efficacy, lack of safety.

Different endpoints in phase 3 trials: Although this targeted medication showed promising clinical outcomes in phase 2, introducing figitumumab to well-established chemotherapy schedules did not enhance survival and, when combined with one regimen, increased significant adverse events and mortality.

Lung cancer often arises in two types: small cell and NSCLC. Radiation therapy, chemotherapy, and laparoscopic removal of the tumor are currently used to treat NSCLC, although long-term survival chances are still low. A particular growth factor (IGF-1R) suspected to play a role in the initiation and progress of NSCLC, as well as other malignancies, was the focus of the development of the drug figitumumab. In phase 1 trials, figitumumab appeared to block the target pathway and shown anticancer effectiveness against numerous types of malignancies, including NSCLC. It also improved the anti-tumor actions of conventional chemotherapies in animal studies. Figitumumab in conjunction with the usual chemotherapy regimen of carboplatin and paclitaxel seemed to have a better response rate than carboplatin and paclitaxel monotherapy in a phase 2 research on NSCLC patients [63].

Based on these findings, two phase 3 studies were carried out in a combination of 1264 NSCLC patients, comparing figitumumab with various standard treatment versus the standard treatments alone. Due to the fact that figitumumab did not increase overall survival, both studies were stopped early [64]. The incidence of deaths and serious adverse events (SAEs) associated with treatment increased when figitumumab was added to one of these standard chemotherapy regimens, with 21% of individuals getting figitumumab experiencing SAEs as opposed to 12% of patients getting the conventional chemotherapy protocol alone. In patients getting figitumumab, the frequency of treatment-related mortality was 5% compared to 1% in those getting the usual regimen.

Pfizer withdrew the publication detailing the phase 2 findings when the phase 3 studies were prematurely stopped due to ineffectiveness and safety issues. The business learned that, contrary to Pfizer's usual operating practices, tumor reduction had not been verified in all responding cases. The data after correction revealed a decreased response rate.

12.4.3 MAGE-A3 VACCINE

Goal: After surgery, non-small cell lung cancer (NSCLC) patients' care.

Sponsor: GlaxoSmithKline.

Difficulty revealed at the time of Phase-3 clinical studies: Lack of efficacy.

Different endpoints in phase 3 trials: The MAGE-A3 vaccine did not offer any clinical advantage when contrasted to a placebo in the phase 3 study, although a positive evidence of concept trial for this specific immune treatment.

Lung cancer often arises in two types: small cell and NSCLC. Radiation therapy, chemotherapy, and laparoscopic removal of the tumor are currently used to treat NSCLC, although long-term survival chances are still low. Recent developments in cancer research suggest the possibility of using the body's immunity to treat NSCLC. Therapeutic cancer vaccines can target surface chemicals on some tumor cells (antigens), perhaps protecting healthy cells. MAGE-A3, a tumor-specific protein found on the outside of specific tumor cells, is one illustration of various cell surface antigens. MAGE-A3, that is not present in healthy lung cells, is expressed by about 33% of NSCLCs, making it a viable target for NSCLC treatments [65].

A MAGE-A3 vaccination was examined in a phase 2 research as a potential therapy for people with MAGE-A3-positive NSCLC. 182 participants were randomly assigned to take between the MAGE-A3 vaccine or a placebo, 13 times over the course of 27 months after operation to eradicate a large portion of the tumor as feasible. The findings revealed that individuals who received this cancer vaccine had improved illness and overall survival rates but that were not statistically significant. Only enough people participated in the study for it to serve as evidence of concept. The vaccination was entered into the largest phase 3 study of an NSCLC therapy yet conducted because the sponsor believed the results were encouraging enough. Following the same protocol as the phase 2 study, researchers randomly assigned 2,272 patients with fully excised MAGEA3-positive NSCLC to get 13 intramuscular doses of either the placebo or a vaccine. However, the study failed to show that administration of the MAGE-A3 cancer vaccine improved patients' survival without disease (60.5 months versus 57.9 months, a data which statistically has no significant difference) [66]. The study's findings lead the researchers to the conclusion that this cancer vaccination does not provide any clinical benefits to NSCLC patients.

12.4.4 INIPARIB

Goal: An extra treatment approach for 'triple negative' breast cancer.

Sponsor: Sanofi.

Difficulty revealed at the time of Phase-3 clinical studies: Lack of efficacy.

Different endpoints in phase 3 trials: Despite encouraging phase 2 tumor reaction and survival outcomes, introducing iniparib to a preexisting chemotherapy treatment did not increase survival in the phase 3 study.

The most frequent malignancy in women is breast cancer. An aggressive and challenging to treat form of breast cancer is triple-negative breast cancer. Since the cancerous tissues don't really overexpress three distinct receptors, it is called triple-negative; if it did, the cancer may be treated with chemotherapies and/or medicines that target the receptors. In preclinical trials, iniparib shown significant activity, boosting the effects of conventional chemotherapy on triple-negative malignant breast cancer cells. 123 malignant triple-negative breast cancer patients were randomly assigned to receive conventional chemotherapy alone or standard chemotherapy with iniparib in phase 2 trial. Without enhancing toxicity, introducing iniparib to a typical chemotherapy course greatly increased patient survival and tumor treatment. Iniparib failed to demonstrate efficacy in phase 3 trials amidst impressive phase 2 results. The usual chemotherapy protocol or the usual chemotherapy protocol with iniparib was assigned randomly to 519 individuals with advanced triple-negative breast cancer. However, the inclusion of iniparib to the conventional protocol did not show any benefit in ultimate or advancement survival. The phase 3 study did not reveal any substantial safety issues. Individuals undergoing conventional treatment had an average survival of 11.1 months compared to 11.8 months for patients also receiving iniparib [67].

12.5 CONCLUSION

Flexibility is required from sponsors and regulatory bodies due to the wide variety of scientific methodologies employed in the creation of cancer vaccines. As of now, certain guideline publications sporadically contain detailed recommendations on individual vaccine kinds. This adaptive regulatory structure enables the establishment of a non-clinical research program that requires factual rationale for each experiment and parameter investigated instead of solely relying on the suggestions of the authorities. Therefore, program design requires knowledge from a wide range of fields, including regulatory affairs, biotechnology, toxicity, and fundamental and applied immunology. Prior to the vaccine's first test on humans, each program also necessitates a thorough risk-benefit analysis that takes into account (a) the cancer evidence and phase of the disease, (b) the planned patient percentage, (c) alternative therapies, and (d) preventative versus therapeutic implementation; Finally, (e) that an innovative pharmaceutical concept for one cancer vaccine (which may be specific to a particular cancer vaccine) does not totally depend on the risk evaluation of other cancer vaccines.

Experience to date suggests that in comparison to small compound drugs or conventional vaccines, the nonclinical development of cancer vaccines requires a smaller, more highly specialized, and commonly more sophisticated program due to the diverse scientific approaches as well as the life-threatening character of the ailments in question. Due to large variations in the field of vaccines, sponsors should engage with the regulatory bodies from time to time to discuss about the progress in the pre-clinical as well as clinical phases before submitting their results. Moreover, they should also give their inputs whenever the draft guidelines related to their products are made. European Medicines Agency has given a choice for an informal meeting with the Innovation Task Force, where an open discussion can be made regarding the regulatory challenges faced by the developers for a particular type or category of vaccines.

These interactions will prove to be valuable, fruitful and will help in aligning the scientific advances and regulatory requirements to enable smooth processing of vaccines. Also, the guidelines for vaccines development should be harmonized so that the queries arising during submission in various countries do not rise, but there should also be a scope to change few things depending on the category of the vaccine keeping in mind the quality, safety and efficacy of the product developed.

KEYWORDS

- **cancer vaccine**
- **clinical trials**
- **personalized vaccines**
- **regulatory guidelines**
- **submission**

REFERENCES

1. Jessy, T., (2011). Immunity over inability: The spontaneous regression of cancer. *Journal of Natural Science, Biology, and Medicine, 2*(1), 43–49.
2. Cann, S. H., Van, N. J. P., & Van, N. C., (2003). Dr William Coley and tumor regression: A place in history or in the future. *Postgraduate Medical Journal, 79*(938), 672–680.
3. Tran, T., Burt, D., Eapen, L., & Keller, O. R., (2013). Spontaneous regression of metastatic melanoma after inoculation with tetanus–diphtheria–pertussis vaccine. *Current Oncology, 20*(3), 270–273
4. Von Mensdorff-Pouilly, S., Snijdewint, F. G., Verstraeten, A. A., Verheijen, R. H., & Kenemans, P., (2000). Human MUC1 mucin: A multifaceted glycoprotein. *The International Journal of Biological Markers, 15*(4), 343–356.
5. Von Mensdorff-Pouilly, S., Moreno, M., & Verheijen, R. H., (2011). Natural and induced humoral responses to MUC1. *Cancers, 3*(3), 3073–3103.
6. Cramer, D. W., Vitonis, A. F., Pinheiro, S. P., McKolanis, J. R., Fichorova, R. N., Brown, K. E., Hatchette, T. F., & Finn, O. J., (2010). Mumps and ovarian cancer: Modern interpretation of an historic association. *Cancer Causes & Control, 21*(8), 1193–1201.
7. Burnet, F. M., (1971). Immunological surveillance in neoplasia. *Immunological Reviews, 7*(1), 3–25.
8. European Medicines Agency, (2019). *Guideline on the Clinical Evaluation of Anticancer Medicinal Products.* EMA/CHMP/205/95/Rev.6. Accessible at: https://www.ema.europa.eu/en/documents/scientific-guideline/draft-guideline-evaluationanticancer-medicinal-products-man-revision-6_en.pdf (accessed on 11 March 2024).
9. Salas-Vega, S., Iliopoulos, O., & Mossialos, E., (2017). Assessment of overall survival, quality of life, and safety benefits associated with new cancer medicines. *JAMA Oncology, 3*(3), 382–390.
10. Health Canada Guidance Document, (2018). '*Harmonized Requirements for the Licensing of Vaccines and Guidelines for the Preparation of an Application.*' Accessible at: https://www.canada.ca/en/health-canada/services/drugs-health-products/biologicsradiopharmaceuticals-genetic-therapies/applications-submissions/guidanced-ocuments/drug/harmonized-requirements-licensing-vaccines-guidelines-preparationap-plication.html (accessed on 11 March 2024).
11. Van, D. B. P., Stroobant, V., Vigneron, N., & Van, D. E. B., (2013). Peptide database of T-cell defined tumor antigens. *Cancer Immunity, 2013.*
12. Dranoff, G., (2012). Experimental mouse tumor models: What can be learnt about human cancer immunology? *Nature Reviews Immunology, 12*(1), 61–66.
13. Sathish, J. G., Sethu, S., Bielsky, M. C., De Haan, L., French, N. S., Govindappa, K., Green, J., et al., (2013). Challenges and approaches for the development of safer immunomodulatory biologics. *Nature Reviews Drug Discovery, 12*(4), 306–324.
14. Heelan, B. T., (2014). Regulatory considerations for clinical development of cancer vaccines. *Human Vaccines & Immunotherapeutics, 10*(11), 3409–3414.
15. *Vaccines Licensed for Use in the United States*. Accessible at: https://www.fda.gov/vaccinesblood-biologics/vaccines/vaccines-licensed-use-united-states (accessed on 11 March 2024).
16. Samy, K. P., & Brennan, T. V., (2018). Dendritic cell therapy in transplantation, phenotype governs destination and function. *Transplantation, 102*(10), 1593.

17. Ogi, C., & Aruga, A., (2015). Approaches to improve development methods for therapeutic cancer vaccines. *Immunology Letters, 164*(2), 100–108.
18. Saxena, M., Van, D. B. S. H., Melief, C. J., & Bhardwaj, N., (2021). Therapeutic cancer vaccines. *Nature Reviews Cancer, 21*(6), 360–378.
19. *WHO 'Guidelines on Nonclinical Evaluation of Vaccines.' WHO Technical Report Series, No. 927*, (2005). Accessible at: https://www.who.int/publications/m/item/nonclinicalevaluation-of-vaccines-annex-1-trs-no-927 (accessed on 11 March 2024).
20. Robert, C., (2020). A decade of immune-checkpoint inhibitors in cancer therapy. *Nature Communications, 11*(1), 1–3.
21. Pardi, N., Hogan, M. J., Porter, F. W., & Weissman, D., (2018). mRNA vaccines—A new era in vaccinology. *Nature Reviews Drug Discovery, 17*(4), 261–279.
22. Haen, S. P., Löffler, M. W., Rammensee, H. G., & Brossart, P., (2020). Towards new horizons: Characterization, classification, and implications of the tumor antigenic repertoire. *Nature Reviews Clinical Oncology, 17*(10), 595–610.
23. *WHO 'Guidelines on Clinical Evaluation of Vaccines: Regulatory Expectations*,' (2017). WHO Technical Report Series 1004, Annex 9. Accessible at: https://www.who.int/publications/m/item/WHO-TRS-1004-web-annex-9 (accessed on 11 March 2024).
24. Mittendorf, E. A., Lu, B., Melisko, M., Price, H. J., Bondarenko, I., Brunt, A. M., Sergii, G., et al., (2019). Efficacy and safety analysis of nelipepimut-S vaccine to prevent breast cancer recurrence: A randomized, multicenter, phase III clinical trial randomized phase III trial of nelipepimut-S in breast cancer. *Clinical Cancer Research, 25*(14), 4248–4254.
25. Lilleby, W., Gaudernack, G., Brunsvig, P. F., Vlatkovic, L., Schulz, M., Mills, K., Hole, K. H., & Inderberg, E. M., (2017). Phase I/IIa clinical trial of a novel hTERT peptide vaccine in men with metastatic hormone-naive prostate cancer. *Cancer Immunology, Immunotherapy, 66*(7), 891–901.
26. Gridelli, C., Ciuleanu, T., Domine, M., Szczesna, A., Bover, I., Cobo, M., Kentepozidis, N., et al., (2020). Clinical activity of a hTERT (vx-001) cancer vaccine as post-chemotherapy maintenance immunotherapy in patients with stage IV non-small cell lung cancer: Final results of a randomized phase 2 clinical trial. *British Journal of Cancer, 122*(10), 1461–1466.
27. Guidance for Industry, (2011). '*Clinical Considerations for Therapeutic Cancer Vaccines*.' Accessible at: https://www.fda.gov/media/82312/download (accessed on 11 March 2024).
28. CDSCO guidance for industry, (2019). *'Submission of Clinical Trial Application for Evaluating Safety and Efficacy*.'Document No. – CT/71108. Accessible at: https://www.kem.edu/wp-content/uploads/2019/12/CDSCO-Guidance-For-Industry.pdf (accessed on 11 March 2024).
29. Tay, B. Q., Wright, Q., Ladwa, R., Perry, C., Leggatt, G., Simpson, F., Wells, J. W., et al., (2021). Evolution of cancer vaccines—Challenges, achievements, and future directions. *Vaccines, 9*(5), 535.
30. Iglesias-Lopez, C., Agustí, A., Obach, M., & Vallano, A., (2019). Regulatory framework for advanced therapy medicinal products in Europe and United States. *Frontiers in Pharmacology. 2019*, 921.
31. Parks, R. J., & Gussoni, E., (2018). Building immune tolerance through DNA vaccination. *Proceedings of the National Academy of Sciences, 115*(39), 9652–9654.
32. Health Canada Guidance Document, (2018). '*Harmonized Requirements for the Licensing of Vaccines and Guidelines for the Preparation of an Application.*'

Accessible at: https://www.canada.ca/en/health-canada/services/drugs-health-products/biologicsradiopharmaceuticals-genetic-therapies/applications-submissions/guidancedocuments/drug/harmonized-requirements-licensing-vaccines-guidelines-preparationapplication.html (accessed on 11 March 2024).
33. Stern, P. L., Beverley, P. C., & Carroll, M., (2000). *Cancer Vaccines and Immunotherapy*. Cambridge University Press.
34. *Guidance for Industry: Data Standards for Drug and Biological Product Submissions Containing Real-World Data.* FDA/USFDA Draft Guidance Document Oct 2021. Accessible at: https://www.fda.gov/media/153341/download (accessed on 11 March 2024).
35. Dinda, A. K., Tripathi, S. K., & John, B., (2020). Revisiting regulatory framework in India for accelerated vaccine development in pandemics with an evidence-based fast-tracking strategy. *The Indian Journal of Medical Research, 152*(1, 2), 156.
36. European Medicines Agency, (2018). '*Guideline on Clinical Evaluation of Vaccines*.' EMEA/CHMP/VWP/164653/05 Rev. 1. Accessible at: https://www.ema.europa.eu/en/documents/scientific-guideline/draft-guideline-clinicalevaluation-vaccines-revision-1_en.pdf (accessed on 11 March 2024).
37. Kudrin, A., (2012). Overview of cancer vaccines: Considerations for development. *Human Vaccines & Immunotherapeutics, 8*(9), 1335–1353.
38. Hinz, T., Kallen, K., Britten, C. M., Flamion, B., Granzer, U., Hoos, A., Huber, C., et al., (2017). The European regulatory environment of RNA-based vaccines. *RNA Vaccines, 2017*, 203–222.
39. Grunwitz, C., & Kranz, L. M., (2017). mRNA cancer vaccines—Messages that prevail. *Cancer Vaccines, 2017*, 145–164.
40. Grabbe, S., Haas, H., Diken, M., Kranz, L. M., Langguth, P., & Sahin, U., (2016). Translating nanoparticulate-personalized cancer vaccines into clinical applications: Case study with RNA-lipoplexes for the treatment of melanoma. *Nanomedicine, 11*(20), 2723–2734.
41. Romero, P., Banchereau, J., Bhardwaj, N., Cockett, M., Disis, M. L., Dranoff, G., Gilboa, E., et al., (2016). The human vaccines project: A roadmap for cancer vaccine development. *Science Translational Medicine, 8*(334), 334ps9-.
42. Kauffmann, F., Van, D. P., Leroux-Roels, G., Vandermeulen, C., Berthels, N., Beuneu, C., & Mali, S., (2019). Clinical trials with GMO-containing vaccines in Europe: Status and regulatory framework. *Vaccine, 37*(42), 6144–6153.
43. Mougel, A., Terme, M., & Tanchot, C., (2019). Therapeutic cancer vaccine and combinations with antiangiogenic therapies and immune checkpoint blockade. *Frontiers in Immunology, 10*, 467.
44. Májek, O., Anttila, A., Arbyn, M., Van, V. E. B., Engesæter, B., & Lönnberg, S., (2019). The legal framework for European cervical cancer screening programs. *European Journal of Public Health, 29*(2), 345–350.
45. Tardón, M. C., Allard, M., Dutoit, V., Dietrich, P. Y., & Walker, P. R., (2019). Peptides as cancer vaccines. *Current Opinion in Pharmacology, 47*, 20–26.
46. Rutschman, A. S., (2019). The vaccine race in the 21st century. *Ariz. L. Rev., 61*, 729.
47. Sharma, P., Jhawat, V., Mathur, P., & Dutt, R., (2022). Innovation in cancer therapeutics and regulatory perspectives. *Medical Oncology, 39*(5), 1–2.
48. Mangla, B., Kumar, P., Goyal, K., Kohli, K., & Jindal, S., (2022). Product development and scale-up challenges in cancer vaccine development. *Nanotherapeutics in Cancer Vaccination and Challenges* (pp. 313–324). Academic Press.

49. Nappi, F., & Galli, M. C., (2022). Regulatory aspects of cancer immunotherapy in Europe. *Cancer Vaccines as Immunotherapy of Cancer* (pp. 75–84). Academic Press.
50. EMEA, Committee for Proprietary Medicinal Products, (2001). '*Note for Guidance on the Quality, Preclinical and Clinical Aspects of Gene Transfer Medicinal Products.*' CPMP/BPW/3088/99. Accessible at: https://www.ema.europa.eu/en/documents/scientificguideline/note-guidance-quality-preclinical-clinical-aspects-gene-transfer-medicinalproducts_en.pdf (accessed on 11 March 2024).
51. Pan American Network on Drug Regulatory Harmonization, (2010). 'HarmonizedRequirements for the Licensing of Vaccines in the Americas and Guidelines for Application.' PANDRH Technical Document No. 1, ISBN 978 92 75 13083 4, Accessible at: https://iris.paho.org/bitstream/handle/10665.2/31215/9789275130834eng.PDF?sequence=1&isAllowed=y (accessed on 11 March 2024).
52. Pallerla, S., Abdul, A. U., Comeau, J., & Jois, S., (2021). Cancer vaccines, treatment of the future: With emphasis on her2-positive breast cancer. *International Journal of Molecular Sciences, 22*(2), 779.
53. 22 Case Studies Where Phase 2 and Phase 3 Trials Had Divergent Results, USFDA, (2017). Accessible at: https://www.fda.gov/media/102332/download (accessed on 11 March 2024).
54. Shemesh, C. S., Hsu, J. C., Hosseini, I., Shen, B. Q., Rotte, A., Twomey, P., Girish, S., & Wu, B., (2021). Personalized cancer vaccines: Clinical landscape, challenges, and opportunities. *Molecular Therapy, 29*(2), 555–570.
55. Tony, D. A., & Yan-Ping, Y., (2019). Advances and challenges in vaccine development and manufacture. Bioprocess International. Accessible at: https://bioprocessintl.com/manufacturing/vaccines/advances-and-challenges-in-vaccinedevelopment-and-manufacture/ (accessed on 11 March 2024).
56. D'Amore, T., (2019). Advances and challenges in vaccine development and manufacture. Bioprocess International. https://bioprocessintl.com/manufacturing/vaccines/advances-andchallenges-in-vaccine-development-and-manufacture/ (accessed on 11 March 2024).
57. Goyal, K., Mangla, B., Singh, N., Goel, H., Goyal, K., Jindal, S., et al., (2022). Regulatory landscape in the approval of cancer vaccine. Nanotherapeutics in Cancer Vaccination and Challenges (pp. 325–348). Elsevier.
58. Future of Health: Cancer Vaccines, (n.d.). UW School of Medicine and Public Health. Retrieved from: https://www.med.wisc.edu/news-and-events/2022/may/future-of-healthcancer-vaccines/ (accessed on 11 March 2024).
59. Vigneron, N., Stroobant, V., Van, D. Eynde, B. J., & Van, D. B. P., (2013). Database of T cell-defined human tumor antigens: The 2013 update. Cancer Immunity, 13, 15.
60. Research CfBEa, (2022). Vaccines Licensed for Use in the United States. FDA.
61. Patel, V. M., Prajapati, B. G., & Patel, M. M., (2007). Formulation, evaluation, and comparison of bilayered and multilayered mucoadhesive buccal devices of propranolol hydrochloride. *AAPS Pharmscitech, 8*(1), E147–E54.
62. Patel, V. M., Prajapati, B. G., & Patel, M. M., (2007). Design and characterization of chitosan-containing mucoadhesive buccal patches of propranolol hydrochloride. *Acta Pharm., 57*(1), 61–72.
63. Singh, J., Bowne, W. B., & Snook, A. E., (2021). Cancer vaccines and immunotherapy for tumor prevention and treatment. *Vaccines, 9*(11).

64. Prajapati, B. G., & Ratnakar, N., (2009). A review on recent patents on fast dissolving drug delivery system. *International Journal of PharmTech Research, 1*(3), 790–798.
65. Patel, V. M., Prajapati, B. G., & Patel, M. M., (2007). Effect of hydrophilic polymers on buccoadhesive Eudragit patches of propranolol hydrochloride using factorial design. *AAPS Pharmscitech, 8*(2), E119–E26.
66. Calvo, T. M., Allard, M., Dutoit, V., Dietrich, P. Y., & Walker, P. R., (2019). Peptides as cancer vaccines. *Current Opinion in Pharmacology, 47*, 20–26.
67. Donninger, H., Li, C., Eaton, J. W., & Yaddanapudi, K., (2021). Cancer vaccines: Promising therapeutics or an unattainable dream. *Vaccines, 9*(6).
68. Advances and challenges in vaccine development and manufacture," by D'Amore, T., Yang, Y. P., 2019, *BioProc Int.* (https://bioprocessintl.com/manufacturing/vaccines/advances-and-challenges-in-vaccine-development-and-manufacture/).
69. Regulatory landscape in the approval of cancer vaccine," by Goyal, K., Mangla, B., Singh, N., Goel, H., Goyal, K., Jindal, S., Ansari, M. J., Almalki, W. H., 2022, *Nanotherapeutics in Cancer Vaccination and Challenges*, Academic Press, pp. 325–48 (https://www.sciencedirect.com/science/article/pii/B978012823686400001X).
70. Personalized cancer vaccines: clinical landscape, challenges, and opportunities," by Shemesh, C. S., Hsu, J. C., Hosseini, I., Shen, B. Q., Rotte, A., Twomey, P., Girish, S., Wu, B., 2021, *Molecular Therapy*, 29(2), pp. 555–70 (https://www.ncbi.nlm.nih.gov/pmc/articles/PMC7828795/).

Index

D

E

I

T

Y

Z